Gene Therapy Protocols

METHODS IN MOLECULAR BIOLOGY™

John M. Walker, SERIES EDITOR

459. **Prion Protein Protocols**, edited by *Andrew F. Hill, 2008*
458. **Artificial Neural Networks:** *Methods and Applications,* edited by *David S. Livingstone, 2008*
457. **Membrane Trafficking**, edited by *Ales Vancura, 2008*
456. **Adipose Tissue Protocols, Second Edition**, edited by *Kaiping Yang, 2008*
455. **Osteoporosis**, edited by *Jennifer J. Westendorf, 2008*
454. **SARS- and Other Coronaviruses:** *Laboratory Protocols,* edited by *Dave Cavanagh, 2008*
453. **Bioinformatics, Volume 2:** *Structure, Function, and Applications,* edited by *Jonathan M. Keith, 2008*
452. **Bioinformatics, Volume 1:** *Data, Sequence Analysis, and Evolution,* edited by *Jonathan M. Keith, 2008*
451. **Plant Virology Protocols:** *From Viral Sequence to Protein Function,* edited by *Gary Foster, Elisabeth Johansen, Yiguo Hong, and Peter Nagy, 2008*
450. **Germline Stem Cells**, edited by *Steven X. Hou and Shree Ram Singh, 2008*
449. **Mesenchymal Stem Cells:** *Methods and Protocols,* edited by *Darwin J. Prockop, Douglas G. Phinney, and Bruce A. Brunnell, 2008*
448. **Pharmacogenomics in Drug Discovery and Development**, edited by *Qing Yan, 2008*
447. **Alcohol:** *Methods and Protocols,* edited by *Laura E. Nagy, 2008*
446. **Post-translational Modification of Proteins:** *Tools for Functional Proteomics, Second Edition,* edited by *Christoph Kannicht, 2008*
445. **Autophagosome and Phagosome**, edited by *Vojo Deretic, 2008*
444. **Prenatal Diagnosis**, edited by *Sinhue Hahn and Laird G. Jackson, 2008*
443. **Molecular Modeling of Proteins**, edited by *Andreas Kukol, 2008*
439. **Genomics Protocols:** *Second Edition,* edited by *Mike Starkey and Ramnanth Elaswarapu, 2008*
438. **Neural Stem Cells:** *Methods and Protocols, Second Edition,* edited by *Leslie P. Weiner, 2008*
437. **Drug Delivery Systems**, edited by *Kewal K. Jain, 2008*
436. **Avian Influenza Virus**, edited by *Erica Spackman, 2008*
435. **Chromosomal Mutagenesis**, edited by *Greg Davis and Kevin J. Kayser, 2008*
434. **Gene Therapy Protocols: Volume 2:** *Design and Characterization of Gene Transfer Vectors,* edited by *Joseph M. Le Doux, 2008*
433. **Gene Therapy Protocols: Volume 1:** *Production and In Vivo Applications of Gene Transfer Vectors,* edited by *Joseph M. Le Doux, 2008*
432. **Organelle Proteomics**, edited by *Delphine Pflieger and Jean Rossier, 2008*
431. **Bacterial Pathogenesis:** *Methods and Protocols,* edited by *Frank DeLeo and Michael Otto, 2008*
430. **Hematopoietic Stem Cell Protocols**, edited by *Kevin D. Bunting, 2008*
429. **Molecular Beacons:** *Signalling Nucleic Acid Probes, Methods and Protocols,* edited by *Andreas Marx and Oliver Seitz, 2008*
428. **Clinical Proteomics:** *Methods and Protocols,* edited by *Antonio Vlahou, 2008*
427. **Plant Embryogenesis**, edited by *Maria Fernanda Suarez and Peter Bozhkov, 2008*
426. **Structural Proteomics:** *High-Throughput Methods,* edited by *Bostjan Kobe, Mitchell Guss, and Huber Thomas, 2008*
425. **2D PAGE: Volume 2:** *Applications and Protocols,* edited by *Anton Posch, 2008*
424. **2D PAGE: Volume 1:** *Sample Preparation and Pre-Fractionation,* edited by *Anton Posch, 2008*
423. **Electroporation Protocols**, edited by *Shulin Li, 2008*
422. **Phylogenomics**, edited by *William J. Murphy, 2008*
421. **Affinity Chromatography:** *Methods and Protocols, Second Edition,* edited by *Michael Zachariou, 2008*
420. **Drosophila:** *Methods and Protocols,* edited by *Christian Dahmann, 2008*
419. **Post-Transcriptional Gene Regulation**, edited by *Jeffrey Wilusz, 2008*
418. **Avidin-Biotin Interactions:** *Methods and Applications,* edited by *Robert J. McMahon, 2008*
417. **Tissue Engineering**, *Second Edition,* edited by *Hannsjörg Hauser and Martin Fussenegger, 2007*
416. **Gene Essentiality:** *Protocols and Bioinformatics,* edited by *Svetlana Gerdes and Andrei L. Osterman, 2008*
415. **Innate Immunity**, edited by *Jonathan Ewbank and Eric Vivier, 2007*
414. **Apoptosis in Cancer:** *Methods and Protocols,* edited by *Gil Mor and Ayesha Alvero, 2008*
413. **Protein Structure Prediction**, *Second Edition,* edited by *Mohammed Zaki and Chris Bystroff, 2008*
412. **Neutrophil Methods and Protocols**, edited by *Mark T. Quinn, Frank R. DeLeo, and Gary M. Bokoch, 2007*
411. **Reporter Genes for Mammalian Systems**, edited by *Don Anson, 2007*
410. **Environmental Genomics**, edited by *Cristofre C. Martin, 2007*
409. **Immunoinformatics:** *Predicting Immunogenicity In Silico,* edited by *Darren R. Flower, 2007*
408. **Gene Function Analysis**, edited by *Michael Ochs, 2007*
407. **Stem Cell Assays**, edited by *Vemuri C. Mohan, 2007*
406. **Plant Bioinformatics:** *Methods and Protocols,* edited by *David Edwards, 2007*
405. **Telomerase Inhibition:** *Strategies and Protocols,* edited by *Lucy Andrews and Trygve O. Tollefsbol, 2007*
404. **Topics in Biostatistics**, edited by *Walter T. Ambrosius, 2007*
403. **Patch-Clamp Methods and Protocols**, edited by *Peter Molnar and James J. Hickman 2007*
402. **PCR Primer Design**, edited by *Anton Yuryev, 2007*

METHODS IN MOLECULAR BIOLOGY™

Gene Therapy Protocols

Volume 1: Production and In Vivo Applications of Gene Transfer Vectors

Third Edition

Edited by

Joseph M. Le Doux

Department of Biomedical Engineering, Georgia Institute of Technology and Emory University, Atlanta, GA

Humana Press

Editor
Joseph M. Le Doux
Department of Biomedical Engineering
Georgia Institute of Technology and Emory University
Atlanta, Georgia 30322-0355
joe.ledoux@bme.gatech.edu

Series Editor
John M. Walker
School of Life Sciences
University of Hertfordshire
College Lane
Hatfield Campus
Hatfield, Herts., UK AL10 9AB
john.walker543@ntlworld.com

ISBN 978-1-62703-956-7 ISBN 978-1-59745-237-3 (eBook)
ISSN 1064-3745 ISSN 1940-6029 (eBook)

Cover illustration: Background image: Photo provided courtesy of Dongsheng Duan. Modified by Nancy Fallatt with permission. Cover image: Figure 1, Chapter 10. "Non-Viral Gene Delivery with Cationic Liposome-DNA Complexes," by Kai K. Ewert, Ayesha Ahmad, Nathan F. Bouxsein, Heather M. Evans, and Cyrus R. Safinya. Reprinted (abstracted/excerpted) with permission from J. Am. Chem. Soc. 2006, 128(12), 4002. Copyright 2006 American Chemical Society.

Printed on acid-free paper

9 8 7 6 5 4 3 2 1

springer.com

Preface

Gene therapy has the potential to significantly impact human healthcare in the twenty-first century. The idea behind gene therapy is simple: to deliver genetic material to cells that will slow down or halt the progression of disease, or to help repair or regenerate damaged or lost tissues. To successfully implement this simple idea, however, we must first address a number of technological challenges. One such challenge is the production of gene transfer vectors that can safely and effectively transfer genes to a wide variety of cells, tissues, and organs. A number of vectors have been developed, but none are ideal. The choice of which system to use depends on the specific gene therapy application being addressed, and is often influenced by its specific functional characteristics, such as the types of cells it can transduce, the efficiency with which it transfers genes, the strength and persistence of transgene expression, and the extent to which it activates the immune response of the host. For all gene transfer vector systems, it is critical that production methods are developed that are economical, that can be conducted on a large scale, and that yield a high-titer product that does not require extensive postproduction processing or purification. Another critical issue is the development of clinically relevant protocols for using these gene transfer vector systems. Numerous experimental protocols have been developed over the past few years. These protocols share a number of common elements, but they also each harbor their own share of unique technical challenges and issues that were addressed and overcome through years of hard work, creativity, and perseverance. These hard-earned protocols can sometimes be used "as is" by others if they are addressing the same or similar gene therapy application, or they can serve as the basis for the development of new protocols.

The field of gene therapy is rapidly advancing on all of these fronts. Significant improvements have been made in the methods used to produce the core gene transfer technologies. In parallel, novel gene transfer protocols have been developed that have helped to expand the list of potential applications of gene therapy. In this new and entirely revised third edition, *Gene Therapy Protocols, Volumes 1 and 2* present a comprehensive collection of detailed methods and protocols used by the leaders in the field of gene therapy. The first volume covers current and emerging methods for the production of major viral and nonviral gene transfer vectors. The second volume provides detailed protocols for some of the most important applications of gene therapy, including the areas of stem cell biology, cancer, diabetes, HIV, and tissue engineering. This compilation of protocols is expected to serve as a valuable resource for graduate students and postdoctoral fellows, as well as for basic and clinical researchers in the industry and academia.

Joseph M. LeDoux

Contents

Contributors

AYESHA AHMAD • *Materials Department, Physics Department and Molecular, Cellular, and Developmental Biology Department, University of California, Santa Barbara, CA*

ERIC W. F. W. ALTON • *Department of Gene Therapy, Faculty of Medicine, Imperial College London, UK, and Medical Genetics, Molecular Medicine Centre, University of Edinburgh, Edinburgh, UK*

STELIOS T. ANDREADIS • *Department of Chemical and Biological Engineering, University of Buffalo, The State University of New York, Buffalo, NY*

JESSAMYN BAGLEY • *Transplantation Research Center, Renal Division, Brigham and Women's Hospital and Children's Hospital Boston, Harvard Medical School, Boston, MA*

NATHAN F. BOUXSEIN • *Materials Department, Physics Department and Molecular, Cellular, and Developmental Biology Department, University of California, Santa Barbara, CA*

A. CHRISTOPHER BOYD • *The UK Cystic Fibrosis Gene Therapy Consortium and Medical Genetics, Molecular Medicine Centre, University of Edinburgh, Edinburgh, UK*

KENNETH CORNETTA • *Department of Medical and Molecular Genetics, Indiana University School of Medicine, Indianapolis, IN*

SCOTT CROSS • *Department of Medical and Molecular Genetics, Indiana University School of Medicine, Indianapolis, IN*

ABHIJIT DE • *Molecular Imaging Program at Stanford (MIPS) and Bio-X Program, Departments of Radiology and Bioengineering, School of Medicine, Stanford University, Stanford, CA*

STÉPHANIE DECHERF • *Laboratoire d'Evolution des Régulations Endocriniennes, Muséum National d'Histoire, Naturelle, Paris, France*

BARBARA A. DEMENEIX • *Laboratoire d'Evolution des Régulations Endocriniennes, Muséum National d'Histoire, Naturelle, Paris, France*

CHRISTOPHER B. DOERING • *Aflac Cancer Center and Blood Disorders Service, Department of Pediatrics, Emory University, Atlanta, GA*

DONGSHENG DUAN • *Department of Molecular Microbiology and Immunology, University of Missouri, Columbia, MO*

HEATHER M. EVANS • *Materials Department, Physics Department and Molecular, Cellular, and Developmental Biology Department, University of California, Santa Barbara, CA*

KAI K. EWERT • *Materials Department, Physics Department and Molecular, Cellular, and Developmental Biology Department, University of California, Santa Barbara, CA*

SANJIV SAM GAMBHIR • *Molecular Imaging Program at Stanford (MIPS) and Bio-X Program, Departments of Radiology and Bioengineering, School of Medicine, Stanford University, Stanford, CA*

ANDRÉS J. GARCÍA • *Petit Institute for Bioengineering and Biosciences and Georgia Tech/Emory Center for the Engineering of Living Tissues, Georgia Institute of Technology, Atlanta, GA*

DEBORAH R. GILL • *The UK Cystic Fibrosis Gene Therapy Consortium and Gene Medicine Group, University of Oxford, Oxford, UK*

JOSEPH C. GLORIOSO • *Department of Molecular Genetics and Biochemistry, University of Pittsburgh School of Medicine, Pittsburgh, PA*

WILLIAM F. GOINS • *Department of Molecular Genetics and Biochemistry, University of Pittsburgh School of Medicine, Pittsburgh, PA*

UTA GRIESENBACH • *Department of Gene Therapy, Faculty of Medicine, Imperial College London, UK, and The UK Cystic Fibrosis Gene Therapy Consortium, London, UK*

ZAHRA HASSANI • *Laboratoire d'Evolution des Régulations Endocriniennes, Muséum National d'Histoire Naturelle, CNRS, Paris, France*

EMMA HOLDER • *Medical Genetics, Molecular Medicine Centre, University of Edinburgh, Edinburgh, UK, and The UK Cystic Fibrosis Gene Therapy Consortium, London, UK*

SHAOHUA HUANG • *Department of Molecular Genetics and Biochemistry, University of Pittsburgh School of Medicine, Pittsburgh, PA*

STEPHEN C. HYDE • *The UK Cystic Fibrosis Gene Therapy Consortium and Gene Medicine Group, University of Oxford, Oxford, UK*

JOHN IACOMINI • *Transplantation Research Center, Renal Division Brigham and Women's Hospital and Children's Hospital Boston, Harvard Medical School, Boston, MA*

LUCIENNE M. IDE • *Aflac Cancer Center and Blood Disorders Service, Division of Hematology/Oncology and Bone Marrow Transplantation, Department of Pediatrics, Emory University School of Medicine, Atlanta, GA*

ELISABETH JAVAZON • *Aflac Cancer Center and Blood Disorders Service, Division of Hematology/Oncology and Bone Marrow Transplantation, Department of Pediatrics, Emory University School of Medicine, Atlanta, GA*

DAVID H. KIRN • *Jennerex Biotherapeutics Inc., San Francisco, CA, and Department of Clinical Pharmacology, University of Oxford, Oxford, UK*
ROBERT M. KOTIN • *Laboratory of Biochemical Genetics, National Heart, Lung, and Blood Institute, Bethesda, MD*
DAVID M. KRISKY • *Department of Molecular Genetics and Biochemistry, University of Pittsburgh School of Medicine, Pittsburgh, PA*
YOUNG JIK KWON • *Department of Biomedical Engineering, Case Western Reserve University, Cleveland, OH*
YI LAI • *Department of Molecular Microbiology and Immunology, University of Missouri, Columbia, MO*
JOSEPH M. LE DOUX • *Department of Biomedical Engineering, Georgia Institute of Technology and Emory University, Atlanta, GA*
DONGWON LEE • *College of Medicine, Department of Internal Medicine, University of South Florida, Tampa, FL*
PEDRO LEI • *Department of Chemical and Biological Engineering, University at Buffalo, The State University of New York, Buffalo, NY*
DEJIA LI • *Department of Molecular Microbiology and Immunology, University of Missouri, Columbia, MO*
MINGJIE LI • *Division of Molecular Biology, Beckman Research Institute of the City of Hope, Duarte, CA*
TA-CHIANG LIU • *Brain Tumor Research Center, Massachusetts General Hospital and Harvard Medical School, Boston, MA*
GORDON J. LUTZ • *Basic and Clinical Myology Laboratory, Department of Physiology, The University of Melbourne, Victoria, Australia*
GORDON S. LYNCH • *Basic and Clinical Myology Laboratory, Department of Physiology, The University of Melbourne, Victoria, Australia*
SHYAM S. MOHAPATRA • *College of Medicine, Department of Internal Medicine, University of South Florida, Tampa, FL*
FELIX M. MUNKONGE • *Department of Gene Therapy, Faculty of Medicine, Imperial College London, UK, and The UK Cystic Fibrosis Gene Therapy Consortium, London, UK*
ALEJANDRO NEGRETE • *Laboratory of Biochemical Genetics, National Heart, Lung, and Blood Institute, Bethesda, MD*
PHILIP NG • *Department of Molecular and Human Genetics, Baylor College of Medicine, Houston, TX*
DONNA J. PALMER • *Department of Molecular and Human Genetics, Baylor College of Medicine, Houston, TX*
JENNIFER E. PHILLIPS • *Coulter Department of Biomedical Engineering, Georgia Institute of Technology and Emory University, Atlanta, GA*

DAVID PORTEOUS • *The UK Cystic Fibrosis Gene Therapy Consortium and Medical Genetics, Molecular Medicine Centre, University of Edinburgh, Edinburgh, UK*
RAJAGOPAL RAMESH • *Department of Thoracic and Cardiovascular Surgery, The University of Texas M.D. Anderson Cancer Center, Houston, TX*
LILITH REEVES • *Division of Experimental Hematology, Children's Hospital Medical Center, University of Cincinnati, Cincinnati, OH*
JOHN J. ROSSI • *Division of Molecular Biology, Beckman Research Institute of the City of Hope, Duarte, CA*
CYRUS R. SAFINYA • *Materials Department, Physics Department and Molecular, Cellular, and Developmental Biology Department, University of California, Santa Barbara, CA*
JONATHAN D. SCHERTZER • *Basic and Clinical Myology Laboratory, Department of Physiology, The University of Melbourne, Victoria, Australia*
SHASHANK R. SIRSI • *Department of Pharmacology and Physiology, Drexel University College of Medicine, Philadelphia, PA*
STEPHEN N. SMITH • *Department of Gene Therapy, Faculty of Medicine, Imperial College London, UK, and Gene Medicine Group, University of Oxford, Oxford, UK*
H. TRENT SPENCER • *Aflac Cancer Center and Blood Disorders Service, Division of Hematology/Oncology and Bone Marrow Transplantation, Department of Pediatrics, Emory University School of Medicine, Atlanta, GA*
STEPHANIE SUMNER-JONES • *The UK Cystic Fibrosis Gene Therapy Consortium and Gene Medicine Group, University of Oxford, Oxford, UK*
STEPHEN H. THORNE • *Department of Pediatrics, Stanford University School of Medicine, Stanford, CA*
CHAORUI TIAN • *Transplantation Research Center, Renal Division, Brigham and Women's Hospital and Children's Hospital Boston, Harvard Medical School, Boston, MA*
JAMES B. WECHUCK • *Department of Molecular Genetics and Biochemistry, University of Pittsburgh School of Medicine, Pittsburgh, PA*
JASON H. WILLIAMS • *Basic and Clinical Myology Laboratory, Department of Physiology, The University of Melbourne, Victoria, Australia*
SHAHRIAR SHAH YAGHOUBI • *Molecular Imaging Program at Stanford (MIPS) and Bio-X Program, Departments of Radiology and Bioengineering, School of Medicine, Stanford University, Stanford, CA*
HONG YU • *Department of Surgery, Vascular Biology Institute, University of Miami School of Medicine, Miami, FL*
YONGPING YUE • *Department of Molecular Microbiology and Immunology, University of Missouri, Columbia, MO*

1

Preparation and Quantification of Pseudotyped Retroviral Vector

Hong Yu and Young Jik Kwon

Summary

Retroviral vectors have been widely used for research and clinical trials in gene therapy because of their high transduction efficiency. Retroviruses interact with target cells through their surface molecules (i.e., envelope proteins) and cellular receptors, which limit the susceptibility of target cells to retroviral vectors. Murine leukemia retrovirus (MuLV) pseudotyped with vesicular stomatitis virus G glycoprotein (VSV-G) overcomes the species barrier and is more resistant to mechanical and biochemical inactivation. A cell line producing VSV-G pseudotyped MuLV vector can be established by transfecting 293T cells expressing Gag, Pol, and VSV-G (293 GPG cell line) with a retroviral vector plasmid. Transduction potency of the resulting VSV-G pseudotyped MuLV retroviral supernatant can be quantified by titration, electron microscopy (EM), and the reverse transcriptase (RT) assay. These protocols provide methods to prepare and quantify a pseudotyped retroviral vector with high transduction rates for most types of target cells.

Key Words: Viral vector; pseudotyped retrovirus; vesicular stomatitis virus G protein (VSV-G); titration; electron microscopy; RT assay.

1. Introduction

Retroviral vectors derived from murine leukemia retrovirus (MuLV) have been widely used for efficient gene transfer to achieve long-term expression of a chosen therapeutic gene in mammalian cells *(1)*. Disadvantages of this vector are the instability and low viral titers generated from packaging cells, low efficiency gene transfer into human cells, especially in vivo, and the requirement for dividing of target cells. Some studies have attempted to increase

From: *Methods in Molecular Biology, vol. 433: Volume 1: Production and In Vivo Applications of Gene Transfer Vectors*
Edited by: J. M. Le Doux © Humana Press, Totowa, NJ

the transduction efficiency by using strategies such as low-speed centrifugation of viral supernatant with cells, multiple viral exposures *(2)*, or increasing viral titers by ultracentrifugation *(3)* and were able to produce an average transduction efficiency of 10–60%. However, all these improvements of transduction efficiency require additional and practically inefficient procedures.

The envelope G glycoprotein from the vesicular stomatitis virus (VSV-G) has been used to construct a pseudotyped MuLV with significant improvement in stability and transduction efficiency *(4,5)*. The VSV-G pseudotyped MuLV (VSV-G/MuLV) can be concentrated to titers exceeding 10^9 colony formation units (cfu)/ml through ultracentrifugation with minimal loss of infectivity *(6)*. This pseudotyped vector has a much broader host range than the vectors having the conventional amphotropic Env and been successfully used to transfer genes into human peripheral blood lymphocytes *(7,8)*, leukocytes *(9)*, hepatocytes *(10)*, vascular tissues *(11)*, and stem cells *(12)*. Recently, VSV-G envelope protein has been modified to achieve easy purification *(13)* and temperature-controlled replication *(14)*. This chapter will discuss the principles and procedures to generate and to quantify such pseudotyped retroviral vectors using methods such as titration of transduction activity, direct physical particle counting under electron microscopy (EM), and the reverse transcriptase (RT) assay.

1.1. MuLV Pseudotyped with VSV-G Protein

Retrovirus contains two single-stranded RNA molecules associated with *gag* proteins and *pol* protein in a core structure. Estimated 100–300 *env* proteins protrude from the lipid bilayer of the particles. The packaging cells are genetically engineered to supply the *env*, *pol*, and *gag* sequences that encode the structural proteins necessary for the formation of the viral particle. Recombinant retroviral vectors are usually produced from a packaging cell line transfected with a retroviral vector DNA that is transcribed inside the cell (producer cell line), and the resulting RNA is recognized by the structural proteins and packaged into retroviral particles that are bud off from the plasma membrane. More detail of retroviral vector can be obtained from the other publications *(1,15,16)*.

The MuLV virus has low transduction efficiency and is unstable. The envelope protein, a key component for efficient transduction by specific virus/cell interactions, is composed of two subunits: soluble unit and transmembrane unit. The Env proteins interact with protein receptors on the cell surface to initiate internalization of the virus. The soluble unit of the envelope is easy to be shed off from the viral particle resulting in inactive Env protein. The labile structural characteristics of the envelope protein are thought to be a main reason for the unstable retrovirus ***(17)***. Therefore, modification of envelope components might result in a more stable particle.

VSV-G, an envelope protein from VSV, is a single-chain protein. Emi et al. ***(18)*** demonstrated that the VSV-G protein could completely replace the env protein of MuLV and produce an infectious MuLV-based viral vector, which is called VSV-G pseudotyped MuLV. VSV-G has been demonstrated to interact with phospholipid components of the plasma membrane ***(19–21)***. As virus entry seems not to be dependent on the presence of specific protein receptors, VSV-G pseudotyped MuLV has broader host range than that of traditional retrovirus.

1.2. Packaging Cell for VSV-G Pseudotyped MuLV

The production of VSV-G pseudotyped MuLV was originally produced from transient transfection of 293T cell line with three plasmids: gag-pol expression plasmid, VSV-G expression plasmid, and retroviral vector plasmid (*see* **Fig. 1**). Human 293T, a subline of Ad5-transformed embryonic kidney cell line 293, is used because they are more transfectable than NIH 3T3 cells from which the original retroviral producer cell lines were derived. In addition, the 293T cell line contains the simian virus 40 (SV40) large tumor antigen whose expression may increase the replication of vectors containing the SV40 origin of replication.

Because VSV-G protein is toxic to cells, stable expression of the protein was not achieved, and a packaging cell line had not been available until an inducible promoter was applied. The stable packaging cell lines for the production of VSV-G pseudotyped MuLV have been developed by applying a tetracycline-modulated promoter ***(22,23)*** or a recombination required system ***(24)*** for the controlled expression of VSV-G to minimize its associated toxicity.

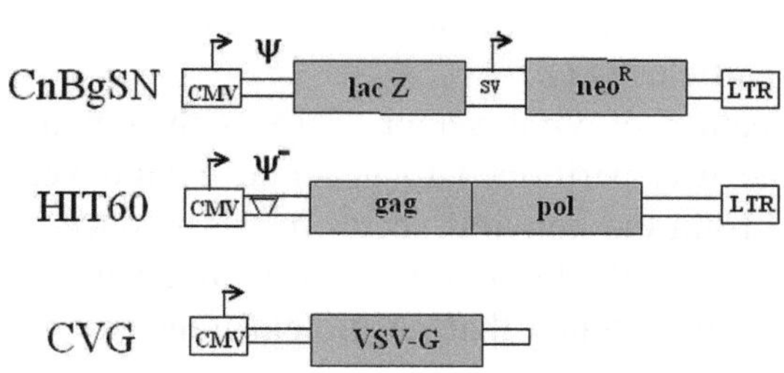

Fig. 1. Schematic representation of plasmids used for the production of viral vectors. CnBgSN is a viral vector plasmid carrying the *lacZ* report gene and *neo*^{r} drug-resistant gene. HIT60 is plasmid carrying *gag* and *pol* viral genes. CVG is the plasmid for VSV-G envelope expression.

In the packaging cell lines, the gag-pol genes are constantly expressed. In the tetracycline-regulatable promoter system, the expression of *VSV-G* gene is inhibited in the presence of tetracycline. Removal of tetracycline from the culture medium allows high expression of VSV-G and production of infectious viral particles. In the recombination-required system, the expression of VSV-G is completely silent because the RNA transcript terminates before VSV-G coding sequence. By introducing Cre recombinase through adenovirus vector, the RNA transcript terminates are excised through a site-specific recombination, and VSV-G can be transcribed ***(24)***. Here we have only used the tetracycline-modulated packaging cell line to generate stable producer cells for the production of VSV-G pseudotyped MuLV vectors.

1.3. Transduction Efficiency

Transduction efficiency of VSV-G/MuLV is ranged 70–95%, much higher than that of the traditional amphotropic MuLV vector (10–50%). At the same multiplicity of infection (MOI), the transduction efficiency of VSV-G/MuLV still remained about fourfold higher than that of amphotropic MuLV ***(11)***. This high transduction efficiency could be a result of several factors: more stable VSV-G glycoprotein and more receptors available for virus binding that potentiates binding efficiency of the VSV-G pseudotyped virus to the cells.

By using the VSV-G pseudotyped vector, we may omit the step of selection of the transduced cells because most cells will be transduced with the vector. This can significantly reduce the amount of time required for the application of gene therapy.

1.4. Pseudotransduction

VSV-G pseudotyped MuLV-mediated gene transfer has been reported to cause pseudotransduction (protein transfer) ***(7,10)***. However, we have demonstrated through four different experiments that there was no protein transfer through the viral particles ***(11)***.

First, co-transfection of viral vector DNA with a nonviral vector DNA carrying *lacZ* gene was performed to generate a viral vector. The proteins (β-Gal) expressed from the nonviral vector in the producer cells could not be transferred to recipient cells through the viral particles generated from the co-transfection. This demonstrated that no protein transfer occurred through the viral vector.

Second, transduction was correlated with viral RT activity. When the activity of RT was blocked during transduction, targeted protein expression was not detected in the recipient cells, therefore indicating that the transduction is a result of reverse transcription from RNA into DNA.

Third, gene transfer using the MuLV vector requires cells to be dividing *(25)*, while protein transfer is independent of cell proliferation. When the proliferation of recipient cells was arrested with γ-irradiation, no cells were able to be transduced. The correlation of transduction and cell proliferation further supports the conclusion that no protein transfer occurs with the VSV-G pseudotyped vector.

Fourth, gene expression resulted from the transduction was stable. For pseudotransduction, the amount of transferred protein would be limited, and its protein activity is transient. Thus, after several passages, the population of cells retaining the transferred protein would decrease. On the other hand, the expression of the protein resulting from the transferred gene would be stable, and the population of the cells possessing the gene would not decrease with passage. High protein expression of the transduced gene was detected following nine passages after transduction, indicating that the protein detected in the transduced cells truly arose from gene expression.

The viral supernatant reported with pseudotransduction was processed through ultracentrifugation while we used the supernatant directly collected from cell culture without concentration. Perhaps the ultracentrifugation altered the pseudotyped vector leading to pseudotransduction.

1.5. Quantification of VSV-G Pseudotyped MuLV

Quantitation of viral vectors can be carried out by measuring transduction efficiency (titration), viral components (matrix protein or RT), or counting physical particles. Currently, the titration method, which is performed by overlaying viral supernatant onto target cells (e.g., NIH 3T3 cells) after serial dilutions, is widely used to represent infectious viral concentration expressed as number of colonies formation per volume (i.e., cfu/ml). However, this method represents only the retroviral particles that successfully transduce target cells *(26)* and usually underestimates the true viral concentration. Unlike the titration method, counting viral particles through EM and measurement of a viral component (e.g., RT) can provide the number of virus particles in the supernatant *(27)*. However, the total number of viral particles derived from these methods cannot distinguish infectious from noninfectious viral particles. Direct counting using EM cannot be used to quantify viral concentration of directly harvested supernatant before the supernatant is concentrated and purified because viral concentration of directly collected supernatant is too low to have enough particles to be viewed on electron micrographs. RT assay can be used to quantify concentration of unconcentrated supernatant but only after RT activity of a standard viral particle solution is correlated with EM counting of a concentrated and purified viral stock solution. Therefore, it is important to quantify infectious retroviral concentration using multiple methods. For example, we have concentrated a batch of supernatant (1000 ml) of VSV-G pseudotyped MuLV vector collected from the producer cells to 1 ml through centrifugation and purified it through a gradient centrifugation ***(1)***. The particle number per titer determined by EM counting and RT activity assay decreased from 847 to 128, which indicates less inactive particle present in concentrated and purified viral solutions (*see* **Note 1**).

2. Materials

2.1. Cell Culture

1. Human 293 cells. Human 293T/17 cell line can be obtained from America Type Culture Collection (CRL11268) ***(28)***. They are maintained in Dulbecco's modified Eagle's medium (DMEM) (Gibco BRL, Gaithersburg, MD) supplemented with 10% fetal bovine serum (FBS) and 2 mM glutamine (Gibco BRL).
2. 293/GPG cell line, a VSV-G pseudotyped MuLV packaging cell line ***(23)*** is maintained in DMEM described above with additional 1 μg/ml tetracycline (Sigma, St. Louis, MO), 2 μg/ml puromycin (Sigma), 0.3 mg/ml G418 (Gibco BRL), and 1 mM sodium pyruvate.
3. VSV-G pseudotyped MuLV producer cells are also maintained in the same medium as for the packaging cells with tetracycline. To produce virus, the cells are cultured in the production medium, which is the same as the maintain medium but without tetracycline, puromycin, nor G418. All cells were maintained in a humidified 37 °C incubator with 5% CO_2.
4. Phosphate-buffered saline (PBS): 0.9% NaCl, 0.0144% KH_2PO_4, 0.0795% $K_2HPO_4 \cdot 7H_2O$ (all w/v), pH 7.2.
5. Trypsin–EDTA: 0.05% Trypsin, 0.53 mM EDTA·4Na, stored at 4°C.
6. Tissue culture dishes (100-mm diameter).

2.2. Plasmids

1. The retroviral vector pCnBgSN ***(29)*** contains a *lacZ* gene encoding nuclear-localized β-galactosidase and a neomycin resistance gene (*neo*^{r}) encoding for neomycin phosphotransferase. The expression of *lacZ* is driven by the hybrid 5′ cytomegalovirus (CMV) long terminal repeat (LTR) promoter, while the *neo*^{r} is controlled by a SV40 promoter (*see* **Fig. 1**).
2. Plasmid pHIT60 is a MoMuLV Gag and Pol expressing plasmid, whose expression is driven by a CMV promoter ***(30)***. Use only JM109 cells to grow this plasmid.
3. Plasmid pCVG is a CMV-driven VSV-G expression vector ***(11)***.
4. All plasmid DNA can be purified from *Escherichia coli* using Qiagen Plasmid Kit (Valencia, CA).

2.3. DNA Transfection

1. 2× HEPES-buffered saline (2× HBS): 50 mM HEPES, pH7.1, 280 mM NaCl, 1.5 mM Na_2HPO_4, 10 mM KCl.
2. 2 M $CaCl_2$.
3. 0.5 M, Sodium butyrate, filtered and stored at –20 °C.
4. Sterile distilled water.
5. Sterile 5-ml conical tubes, 1-ml pipettes.

2.4. Virus Transduction

1. Polybrene (hexadimethrine bromide, H-9268; Sigma) stock solution (8 mg/ml).
2. G418 (neomycin phosphotransferase or Geneticin, GIBCO/BRL) stock solution (50 mg/ml).
3. X-gal staining solution: prepare X-gal in stock (40 mg/ml, Gibco BRL) and dilute just before use in 1:40 with X-gal staining buffer (2 mM $MgCl_2$, 5 mM $K_3Fe(CN)_6$, and 5 mM $K_4Fe(CN)_6$ in PBS).
4. 10% formaldehyde.

2.5. Virus Concetration, Purification, EM, and RT Assay

1. TNE buffer: 0.01 M Tris–HCl, pH 7.2, 0.1 M NaCl, 0.001 EDTA.
2. 10 and 60% sucrose solutions, with a linear gradient making apparatus.
3. Polystyrene particle solution; diameter, 102 ± 3 nm (Duke Scientific Corp., Palo Alto, CA).
4. A copper mesh with 300 grids (Ted Pella, Redding, CA) precoated with Formvar (Fluka, Milwaukee, WI).
5. 1% Uracyl acetate (Fluka).
6. RT cocktail: 15 μg of poly(A)/ml, 7.5 μg of oligo(dT)/ml, 15 μM dTTP, 75 mM Tris (pH 8.3), 30 mM dithiothreitol, 0.9 mM $MnCl_2$, 94 mM NaCl, 0.075% Nonidet P-40, and 15 μCi of α-S^{35}-dTTP (1 Ci/μmol; ICN, Costa Mesa, CA).
7. 2× SSC: 0.3 M NaCl plus 0.03 M sodium citrate.

3. Methods

3.1. Generation of VSV-G Pseudotyped MuLV from a Transient Transfection

VSV-G pseudotyped MuLV (VSV-G/MuLV) vectors can be generated from a transient three-plasmid transfection system ***(30)***. 293T/17 cells are transfected by calcium phosphate precipitation with plasmids pCnBgSN (or any viral vector you have), pHIT60 (for gal-pol), and pCVG (for VSV-G). The viral titers are ranged from 10^6 to 10^7 cfu/ml.

1. Day 1. Prepare DNA and cells. For transfection in one 100-mm plate, 10 μg of each of the three plasmids is needed: pCnBgSN, pHIT60, and pCVG. Calculate

the volume and pipette all required volume into a sterile microfuge tube. Add H_2O to make the total volume 100 µl. Add 1/10 volume (10 µl) of 3 M sodium acetate, 2.6 times volume (260 µl) of absolute ethanol. Place at –20 °C overnight.

2. Trypsinize the 293T/17 cells (80% confluent) on a 100-mm tissue culture dish and pass them in 1:4 dilution. Culture for overnight (15–18 h) at 37 °C.
3. Day 2. Cells should be about 60% confluent. Change medium on the plate with fresh 10 ml medium and keep culturing. Perform transfection within 4 h after the medium change.
4. About 2 h after medium change, centrifuge DNA at 12,000 × *g* for 20 min at 4 °C in a microfuge. The pellet of DNA may be washed with 70% ethanol. Remove the supernatant, and air dry the pellet for about 10 min inside a tissue culture hood. The pellet should appear clear at this point.
5. Add 438 µl of H_2O to completely dissolve the DNA pellet. Transfer the solution to a 5-ml sterile polypropylene tube.
6. Add 62 µl of 2 M $CaCl_2$ into the tube containing the DNA. This should be done slowly drop by drop. Do not mix. Leave it in the hood 5 min.
7. Using a sterile 1-ml plastic pipette, take up 500 µl of 2× HBS. Insert the pipette into the DNA/$CaCl_2$ mixture until the tip is just above the bottom of the tube but not touching it. Very gently, add the HBS to the bottom, followed by blowing 30 bubbles. Do not mix. Cap the tube and leave in the hood for 30 min for precipitate to form. A fine milky precipitate should be seen.
8. Using a 1-ml pipette, gently pipette up and down the DNA precipitate a few times. Add the precipitate to the 293T cell plate drop by drop while rocking the plates. Incubate at 37 °C for overnight.
9. Day 3 morning. Wash the cells with 5 ml pre-warmed PBS to remove the precipitate. The cells are cultured in 6 ml fresh medium containing 10 mM sodium butyrate. In the evening, after 8–12-h incubation with sodium butyrate, the supernatant is replaced with 6 ml fresh DMEM medium, and the cells are incubated overnight.
10. Day 4 morning. The supernatant is collected, filtered with 0.45-µm filter, and aliquoted. The cell plate can be continually cultured with 6 ml fresh medium and the supernatant be harvested after about 24 h on day 4. The supernatants can be used for various assays or stored at –70 °C. The supernatant should have virus about 10^6–10^7 cfu/ml (*see* **Note 2**).

3.2. Generation of VSV-G Pseudotyped MuLV Producer Cell Line

The viral supernatants from the transient transfection are used to transduce packaging cell line 293/GPG to generate VSV-G/MuLV producer cell line. Because 293/GPG has already the G418 resistant marker gene, G418 selection cannot be used to select transduced cells. However, as the transduction efficiency of VSV-G pseudotyped MLV vector is higher than 90%, the pools of the transduced 293/GPG, without selection, can be used directly as stable producer cells.

1. Day 1. Grow 293/GPG cells on 100-mm tissue culture dishes in DMEM with tetracycline, until 80% confluent. Trypsinize the cells and pass them in 1:5 dilution. Culture for overnight (15–18 h) at 37 °C.
2. Day 2. Cells should be about 40–50% confluent. Thaw VSV-G/MuLV supernatant from transient transfection at 37 °C water bath (require 3 ml per 100-mm plate), add polybrene (1:1000) to final concentration 8 μg/ml. Remove medium in the cell culture plate. Add the 3 ml supernatant into the plate and culture for 2 h at 37 °C.
3. Add 10 ml DMEM with tetracycline into the virus–cell mixture plate. Culture for overnight.
4. Day 3. Replace the medium with 10 ml fresh medium with tetracycline. Culture for 2–3 days till confluent. This is a pool of producer cells. To obtain pure population, see **Note 3** for single colony isolation.
5. Split the cells into five plates. Freeze the cells for later use or change medium into viral production medium to produce virus.

3.3. Generation of VSV-G Pseudotyped MuLV from a Producer Cell Line

1. Grow the producer cells (293/GPG/CnBgSN) to 80% confluent in maintain medium.
2. Remove the medium and wash the cells with PBS once. Replace the medium with 6 ml production medium (without tetracycline), culture for overnight.
3. Collect the supernatant in 24 h and replace with fresh 6 ml production medium. This can be repeated for 3–4 times until the most of cells are detached and float up. This first collection has low virus titer (10^4–10^5 cfu/ml) and could be discarded. The following collections are filtered through a 0.45-μm filter and aliquoted. The supernatant should have virus about 10^6–10^7 cfu/ml.

3.4. Concentration and Purification of VSV-G Pseudotyped MuLV

1. Centrifuge the viral supernatant (100–500 ml) at 9500 rpm (13,800×*g*) in Beckman rotor JA-14 for 12 h.
2. The pellet is resuspended in less than 1 ml TNE buffer, and this will result in 100–1000-fold concentration. This virus can be directly used.
3. The concentrated virus can be further purified by centrifugation through a 10–60% linear sucrose gradient in a Beckman SW-40 rotor at 30,000 rpm (111,000 × g), for 2 h at 20 °C.
4. The virus forms a white milky band viewable with a black background, which is recovered using a syringe punching through the centrifuge tube.
5. The recovered virus (1–2 ml) is suspended in 10 ml TNE buffer. The solution is pelleted by centrifugation at 17,000 rpm (39,800 × *g*) in Beckman rotor JA-17 for 2 h at 4 °C. The pellet is resuspended in TNE buffer at 1/1000 of original supernatant volume.

3.5. Virus Titer and Transduction Efficiency Assay

The viral titer is analyzed by either G418-resistant colony formation or β-galactosidase activity assay, which generates similar results. The transduction efficiency is measured by comparing β-galactosidase positive and negative cells among transduced cells. See **Note 4** for the effect of freezing-thawing processes on vital titer.

1. Day 1. Plate NIH 3T3 cells (or your target cells for transduction efficiency assay) in 30-mm diameter wells of a 6-well plate (5×10^4 cells/well).
2. Day 2. Mix viral supernatant with polybrene (final concentration 8 μg/ml). For titration, do series 1:10 dilution (up to 10^{-6}) of the viral supernatant with DMEM containing polybrene.
3. Remove culture medium on cells plate. Add 1 ml viral supernatant to a well. For titration, use 10^{-1}–10^{-6} dilutions to six corresponding wells. For transduction efficiency, use the undiluted supernatant. Culture for 2 h at 37 °C, followed by the addition of 2 ml of fresh medium.
4. Day 3. Change the medium with 3 ml DMEM containing 0.8 mg/ml G418 for G418 selection assay or without G418 for β-galactosidase assay.
5. For X-Gal staining, the cells are cultured for another 48 h and then fixed with 0.5% glutaraldehyde in PBS followed by three washes with PBS for 10 min each time at room temperature. The cells are then stained for β-galactosidase activity by incubating overnight at 37 °C with 2 ml X-Gal staining solution. A cluster of blue cells is counted as one colony under an inverted microscope.
6. For G418 selection, the cells are cultured with medium containing 0.8 mg/ml G418 for 7 days. After 7 days, the media are removed, and cells are fixed with 10% formaldehyde followed by three times PBS wash. The cells on plate are stained with 1 ml of 1% methylene blue (Sigma) in methanol for 5 min and washed with tap water. The resistant colonies in blue are counted. The titer (cfu/ml) is calculated as numbers of colonies/(dilution factor × virus volume added).
7. For transduction efficiency assay, the cells are stained with X-gal as described above. Transduction efficiency was calculated by the ratio of blue cells (transduced) to the total cells contained in 10 randomly picked high power fields (200×).

3.6. Viral Particle Counting using Electron Microscopy

1. Mix 50 μl of the purified virus stock obtained in 3.4 with an equal volume of a polystyrene particle solution with a known concentration.
2. Drop 10 μl of the mixture on a copper mesh with 300 grids precoated with Formvar and incubate it on parafilm sheet at room temperature for 30 min.
3. The grid was stained with 1% uracyl acetate solution for 5–30 min and washed with water. Darkness of electron micrograph is determined by staining time.

4. The viral particles were examined under Teknai transmission electron microscope (TEM), and both polystyrene beads and viral particles on at least 50 grids were counted (*see* **Fig. 2**).
6. The viral particle concentration of the viral stock was then determined from the ratio of the viral particle number to the polystyrene bead number.

3.7. RT Assay

1. Prepare RT cocktail: 15 μg of poly(A)/ml, 7.5 μg of oligo(dT)/ml, 15 μM dTTP, 75 mM Tris (pH 8.3), 30 mM dithiothreitol, 0.9 mM $MnCl_2$, 94 mM NaCl, 0.075% Nonidet P-40, and 15 μCi of α-S^{35}-dTTP.
2. Mix 10 μl of virus sample with 20 μl of the RT cocktail.
3. Incubate the mixture for 1 h at 37 °C and inactivate the reaction at 95 °C for 5min.
4. Transfer an aliquot (10 μl) of the reaction solution to Whatman DE81 filter paper.
5. Air dry the filter paper, wash it with 2× SSC (0.3 M NaCl plus 0.03 M sodium citrate), rinse it with ethanol, and dry it again.
6. Transfer the filter paper into a vial, add 3 ml of scintillation fluid, and count the radioactivity with a Beckman LS 8000 scintillation counter.
7. Linearly proportionality of radioactivity and concentration of viral particles determined by EM will be obtained. Concentration of an unknown sample will be quantified by a correlated RT activity.

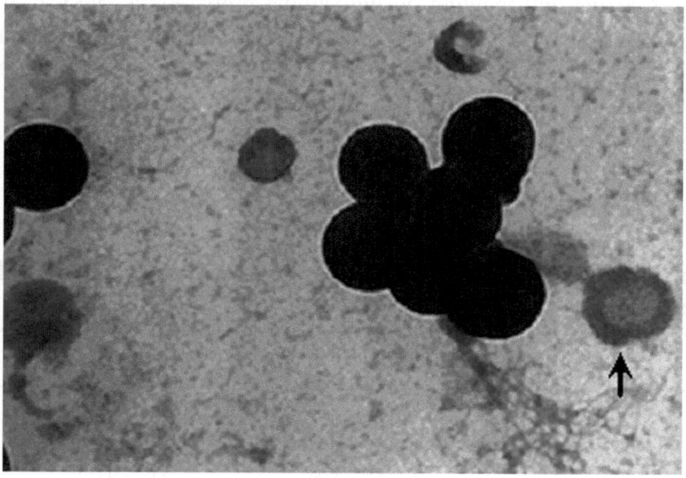

Fig. 2. Electron micrograph of VSV-G pseudotyped MuLV with polystyrene particles. Viral particles with envelope were visualized by transmission electron microscopy (arrow), and the relative concentration of viral particles is determined by comparing the number of polystyrene particles shown as dark spheres.

4. Notes

1. Usually order of retroviral concentrations determined by EM and RT assays is two orders of magnification higher than titer. As free RT, which is not incorporated to viral particles, is counted by RT assay, retroviral concentration of supernatant determined by RT assay is higher than the one by EM. However, there is not significant difference in concentrated and purified viral particles. More detailed quantitative correlations with titer and particles count by EM and RT assays were summarized in **Table 1**.
2. To check the transfection efficiency in **Subheading 3.1**, the 293 cells can be fixed and X-gal stained to measure the percentage of blue cells after collection of supernatant is finished.
3. If the transduction efficiency cannot reach higher than 85% during production of a stable producer cell line, the titer of virus from thus generated stable producer cells could be low. At this situation, a single colony should be isolated. The pool of the transduced cells can be diluted to 1 cell/ml, plate 0.25 ml of the cell suspension on a 96-well plate, and culture it till the single cell in some wells grows into single colony. Pipette up and down the medium to dissociate the cells and pass them into 24-well plate and then duplicate the pass into two 6-well plate. In one 6-well plate, the cells will be cultured in virus production medium, and the supernatant will be collected and the titer be checked. The plate will be discarded after the viral production. The cells that generate the highest titer in the corresponding well of the other plate will be passed to a 100-mm plate and used to produce viral supernatant in large quantity.
4. One of the reasons of the remarkable discrepancy between titer and EM or RT assay is freezing and thawing during sample preparations. VSV-G pseudotyped MuLV lost 30% of its infectious activity after the first two thawing-freezing processes while no further loss of the infectivity was observed after the third freezing-thawing and thereafter (*see* **Fig. 3**). It was reported that ecotropic MuLV lost 80% of its infectivity after three repeats of freezing-thawing *(26)*. This

Table 1
Comparison of Viral Concentration Determined by Titration, EM, and RT Activity Assay

Samples[a]	Volume (ml)	Titer (cfu/ml)	C_p^b (particles/ml)	C_p/titer
FV	1000	4.67×10^6	3.95×10^9	847
CV	1	5.00×10^8	8.25×10^{10}	165
PV	1	1.73×10^8	2.21×10^{10}	128

[a]Fresh viral supernatant (FV), concentrated viral supernatant (CV), and purified viral supernatant (PV).

[b]Retroviral particle number (C_p) of CV and PV was determined by counting viral particles under EM. C_p of FV was determined by RT activity assay and normalized with purified viral suspension with known viral particle number.

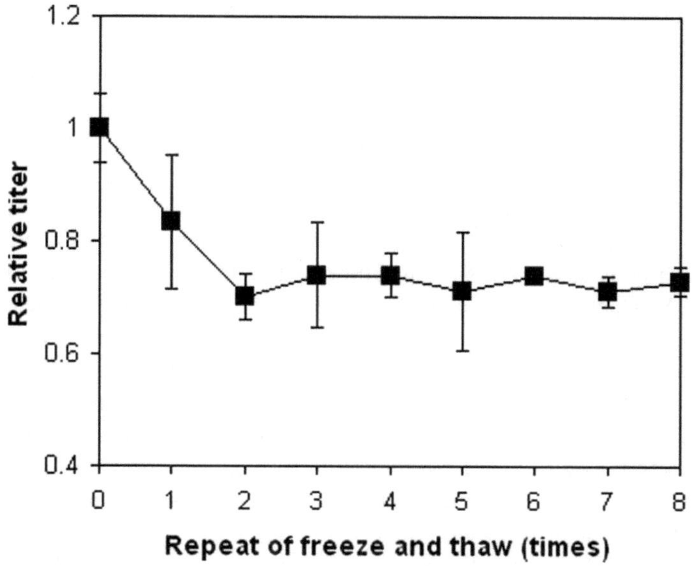

Fig. 3. Effect of freeze-thaw on VSV-G MuLV infectivity. After produced from 293GPG/lacZ cells, the supernatant was stored at –80 °C and thawed. This freezing and thawing process was repeated for additional seven times. Each time after the thaw, the viral supernatant was titered, and the titer was compared to the titer of freshly collected supernatant to generate relative titer.

clarified that pseudotyping with VSV-G increases stability of MuLV against freezing and thawing process.

References

1. Miller, A. D., Miller, D. G., Garcia, J. V., and Lynch, C. M. (1993) Use of retroviral vectors for gene transfer and expression. *Methods Enzymol* **217**, 581–99.
2. Inaba, M., Toninelli, E., Vanmeter, G., Bender, J. R., and Conte, M. S. (1998) Retroviral gene transfer: effects on endothelial cell phenotype. *J Surg Res* **78**, 31–6.
3. Zelenock, J. A., Welling, T. H., Sarkar, R., Gordon, D. G., and Messina, L. M. (1997) Improved retroviral transduction efficiency of vascular cells in vitro and in vivo during clinically relevant incubation periods using centrifugation to increase viral titers. *J Vasc Surg* **26**, 119–27.
4. Friedmann, T., and Yee, J. K. (1995) Pseudotyped retroviral vectors for studies of human gene therapy. *Nat Med* **1**, 275–7.
5. Schnell, M. J., Buonocore, L., Kretzschmar, E., Johnson, E., and Rose, J. K. (1996) Foreign glycoproteins expressed from recombinant vesicular stomatitis viruses are incorporated efficiently into virus particles. *Proc Natl Acad Sci USA* **93**, 11359–65.
6. Burns, J. C., Friedmann, T., Driever, W., Burrascano, M., and Yee, J. K. (1993) Vesicular stomatitis virus G glycoprotein pseudotyped retroviral vectors: concen-

tration to very high titer and efficient gene transfer into mammalian and nonmammalian cells [see comments]. *Proc Natl Acad Sci USA* **90**, 8033–7.

7. Gallardo, H. F., Tan, C., Ory, D., and Sadelain, M. (1997) Recombinant retroviruses pseudotyped with the vesicular stomatitis virus G glycoprotein mediate both stable gene transfer and pseudotransduction in human peripheral blood lymphocytes. *Blood* **90**, 952–7.
8. An, D. S., Koyanagi, Y., Zhao, J. Q., Akkina, R., Bristol, G., Yamamoto, N., Zack, J. A., and Chen, I. S. (1997) High-efficiency transduction of human lymphoid progenitor cells and expression in differentiated T cells. *J Virol* **71**, 1397–404.
9. Sharma, S., Cantwell, M., Kipps, T. J., and Friedmann, T. (1996) Efficient infection of a human T-cell line and of human primary peripheral blood leukocytes with a pseudotyped retrovirus vector. *Proc Natl Acad Sci USA* **93**, 11842–7.
10. Liu, M. L., Winther, B. L., and Kay, M. A. (1996) Pseudotransduction of hepatocytes by using concentrated pseudotyped vesicular stomatitis virus G glycoprotein (VSV-G)-Moloney murine leukemia virus-derived retrovirus vectors: comparison of VSV-G and amphotropic vectors for hepatic gene transfer. *J Virol* **70**, 2497–502.
11. Yu, H., Eton, D., Wang, Y., Kumar, S., Tang, L., Terramani, T., Benedict, C., Hung, G., and Anderson, W. (1999) High efficiency in vitro gene transfer into vascular tissues using a pseudotyped retroviral vector without pseudotransduction. *Gene Ther* **6**, 1876–83.
12. Jang, J. E., Shaw, K., Yu, X. J., Petersen, D., Pepper, K., Lutzko, C., and Kohn, D. B. (2006) Specific and stable gene transfer to human embryonic stem cells using pseudotyped lentiviral vectors. *Stem Cells Dev* **15**, 109–17.
13. Yu, J. H., and Schaffer, D. V. (2006) Selection of novel vesicular stomatitis virus glycoprotein variants from a peptide insertion library for enhanced purification of retroviral and lentiviral vectors. *J Virol* **80**, 3285–92.
14. Guibinga, G. H., Hall, F. L., Gordon, E. M., Ruoslahti, E., and Friedmann, T. (2004) Ligand-modified vesicular stomatitis virus glycoprotein displays a temperature-sensitive intracellular trafficking and virus assembly phenotype. *Mol Ther* **9**, 76–84.
15. Yee, J. K. (1999) Retroviral vectors. In *The Development of Human Gene Therapy* (Friedmann, T., Ed.), pp. 21–45, Cold Spring Harbor Laboratory Press, Cold Spring Harbor.
16. Swaney, W. P., Novelli, E. M., Bahnson, A. B., and Barranger, J. A. (2002) Retrovirus-mediated gene transfer to human hematopoietic stem cells. In: *Gene Therapy Protocols* (J. R. Morgan, Ed.), 2nd Edition, pp. 187–202, Humana Press, Totowa.
17. Gliniak, B. C., Kozak, S. L., Jones, R. T., and Kabat, D. (1991) Disulfide bonding controls the processing of retroviral envelope glycoproteins. *J Biol Chem* **266**, 22991–7.
18. Emi, N., Friedmann, T., and Yee, J. K. (1991) Pseudotype formation of murine leukemia virus with the G protein of vesicular stomatitis virus. *J Virol* **65**, 1202–7.

19. Schlegel, R., Tralka, T. S., Willingham, M. C., and Pastan, I. (1983) Inhibition of VSV binding and infectivity by phosphatidylserine: is phosphatidylserine a VSV-binding site? *Cell* **32**, 639–46.
20. Mastromarino, P., Conti, C., Goldoni, P., Hauttecoeur, B., and Orsi, N. (1987) Characterization of membrane components of the erythrocyte involved in vesicular stomatitis virus attachment and fusion at acidic pH. *J Gen Virol* **68**, 2359–69.
21. Conti, C., Mastromarino, P., Ciuffarella, M. G., and Orsi, N. (1988) Characterization of rat brain cellular membrane components acting as receptors for vesicular stomatitis virus. Brief report. *Arch Virol* **99**, 261–9.
22. Chen, S. T., Iida, A., Guo, L., Friedmann, T., and Yee, J. K. (1996) Generation of packaging cell lines for pseudotyped retroviral vectors of the G protein of vesicular stomatitis virus by using a modified tetracycline inducible system. *Proc Natl Acad Sci USA* **93**, 10057–62.
23. Ory, D. S., Neugeboren, B. A., and Mulligan, R. C. (1996) A stable human-derived packaging cell line for production of high titer retrovirus/vesicular stomatitis virus G pseudotypes. *Proc Natl Acad Sci USA* **93**, 11400–6.
24. Arai, T., Matsumoto, K., Saitoh, K., Ui, M., Ito, T., Murakami, M., Kanegae, Y., Saito, I., Cosset, F. L., Takeuchi, Y., and Iba, H. (1998) A new system for stringent, high-titer vesicular stomatitis virus G protein-pseudotyped retrovirus vector induction by introduction of Cre recombinase into stable prepackaging cell lines. *J Virol* **72**, 1115–21.
25. Miller, D. G., Adam, M. A., and Miller, A. D. (1990) Gene transfer by retrovirus vectors occurs only in cells that are actively replicating at the time of infection [published erratum appears in *Mol Cell Biol* (1992) **12**(1), 433]. *Mol Cell Biol* **10**, 4239–42.
26. Kwon, Y. J., Hung, G., Anderson, W. F., Peng, C.-A., and Yu, H. (2003) Determination of infectious retrovirus concentration from colony-forming assay with quantitative analysis. *J Virol* **77**, 5712–20.
27. Bierley, S. T., Raineri, R., Poiley, J. A., and Morgan, E. M. (1996) A comparison of methods for the estimation of retroviral burden. *Dev Biol Stand* **88**, 163–5.
28. Pear, W. S., Nolan, G. P., Scott, M. L., and Baltimore, D. (1993) Production of high-titer helper-free retroviruses by transient transfection. *Proc Natl Acad Sci USA* **90**, 8392–6.
29. Han, J. Y., Cannon, P. M., Lai, K. M., Zhao, Y., Eiden, M. V., and Anderson, W. F. (1997) Identification of envelope protein residues required for the expended host range of 10A1 murine leukemia virus. *J Virol* **71**, 8103–8.
30. Soneoka, Y., Cannon, P. M., Ramsdale, E. E., Griffiths, J. C., Romano, G., Kingsman, S. M., and Kingsman, A. J. (1995) A transient three-plasmid expression system for the production of high titer retroviral vectors. *Nucleic Acids Res* **23**, 628–33.

2

Production of Retroviral Vectors for Clinical Use

Kenneth Cornetta, Lilith Reeves, and Scott Cross

Summary

Retroviral vectors were the first viral vectors to enter clinical trials and continue to be attractive candidates for applications where integration of the transgene is required. While these vectors are versatile and are used widely in the research setting, large-scale production for human use poses various challenges to insure quality and high titer. Our vector production facility has produced and certified over 20 vectors for clinical use and continues to be challenged to adapt the ever-changing vector technology to a method of production that complies with Good Manufacturing Practice (GMP). We describe two manufacturing methods for producing material for Phase I/II clinical trials and suggest ways for investigators to adapt these methods for multiple applications.

Key Words: Retroviral vectors; gene therapy; packaging cell line; transient transfection; Good Manufacturing Practice.

1. Introduction

Gammaretrovirus-based retroviral vectors (subsequently referred to as "retroviral vectors") are membrane-bound RNA viruses initially derived from murine leukemia viruses. These are the first viral vectors to be developed ***(1–3)*** and the first viral vector to enter clinical trials ***(4)***. The attractive feature of retroviruses is the ability to stably integrate into target cell DNA, becoming a fixed part of the target cell genome and transmitted to all of the cell's progeny. This property makes retroviral vectors attractive gene transfer agents when targeting stem cells, hematopoietic progenitor cells, lymphocytes, and other cells in which a large expansion of the transduced cells is anticipated.

From: *Methods in Molecular Biology, vol. 433: Volume 1: Production and In Vivo Applications of Gene Transfer Vectors*
Edited by: J. M. Le Doux © Humana Press, Totowa, NJ

Compared to other retroviruses, the genome of the murine leukemia viruses are relatively simple and include the *gag* region encoding structural proteins involved with capsid formation, the *pol* region encoding enzymes such as reverse transcriptase (RT) and integrase and the *env* gene encoding a membrane glycoprotein that targets the virions to specific cell receptors. The unique nature of the retroviral life cycle is important when designing vector constructs and generating vector particles. Specifically, retroviral particles package RT and integrase, so that integration occurs in the target without requiring expression of the viral genes. This allows one to delete most of the viral gene sequences and permits substitution with one or more transgenes. Generally, the retroviral long terminal repeats (LTR) required for integration are retained along with a small portion of the *gag* region, because complete deletion of *gag* is associated with a marked decrease in vector titer ***(5)***. Vector design can be a complicated undertaking, and modifications in the LTR can affect the level of gene expression, the risk of replication competent retrovirus (RCR), and the risk of insertional mutagenesis [for review, see Pannell and Ellis ***(6)*** and Hawley *(7)*].

Deletion of the viral genes generates a vector genome that is replication-defective. The challenge for vector production is to generate virion particles that contain the vector genome, RT, integrase, and the other viral proteins without creating contaminating replication-competent virus. The challenge has been met by segregating the vector, the *gag* and *pol* genes, and the viral envelope into three distinct plasmids. The *gag/pol* and *env* genes are also expressed from promoters that lack homology with the LTR. The plasmid constructs have been utilized in two ways to produce clinical grade material through use of the transient transfection method or by generating stable packaging cell lines.

The simplest and quickest method for generating retroviral vectors is to use the transient transfection method. Generally, three plasmids are used to encode the vector, the viral *gag* and *pol* regions, and the envelope gene (*see* **Fig. 1**). These can be introduced using various techniques, such as calcium phosphate transfection, lipofection, or electroporation. Cell lines such as HEK293 and HT1080 cells tend to be the cells of choice due to their high transfection efficiency and ability to produce vector at high titer. An alternative method is to use stable packaging cell lines in which a plasmid expressing *gag* and *pol* along with a plasmid expressing an *env* gene have been incorporated into a cell line (*see* **Fig. 2**). A vector plasmid can then be introduced, and stable transfectants historically were obtained with the aid of a drug selection marker; however, recent trials have moved away from nonessential coding sequences. The population or individual high titer clones can then be used to generate a stable Master Cell Bank (MCB). Over the past 20 years, a number of packaging cell lines have become available that express various envelopes. Options include amphotropic envelopes derived from viruses such as the murine

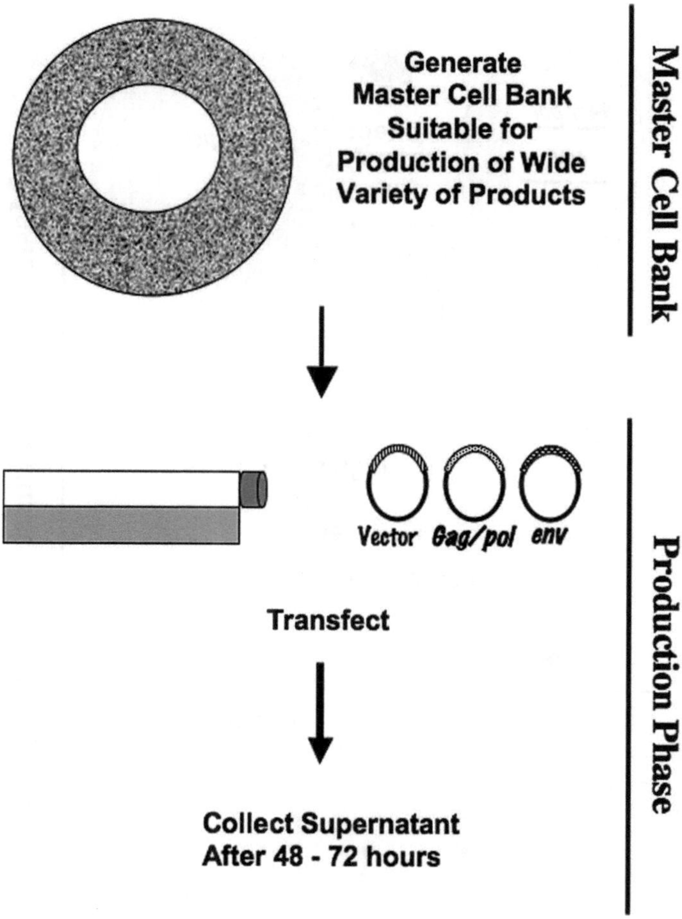

Fig. 1. Vector production by transient transfection and producer cell lines. When producing vector by transient transfection, first a Master Cell Bank (MCB) is generated from a cell that transfects with high efficiency (e.g., HEK293T cells). The MCB is characterized to insure it complies with regulatory requirements. A vial of the cells are then transfected with plasmids expressing the vector, the viral *gag* and *pol* gene, and the viral envelope using calcium phosphate, lipofection or electroporation. Vector supernatant is harvested after 48–72 h later. Although the titer will decrease significantly upon subsequent harvest, it may be appropriate to attempt a second harvest, especially if vector will be concentrated.

4070A and 10A1 viruses, which can bind to receptors on murine, human, and many other cell types. Xenotropic viruses, that do not infect murine cells but can infect human cells, include those derived from the gibbon ape leukemia virus (GALV) and the feline endogenous virus RD114. Vectors pseudotyped with the amphotropic and xenotropic envelopes have been used in clinical trials. More

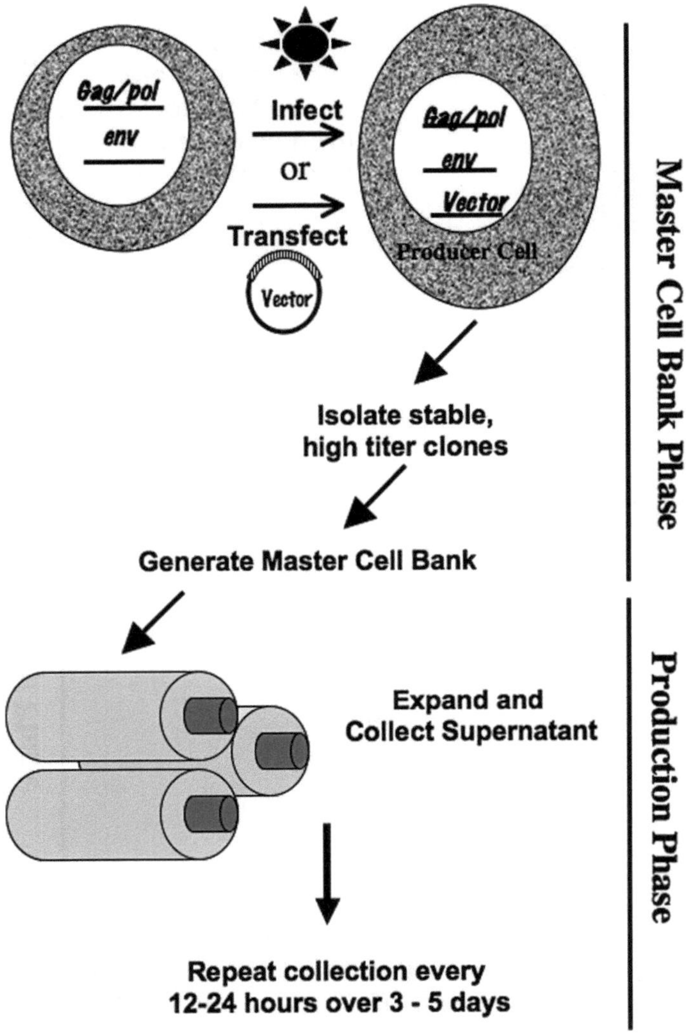

Fig. 2. Vector production using producer cell lines. Producer cell lines are generated from packaging cell lines that are stable cell lines expressing the viral *gag*, *pol*, and *env* genes. A vector plasmid is introduced by transfection or the vector is introduced by infection with vector particles. Stable, high-titer clones are then isolated. The optimal clone is expanded to a MCB that is then certified. A vial of the MCB is expanded to generate vector. The cells generally maintain high titer throughout multiple harvest for 3–5 days.

recently, pseudotyping using nonretroviral envelopes have been successful for the vesicular stomatitis virus G (VSV-G) protein and the alphavirus Ross River Virus (RRV) *(8)*. A summary of packaging cell lines is described in **Table 1**. Initially, packaging cell lines were generated in murine cell lines, but the newer generation of packaging cell lines have used human HEK293 and HT1080 to generate the Phoenix and FLY cell lines, respectively. Murine cell lines do pose the risk of inadvertent transfer of VL30 and other endogenous retroviral sequences *(9)*. The development of packaging cell lines using cells of human origin does address this theoretical safety concern and also generates particles that are less likely to be inactivated by human serum *(10)*. As these human cell lines allow for high transfection efficiency, the vector plasmid can be introduced using transient transfection without the need for the *gag/pol* and

Table 1
Retroviral Packaging Cell Lines

Parent cell line	Target cell species	Cell line	Parent viral envelope	Reference
Murine NIH 3T3				
	Rodent	GPE+86	Mo-MLV	**13**
	Rodent, primate	GP+*env*AM12	4070A MLV	**14**
	Rodent, primate	PA317	4070A MLV	**15**
	Primate	PG13	GALV	**16**
Human HEK293T				
	Rodent	EcoPhoenix	Moloney Ecotropic MLV	[a]
	Rodent, primate	AmphoPhoenix	Human HEK293T	[a]
	PhoenixGP	None		[a]
	Primate	Phoenix RD114	RD114	**17**
Human HT-1080				
	Rodent, primate	FLYA13	4070A MLV	**10**
	Primate	FLYRD18	RD114	**10**

Mo, Moloney Ecotropic; MLV, murine leukemia virus; GALV, gibbon ape leukemia virus; RRV, Ross River virus. The PhoenixGP cells express Moloney *gag* and *pol* and be engineered to express additional pseudotypes by introducing the desired viral envelope. RD114 is derived from the cat endogenous virus RD114.

[a]Information regarding the Phoenix cell lines can be found at www.stanford.edu/group/nolan, cells are also available at the American Tissue Culture Collection, Manassas, VA, USA (www.atcc.org).

Table 2
Regulatory Guidance for the Production of Clinical Gene Therapy Products

Document	Issued	www.fda.gov/cber/gdlns/
Points to Consider in the Characterization of Cell Lines Used to Produce Biologicals	July 12, 1993	ptccell.pdf
ICH Guidance on Quality of Biotechnological/Biological Products: Derivation and Characterization of Cell Substrates Used for Production of Biotechnological/Biological Products (63 FR 50244)	September 21, 1998	qualbiot.pdf
ICH Guideline Q5A. Guidance on Viral Safety Evaluation of Biotechnology Products Derived From Cell Lines of Human or Animal Origin (63 FR 51074)	September 24, 1998	virsafe.pdf
Guidance for Industry: Guidance for Human Somatic Cell Therapy and Gene Therapy (3/30/1998)	March 30, 1998	somgene.htm
Guidance for Industry: Supplemental Guidance on Testing for Replication Competent Retrovirus in Retroviral Vector Based Gene Therapy Products and During Follow-up of Patients in Clinical Trials Using Retroviral Vectors	October 8, 2000	retrogt1000.htm
Draft Guidance for Reviewers: Instructions and Template for Chemistry, Manufacturing, and Control (CMC) Reviewers of Human Somatic Cell Therapy Investigational New Drug Applications (INDs)	2003	cmcsomcell.htm
Draft Guidance for FDA Review Staff and Sponsors: Content and Review of Chemistry, Manufacturing, and Control (CMC) Information for Human Gene Therapy Investigational New Drug Applications (INDs)	2004	gtindcmc.htm
Guidance for Industry: Supplemental Guidance on Testing for Replication Competent Retrovirus in Retroviral Vector Based Gene Therapy Products and During Follow-up of Patients in Clinical Trials Using Retroviral Vectors.	2006	retrogt1000.htm
Guidance for Industry: Gene Therapy Clinical Trials – Observing Subjects for Delayed Adverse Events.	2006	gtclin.htm

Table 3
Common Release Testing Required for Retroviral Vectors Intended for Phase I/II Clinical Trials

	MCB for transient transfection	MCB producer cells	Final product
Purity			
Sterility (21 CFR 610.12)	√	√	√
Mycoplasma	√	√	√
In vivo virus assay	√	√	
Adventitious virus assay	√	√	√
Bovine virus assay	√	√	[b]
Porcine virus assay	√	√	[b]
Transmission Electron Microscopy	√	√	
PCR for Human Viruses[c]	√	√	
MAP/LCM assay	√		
RCR (co-culture and supernatant)[a]	√	√	√
General safety (21 CFR 610.11)		√	
Endotoxin		√	
Identity			
Cell line identity (e.g., murine origin)	√	√	
Vector insert stability (e.g., Southern blot)	√	√	
Sequencing		√	
Potency			
Vector function		√	
Titer		√	

[a]Replication competent retrovirus (RCR) testing is individualized for the envelope glycoprotein specific to the vector construct.

[b]Bovine and Porcine virus testing is not repeated on the final product if the MCB has tested negative for these viruses and all serum and trypsin used in production is certified virus free.

[c]PCR for human viruses: HIV1/2,HTLV I/II,HEP B,HEP C,HHV 6, HHV 7,HHV 8,EBV,CMV ,E1A, SV-40 (not required if all cells used in production are of murine origin).

env plasmids. These cell lines are also amenable to the production of stable cell lines.

The transient transfection method can provide the shortest time to the generation of clinical grade material. It is well suited to the production of vectors pseudotyped with the VSV-G envelope that is generally toxic if expressed at high concentration.

Also, vectors that utilize self-inactivating LTRs are generally produced by the transient transfection method as the integrated forms of these vectors will not generate full-length RNA vector genomes. This method does have a number of disadvantages including the need for high-quality plasmid preparations with each production, a relatively large amount of residual plasmids in the final product, and a greater potential for batch-to-batch variability. A disadvantage of the packaging cell line approach is the time required to generate and analyze a high-titer clone. Nevertheless, once a cell line is obtained, a MCB can be produced, which can provide consistent highly reproducible product.

The focus of this chapter is on the production of bulk material, additional modifications to the preparation may be required. Most notably, concentration of the vector material may be required, and this option will be covered elsewhere in this book. Another major requirement of Good Manufacturing Practice (GMP) is release testing. Generally, there are tests required of the MCB and the Final Vector Product that address the identity, purity, and potency of the material. A list of important regulatory documents that help guide the process and release testing is provided in **Table 2**. While each vector product will require a unique set of release tests, some common assays required for Phase I/II retroviral vector products are listed in **Table 3**.

2. Materials

2.1. Media and Cell Culture Supplies

1. D10 medium for cell expansion: Dulbecco's Modified Eagle Medium (D-MEM) supplemented with 2 mM L-glutamine and 1 mM sodium pyruvate (all from GIBCO/Invitrogen, Carlsbad, CA, USA) and containing 10% fetal bovine serum (HyClone, Logan, UT, USA) (*see* **Note 1**).
2. For cell expansion: 75, 175, and 300 cm^2flasks, calibrated pipettes (various sizes), aspirating pipettes (10 ml), centrifuge tubes (15 and 50 ml) from BD Falcon (Franklin Lakes, NJ, USA) or equivalent. 850 cm^2roller bottles with vent cap from Corning (item #431198, Corning, NY, USA).
3. For cell passage: Dulbecco's phosphate-buffered saline (DPBS) without calcium or magnesium from BioWhittaker (Cambrex Bio Science, East Rutherford, NJ, USA) and Trypsin–EDTA from GIBCO.
4. Glucose monitoring: Accu-Chek Glucose test strips, Accu-Chek Glucose test controls, Level 1 and 2 (Roche; 2030373 and 2030411, Hoffman-La Roche Ltd., Basel, Switzerland).

2.2. Supernatant Filtration and Vialing

1. For filtering cell debris from final a product: Fenwal® 40/150 micron Dual Screen Filter, Transfer Pack Container with Coupler-2000 ml , Vented Y-type Blood/Soln

Set with High Capacity 20 Micron Filter, Sepacell Leukocyte Reduction Filter for RBCs. All from Baxter Healthcare Corporation (Deerfield, IL, USA).

2. For freezing final product: Cryocyte freezing containers with pocket and Cryocyte manifold sets for 50 or 500 ml (depending on Sponsor needs) from Nexell Inc. (Irvine, CA, USA) Frozen Cryocyte containers should be stored in protective canisters (Custom BioGenic Systems/Thermo Forma, Shelby, MI, USA) (*see* **Note 2**).
3. Stedim transfer packs for pooling of final product and dispensing or draining culture medium (sizes 3-, 5-, and 20-l bags). These are the sizes we have used.

2.3. Equipment

1. Biologic safety cabinet (BSC), tissue culture incubator with roller bottle apparatus (if applicable), waterbath, centrifuge, freezer, and balance. An established method for routine equipment monitoring and maintenance should be established.
2. Baxa Repeater™ Pump (Baxa Corp., Englewood, CO, USA). A fluid transfer tube set (Baxa) will also be required.
3. Additional equipment that will be required includes a sterile tubing welder (Terumo SCD® 312) and hand-held tube sealer (SEBRA® model 1090/1105).

3. Methods

3.1. Preparatory Work for GMP Productions

1. The facility in which clinical grade material is produced will vary with the product, with more stringent requirements as a product moves toward licensure. In general, vector production occurs in at least a class 10,000 (ISO class 7) cleanroom with actual manipulation of the product occurring in a BSC at class 100 (ISO 5). All work should be performed under Standard Operating Procedures (SOP) and reviewed by a quality assurance specialist (*see* **Note 3**).
2. Pre-initiation documentation. This generally involves a written "Agreement" with the Sponsor that describes what product will be made, how much material will be generated, and any special requirements for production and testing. On the basis of the Sponsor's request and the facilities SOPs for production, a Batch Production Record (BPR) is generated that is a detailed plan for the work to be done. The BPR serves both as a guide to the procedure and is also used to capture key information such as what was performed and when, along with lot numbers and amounts of reagents and supplies. At a minimum, a production supervisor and the QA specialist should approve the document.
3. Insure that the facility cleaning and environmental monitoring is current and acceptable. Before starting the procedure, make sure all reagents are available and meet pre-established guidelines (e.g., we prepare media that must be tested for sterility and demonstrate no growth for at least 7 days before release).
4. The US FDA has detailed requirements for the labeling of final products to ensure specific information is captured. Final product labels should be approved by the

QA specialist before use. Generally, the labeling procedure would be described in a Labeling of Final and Intermediate Products SOP. Each final product label must contain: product name, study number, lot number, unit (or harvest) number, weight (or volume), formulation, storage instructions, name and address of the production facility, and (as appropriate) that the product is for investigational use only. Labels for release testing containers should also be prepared but are not subject to the same FDA regulations as final product labels.

5. Before production, a MCB should have been generated that will serve as the source of cells for this and future vector productions (MCBs are generally 100–200 vials). As shown in **Tables 2 and 3**, the MCB will require extensive testing per regulatory guidelines (*see* **Note 4**).
6. The MCB cells are evaluated before production to determine the optimal time and temperature for harvest. As previously published, certain cell lines perform optimally at 32°C while others produce optimal titers at 37°C ***(11,12)***. In addition, the time between harvest influences titers, with most cell lines providing the highest titers if allowing 12 or 24 h between harvests. We test these parameters in small scale and define our production schedule based on this data.

3.2. Production from Packaging Cell Lines

1. Thaw one vial of the appropriate MCB producer cells in compatible medium into a 75-cm^2 flask. The laboratory should have an SOP that describes assignment of lot numbers, and subsequent passages should reflect the appropriate nomenclature. The cell viability should be assessed at thawing, and if less than a predetermined amount (e.g., <50%), an investigation whether to continue production should be made by a senior staff member and the QA specialist.
2. Re-feed the flask 12–36 h after thaw. Continue to observe the cells daily. When nearing confluence the cells should be passaged into an increasing number and size of flasks (75, 175, 300 cm^2) until a sufficient number of cells are available to seed the production vessels. To remove cells for passage, the medium should be removed and the monolayer rinsed with D-PBS, which is then replaced with Trypsin–EDTA (approximately 2 ml per 75 cm^2). Cultures should be returned to 37°C for 2–5 min or until cells detach. Subcultures are generally split at a ratio of 1:6 to 1:8, but this may need to be altered depending on the growth characteristics of the cell line. Cells should be re-fed at least every 4 days if they are not ready for passage within that time.
3. For established packaging cell lines, we have obtained the highest titers with roller bottles although other vessels can be used. For a typical production, we use fifty 850-cm^2 roller bottles.
4. Plate the final culture vessels. Harvest and pool cells in a sterile container for the final subcultivation. It is suggested to not increase total surface area more than sixfold for optimal cell plating. Calculate the volume required to equally distribute the cells among the final production vessels. Fill the final vessels with an appropriate volume of tissue culture growth media: 850-cm^2 roller bottles are filled with a minimum of 60 ml media/bottle; 300-cm^2 flasks are filled with

60 ml; and cell factories require a minimum of 100 ml media per chamber. Add the determined volume of cells to each of the final production vessels. Number each vessel as cells are added. Return each to the 37°C 5% CO_2 incubator. If roller bottles are used, adjust the roller unit rotation rate to a setting of 8% of maximum speed to allow cells to adhere to the vessels. After 18–48 h, increase the rotation rate to 15–20% of maximum. Rotation rates can vary depending upon the unit. Our roller bottle unit's maximum rate is approximately 7 rpm.

5. Examine the final production vessels periodically after plating to determine the point at which the cells become confluent. On the day the cells are confluent, re-feed the vessels with an appropriate volume of media (e.g., 60 ml per 850-cm^2 roller bottle). To minimize fetal bovine serum in the final product, serum-free medium can be used in place of D10 medium (*see* **Note 5**).
6. Prepare Collection units, Final Product Unit sets, Final Product labels, and Certification Testing labels for each anticipated harvest.
7. We attempt to harvest a total of 9–18 l of material from 50 roller bottle production, and generally limit harvest to a maximum of 5–6 days.
8. Using the harvest intervals determined before production, harvest supernatant ± 1 h of the previous re-feed time for daily harvests, or ±0.5 h for harvests performed at shorter intervals. Immediately before supernatant harvest, examine each vessel macroscopically and microscopically and record any unusual findings.
9. Before harvest, obtain 1 ml of supernatant from a random vessel to determine glucose concentration. If the glucose is ≥200 mg/dl replace the media with 60 ml of fresh media. If the glucose is <200 mg/dl, replace with 120 ml of fresh media.
10. Harvest supernatant into a collection unit. We use FDA-approved tubing and collection bags, and we weld the parts together using a Sebra sealer. We use a BAXA pump to assist in the removal of media, and the collection bags (Stedim transfer packs) are weighed before collection, so that the final volume of material can be estimated. A sterile aspirating pipette is attached to the collection tubing and used to remove supernatant that is pumped into a transfer pack located outside the BSC. A second operator re-feeds each roller bottle immediately after supernatant removal then returns to the incubator.
11. After completion of the harvest, the pooled supernatant passes through a series of blood banking filters to remove cell debris flowing into the final collection pack. The Sebra sealer is then used to disconnect the transfer pack from the filter set. After subtracting the pre-determined weight of the transfer pack, the actual collection volume is determined by weighing the supernatant filled media pack (assuming 1 ml = 1 g). The percent difference is then determined by subtracting the actual yield from the calculated yield then dividing by the calculated yield. A predetermined acceptable percent difference should be established in the production SOP (we use ≤3%). If the percent difference exceeds acceptable limit, an investigation into the cause is initiated.

3.3. Production Using Transient Transfection Method

1. Thaw a vial of the selected MCB cells and expand through a series of increases in the number and size of flasks as discussed above (*see* **Subheadings 3.2.1.**

and **3.2.2.**) until a sufficient number of cells are obtained for final plating. For transfections, we obtain the highest titer using two-layer cell factories (636 cm^2/chamber) and plate 4.2×10^7 HEK293 cell-based lines per layer. For a 10-l harvest, 50 two-layer cell factories are required (4.2×10^9 cells).

2. Prepare bags or syringes with 100 ml of D10 media for each chamber.
3. In the BSC, replace one of the cell factory vent caps with a Luer lock fill cap. Turn the cell factory on its side, so that the vented port is on the top and the fill port is on the bottom. Introduce both cells and media into the unit and mix, ensuring that each layer contains an equal amount of medium. The media bag/syringe can be removed, and the Luer plug replaced or the bag/syringe can be left attached to the cell factory to be used to remove media during subsequent re-feedings.
4. Allow cells to adhere and grow approximately 10–24 h after plating. Transfection procedure should begin when cells are 40–75% (ideally 50–60%) confluent (*see* **Note 6**). If cells are less than 40% confluent, allow them to grow another 24 h.
5. Approximately 2–4 h before performing the transfection, drain the medium from the cell factories into a bag through the Luer port. Remove the bag and introduce 100 ml per layer of fresh D10 media.
6. Prepare the DNA/$CaCl_2$/water solution. For each two-layer cell factory, use 785 μl of 2 M $CaCl_2$ along with the appropriate amounts of plasmid solution(s); then bring the total volume of this mixture up to 10 ml with nuclease-free water. Multiply the volumes of each individual component times the number of layers to obtain the amount of transfection reagents needed.
7. In a second container, aliquot 2× HEPES-buffered saline, 10 ml per 2 layer cell factory to be transfected.
8. Prepare the calcium phosphate precipitate mixture for each cell factory one at a time. For two-layer cell factories, aliquot 10 ml of the water/DNA/$CaCl_2$ mixture in a 50-ml conical tubes. While gently vortexing the water/DNA/$CaCl_2$ mixture, carefully add 10 ml of HEPES-buffered saline drop wise.
9. Add the transfection mixture to the cell factories by draining the media into a container using the Luer-lock port. Draw up the 20 ml precipitate solution into a syringe and inject into the drained media. Re-introduce the media into the cell factories and ensure it is evenly distributed between both layers before placing it back in the 37°C 5% CO_2 incubator.
10. Re-feed each cell factory 6–24 h after transfection with D10 medium or serum-free medium according to the requests of the investigator.
11. Immediately before supernatant harvest, examine each culture to evaluate for contamination. Obtain a container (e.g., 20-l bag) in which the entire contents of all vessels can be collected.
12. Drain or pump the supernatant from the first vessel through a 0.45 μM clarifying filter into the container (replace the clarifying filter as required). If additional harvests are to be performed, simultaneously re-feed each vessel using appropriate type and volume of media.

3.4. Fill Final Product Units

1. Different clinical applications will require different volume aliquots of the final product. Nevertheless, it is optimal if containers can be found that allow aliquoting to occur within a closed system (*see* **Note 7**). We have found that Cryocyte containers, which are available in different sizes, easily accommodate commonly requested volumes and have ports that allow introduction of the supernatant in an essentially closed system. Follow the manufacturers' recommendations regarding the volume range for each container.
2. Place the Stedim transfer packs containing the pooled filtered supernatant into the BSC. Connect the Cryocyte to the pooled supernatant in the transfer pack(s) by inserting the male spike from both series into ports of separate transfer pack(s). Hang the transfer packs from the bar within the BSC.
3. Place two balances in the BSC. Place the first empty Cryocyte freezing container onto a balance and tare to zero. Open the clamp on the unit to allow supernatant to flow into the first Cryocyte freezing container. Fill to the desired weight, then close the clamp. Simultaneously, a second technician can fill containers to minimize the time between harvest and freezing.
4. To meet FDA requirements for sterility testing of final product, the first, last, and every tenth container is filled with an additional 20 g of supernatant that is subsequently removed for sterility and sterility reserve samples (e.g., in a harvest of 24 bags, #1, 10, 20, and 24 would be tested for sterility).
5. All final product containers from each harvest are checked for complete and accurate labeling by two technicians. Additional samples, as needed, are taken for release testing.
6. Final product containers are stored between <–70°C and ≥–90°C, in a freezer equipped with continuous monitoring and a remote temperature alarm. The product is designated as QUARANTINED until all certification testing is complete and a Certificate of Analysis has been issued. All quarantined final product must be physically separated from certified ("Approved") and research grade products.
7. For packaging cell lines, repeat the harvests at the predetermined intervals for up to 6 days *or* until cells demonstrate excess peeling or overgrowth. For transient transfection harvest, the titer will fall off quickly and some testing will be required to determine whether a second harvest (generally 24 h later) is warranted.

3.5. Final Harvest

1. Harvest supernatant as described above in **Subheading 3.2.1** or **3.2.2**, but do not re-feed the vessels. Instead, wash each vessel with phosphate-buffered saline. Remove the adherent cells using cell scrapers or Trypsin–EDTA, and then pool and count harvested cells from all vessels. Cryopreserve a sufficient number of cells to perform release testing and provide reserve samples (e.g., Current FDA recommendations call for 1% up to 10^8 cells be tested for RCR) (*see* **Table 2**).

3.6. Certification Testing

1. Initiate release testing and complete the changeover decontamination before using the room for another clinical production.
2. Final products may be released for clinical research use when release assays, reports, and Certificate of Analysis are completed satisfactorily and released by the laboratory personnel and the Quality Assurance Unit. Final approval for use is determined by the FDA reviewer of the IND.

4. Notes

1. Vector production should be performed without antibiotics present in the medium. Occasionally, antibiotics are required in the process (e.g., when the tetracycline regulatory system is employed), but this should be discussed with regulators before production to ensure the planned production schema is acceptable. The choice of serum will depend on the requirements of the regulatory agencies overseeing the clinical trial work. In general, serum from Europe is not acceptable because of concerns regarding bovine spongiform encephalitis. For trials in the USA, acceptable serum sources include USA and New Zealand. Currently, European regulators will not accept US serum, and New Zealand sources should be utilized. These requirements do change and regulators should be consulted prior to production. These requirements apply to the final product and to the cell lines. The source of the cell line and documentation of the serum source for all passages in the investigators laboratory should be available for regulators.
2. Cryocyte bags do well at –70 to –90°C but will rupture when thawed if exposed to liquid nitrogen.
3. Complete discussion of GMP quality control is beyond the scope of this chapter, but those generating clinical material should consider select parts of the Code of Federal Regulations: Laboratory record maintenance 21 CFR 58.185 and 58.195 and 21 CFR 211 subpart j and 211.160; Control of components and drug product containers and closures 21 CFR 211.80, 211.82, 211.84, 211.86, 211.87, 211.89, 211.94 and 211.160; Control of reagents used in growth and processing cell lines; 21 CFR 211.80, 211.82, 211.84, 211.86, 211.87, 211.89, 211.94 and 211.160.
4. For the MCB, we generally test sterility and exclude contamination with Mycoplasma and RCR before performing the production. Once the production is complete and titer is satisfactory, all of the remaining release tests are performed.
5. Generally, the titer obtained from packaging cell lines peak within the second or third harvest, plateaus, then falls off with time. The number of harvests one can perform depends on the cell line, and there is also variation within different clones from the same cell line. If serum-free medium is used during harvests, the titer will also fall off more rapidly. For transient transfection, the first harvest provides the highest titer, the second harvest titer is generally 20–50% of the first harvest, and falls further with subsequent harvests. Serum-free medium often contain growth additives that can effect the growth of the target cells. Not all

manufactures list these additives as ingredients. Sponsors should test different serum-free media on target cells before production.

6. We utilize calcium phosphate for transient transfection because GMP grade material is readily available. Certain lipofection reagents can provide improved transfection and higher titer but are prohibitive because of costs and may not be available at GMP grade. Electroporation techniques are difficult to scale up. Both alternatives are being developed for large-scale clinical production and may be available in the near future.
7. Future directions for both transient and stable producer cells productions should move toward more scalable systems with fewer open manipulations. The Cell Factory (Nunclon), as described in the transient production, diminishes the open manipulations compared with roller bottles; however, multiple vessels with open manipulations are still required. Initial work with closed, scalable, single-use bioreactor systems have been found to produce poor viral titers, presumably because of the lack of cell adherence. Current development work includes using growth matrices to overcome this obstacle while retaining the closed system scalable features.

Acknowledgments

The author is supported in part by the Indiana Genomics Initiative (INGEN) and a grant from the National Institutes of Health (NCRR U42 RR11148 and NHLBI PPG 1 P01 HL53586).

References

1. Miller, A. D., Eckner, R. J., and Jolly, D. J. (1984) Expression of a retrovirus encoding human HPRT in mice. *Science* **225**, 630–2.
2. Williams, D. A., Orkin, S. H., and Mulligan, R. C. (1986) Retrovirus-mediated transfer of human adenosine deaminase gene sequences into cells in culture and into murine hematopoietic cells in vivo. *Proceedings of the National Academy of Sciences of the United States of America* **83**, 2566–70.
3. Eglitis, M. A., Kantoff, P., Gilboa, E., and Anderson, W. F. (1985) Gene expression in mice after high efficiency retroviral-mediated gene transfer. *Science* **230**, 1395–98.
4. Rosenberg, S. A., Aebersold, P. M., Cornetta, K., Kasid, A., Morgan, R. A., Moen, R., Karson, E. M., Lotze, M. T., Yang, J. C., Topalian, S. L., Merino, M. H., Culver, K., Miller, A. D., Blaese, M. D., and Anderson, W. F. (1990) Gene transfer into humans-immunotherapy of patients with advanced melanoma, using tumor infiltrating lymphocytes modified by retroviral gene transduction. *New England Journal of Medicine* **323**, 570–78.
5. Armentano, D., Yu, S. F., Kantoff, P. W., von Ruden, T., Anderson, W. F., and Gilboa, E. (1987) Effect of internal viral sequences on the utility of retroviral vectors. *Journal of Virology* **61**, 1647–50.

6. Pannell, D., and Ellis, J. (2001) Silencing of gene expression: implications for design of retrovirus vectors. *Reviews in Medical Virology* **11**, 205–17.
7. Hawley, R. G. (2001) Progress towards vector design for hematopoietic stem cell gene therapy. *Current Gene Therapy* **1**, 1–17.
8. Sharkey, C. M., North, C. L., Kuhn, R. J., and Sanders, D. A. (2001) Ross River virus glycoprotein-pseudotyped retroviruses and stable cell lines for their production. *Journal of Virology* **75**, 2653–59.
9. Hatzoglou, M., Hodgson, C. P., Mularo, F., and R.W., H. (1990) Efficient packaging of a specific VL30 retroelement by psi 2 cells which produce MoMLV recombinant retroviruses. *Human Gene Therapy* **1**, 385–97.
10. Cosset, F., Takeuchi, Y., Battini, J., Weiss, R. A., and Collins, M. K. L. (1995) High-titer packaging cells producing recombinant retroviruses resistant to human serum. *Journal of Virology* **69**, 7430–36.
11. Kotani, H., Newton, P. B., Zhang, S., Chiang, Y. L., Weaver, L., Blaese, R. M., Anderson, W. F., and McGarrity, G. J. (1994) Improved methods of retroviral vector transduction and production for gene therapy. *Human Gene Therapy* **5**, 19–28.
12. Reeves, L., Smucker, P., and Cornetta, K. (2000) Packaging cell line characteristics and optimizing retroviral vector titer: the National Gene Vector Laboratory experience. *Human Gene Therapy* **11**, 2093–103.
13. Markowitz, D., Goff, S., and Bank, A. (1988) A safe packaging line for gene transfer: separating viral genes on two different plasmids. *Journal of Virology* **62**, 1120–24.
14. Markowitz, D., Goff, S., and Bank, A. (1988) Construction and use of a safe and efficient amphotropic packaging cell line. *Virology* **167**, 400–6.
15. Miller, A. D., and Buttimore, C. (1986) Redesign of retrovirus packaging cell lines to avoid recombination leading to helper virus production. *Molecular and Cellular Biology* **6**, 2895–902.
16. Miller, A. D., Garcia, J. V., Von Suhr, N., Lynch, C. M., Wilson, C., and Eiden, M. V. (1991) Construction and properties of retrovirus packaging cells based on gibbon ape leukemia virus. *Journal of Virology* **65**, 2220–24.
17. Neff, T., Peterson, L. J., Morris, J. C., Thompson, J., Zhang, X., Horn, P. A., Thomasson, B. M., and Kiem, H. P. (2004) Efficient gene transfer to hematopoietic repopulating cells using concentrated RD114-pseudotyped vectors produced in human packaging cells. *Molecular Therapy* **9**, 157–59.

3

Methods for the Production of Helper-Dependent Adenoviral Vectors

Donna J. Palmer and Philip Ng

Summary

Helper-dependent adenoviral vectors (HDAd) are deleted of all viral coding sequences and have demonstrated tremendous potential for gene therapy by providing long-term, high-level transgene expression in the absence of chronic toxicity. Thus, HDAd are superior to early generation Ad for gene therapy of genetic diseases where long-term transgene expression is required. This chapter describes in detail the rescue, amplification, and large-scale production of HDAd.

Key Words: Adenovirus; helper-dependent; helper; virus; Cre; suspension; large-scale; spinner; gene therapy; gutless; gutted.

1. Introduction

Helper-dependent adenoviral vector (HDAd) (also referred to as gutless, gutted, mini, fully-deleted, high-capacity, Δ, pseudo) is deleted of all viral coding sequences. HDAds retain the advantages of early generation Ad vectors including high efficiency in vivo transduction and high-level transgene expression. However, the absence of viral gene expression in transduced cells permits stable, long-term transgene expression in the absence of chronic toxicity in both rodents ***(1,2)*** and large animals ***(3,4)***. Moreover, the deletion of the viral sequences permits a large cloning capacity of ~37 kb allowing for the delivery of whole genomic loci, multiple transgenes, and large *cis*-acting elements to enhance, prolong, and regulate transgene expression. In addition, because the vector genome exists episomally in transduced cells, the risks of germline transmission and insertional mutagenesis leading to oncogenic transformation are

From: *Methods in Molecular Biology, vol. 433: Volume 1: Production and In Vivo Applications of Gene Transfer Vectors*
Edited by: J. M. Le Doux © Humana Press, Totowa, NJ

negligible. The reader is referred elsewhere for a more comprehensive review on HDAds *(5)*.

Because HDAds are devoid of viral coding sequences a helper virus (HV) is required for its propagation. **Figure 1** provides an overview for the production of HDAd first developed by Parks et al. *(6)*. The HDAd is first constructed as a bacterial plasmid. The plasmid form of the HDAd (pHDAd) is digested with a restriction endonuclease to release the HDAd genome and transfected into E1-complementing producer cells that express the bacteriophage P1 site-specific recombinase Cre. The transfected cells are then infected with a HV bearing a packaging signal (Ψ) flanked by loxP sites. Cre-mediated recombination between the two loxP sites results in Ψ excision rendering the HV genome unpackagable but still able to replicate and *trans*-complement HDAd replication and encapsidation. The HDAd is propagated by serial coinfections of 293Cre cells with the HDAd and HV and finally purified by CsCl ultracentrifugation. This chapter will focus on the methods used to produce HDAd based on the pΔ28E4 backbone *(7)* using the producer cell line 116 *(8)* and the HV AdNG163R-2 *(8)* or AdNG163 *(9)*, although these methods should be generally applicable with other reagents.

1.1. Helper-Dependent Adenoviral Vector

The HDAd contains the expression cassette of interest and ~500 bp of *cis*-acting Ad sequences necessary for vector DNA replication [inverted terminal repeat (ITRs)] and packaging (Ψ) (*see* **Fig. 1**). In addition to these minimal Ad sequences, Sandig et al. *(10)* have shown that inclusion of a small segment of noncoding Ad sequence from the E4 region adjacent to the right ITR increases vector yields possibly by enhancing packaging of the vector DNA.

The size of the HDAd genome is an important consideration. For efficient packaging, the genome should be between ~75% (>27 kb) *(11)* and 105% (<37.8 kb) *(12)* of wildtype. HDAd outside this range may undergo rearrangements or may not be efficiently packaged, if at all. Thus, the HDAd must often include "stuffer" DNA if the genes or cassettes to be cloned are relatively small. The choice of stuffer DNA is important with regard to vector stability and replication efficiency *(10,13,14)*. In general, noncoding eukaryotic DNA is preferable while repetitive elements and unnecessary homology with the HV should be avoided to ensure vector stability and prolonged transgene expression. We have extensively used vector backbones based on pΔ28E4 *(7,15)*, which is composed of DNA from the human hypoxanthine-guanine phosphoribosyltransferase (*HPRT1*) gene and the C346 cosmid. HDAds based on this backbone are stable and perform well in vivo with respect to level and duration of expression *(2–4,15–17)*.

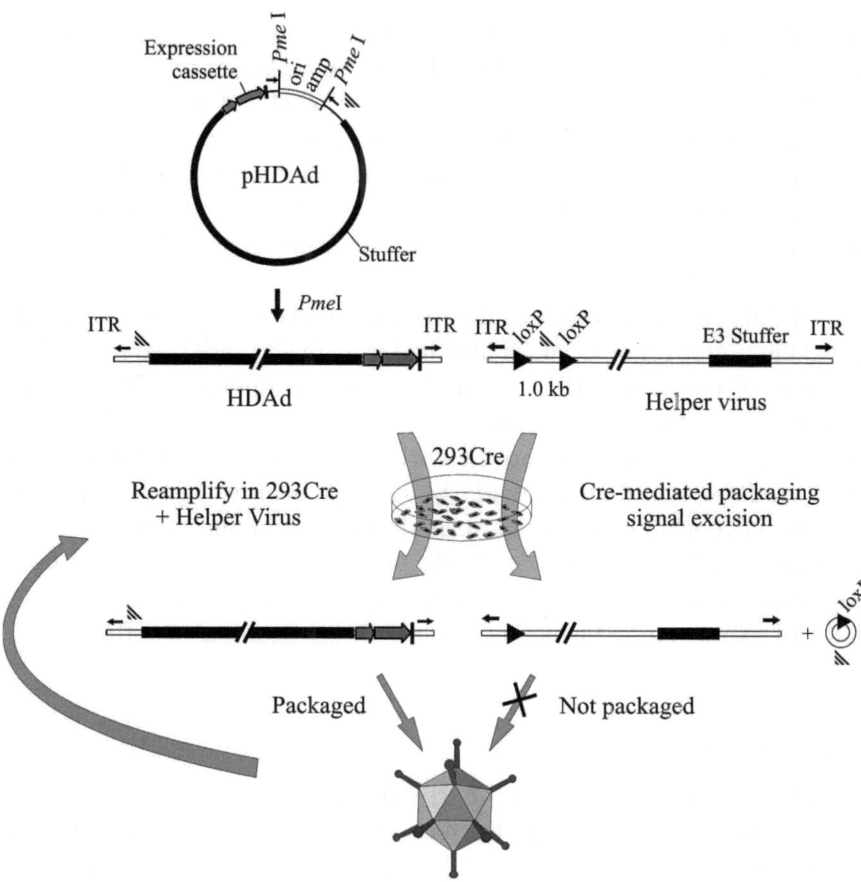

Fig. 1. The Cre/loxP system for producing HDAd. The HDAd contains only ~500 bp of *cis*-acting Ad sequences required for DNA replication [inverted terminal repeat (ITRs)] and packaging (Ψ), the remainder of the genome consists of the desired transgene and non-Ad "stuffer" sequences. The HDAd genome is constructed as a bacterial plasmid (pHDAd) and is liberated by restriction enzyme digestion (e.g., PmeI). To rescue the HDAd, the liberated genome is transfected into 293 cells expressing Cre and infected with a helper virus (HV) bearing a packaging signal (Ψ) flanked by loxP sites. Cre-mediated excision of Ψ renders the HV genome unpackagable, but still able to replicate and provide all of the necessary *trans*-acting factors for propagation of the HDAd. The titer of the HDAd is increased by serial coinfections (called passages, *see* **Fig. 2**) of 293Cre cells with the HDAd and the HV. Finally, the HDAd is purified by CsCl ultracentrifugation (*see* **Fig. 4**).

In addition to the absolute size of the HDAd, its relative size compared to the HV is an important consideration. This is because Cre-mediated selection against the HV, while efficient, is not absolute *(8,18)*. If the genome size of the HDAd is sufficiently different from that of the HV then the two species can be physically separated by CsCl ultracentrifugation based on their different buoyant densities. HDAds between 28 and 31 kb are ideal because they are efficiently packaged and easily separated from residual HV (35–37 kb) following CsCl ultracentrifugation.

The HDAd is initially constructed as a bacterial plasmid (pHDAd, *see* **Fig. 1**) such that the vector genome can be liberated from the bacterial sequences before transfection of producer cells by restriction enzyme digestion. The restriction enzyme sites are located immediately adjacent to the ITRs. Obviously, the restriction enzyme used should not cleave elsewhere in the vector genome. The pHDAd used for transfection should be of high quality, and purification by CsCl gradient centrifugation is the method of choice in our laboratory (see the Chapter entitled *Methods for the Production of First Generation Adenoviral Vectors* in this volume for a detailed protocol) but other commercially available methods of extracting and isolating plasmid DNA should also suffice.

1.2. Helper Virus

The HV is a first generation Ad (E1-deleted) with its packaging signal flanked by loxP sites (*see* **Fig. 1**). As with all first generation Ad vectors propagated in 293 or 293-derived cells (such as 116 cells), the potential exists for the generation of replication competent Ad (RCA; $E1^+$) as a consequence of homologous recombination between the HV and the Ad sequences present in 293 cells. To prevent the formation of RCA, a DNA stuffer sequence is inserted into the E3 region to render any $E1^+$ recombinants too large to be packaged *(6)*. The size of the stuffer should be chosen such that the total size of the HV genome is <105% of wildtype but >105% following homologous recombination with Ad sequences in the 293 cells. In general, noncoding stuffer sequences with no homology to the HDAd are desirable. A number of sequences have been found to be stable and not to adversely affect virus propagation *(6,8,10,19,20)*. We use the human serotype 5 HV AdNG163R-2 *(8)*. This HV has a number of unique features designed to improve HDAd production; it contains a reverse orientation packaging signal to prevent the generation of Cre-resistant mutants, it contains a DNA stuffer, composed of noncontiguous lambda DNA segments with no intact coding sequences, inserted into the wildtype E3 region to prevent the emergence of RCA, and it has a large genome size of 37.2 kb to permit maximum separation from HDAds (we construct HDAds to be 28 –31 kb) following CsCl ultracentrifugation.

It is important that HV stocks be derived from plaque purified isolates because any variants present may be preferentially amplified during serial passage (i.e., no Cre-resistant mutants). High-titer stocks should be purified by CsCl ultracentrifugation, dialyzed against 10 mM Tris–HCl, pH 8.0, supplemented with glycerol to 10%, and stored in aliquots at –80 °C. Working stocks are diluted (to ~2×10^8 pfu/ml) in 10 mM Tris–HCl pH 8.0 supplemented with glycerol to 10% and stored in small aliquots at –80 °C. Accurate determination of the infectious titer of the HV is critical for the successful production of HDAd (see the Chapter entitled *Methods for the Production of First Generation Adenoviral Vectors* in this volume for a detailed protocol).

1.3. Producer Cells

A number of properties are desirable in a producer cell line. First, they must complement E1 functions of the HV. Second, high levels of Cre expression are desirable to minimize HV contamination because it has been shown that incomplete excision of Ψ because of sub-optimal Cre levels is responsible for HV contamination ***(8,18)***. Third, the ability to grow both as adherent cells for vector amplification and in suspension to facilitate large-scale HDAd production is desirable. We use the 293-derived producer cell called 116 *(8)* that expresses very high levels of Cre resulting in exceedingly low levels of HV contamination. In addition, 116 cells can grow as both adherent monolayers and in suspension rendering them ideal for both HDAd amplification and large-scale production.

2. Materials

2.1. Cell Culture

1. 116 cells.
2. Minimal essential media (MEM, Invitrogen, Carlsbad, CA, USA, #61100) supplemented with 100 units/ml penicillin/streptomycin and 2 mM L-glutamine. For adherent growth of 116 cells in monolayers.
3. MEM eagle, Joklik modification for suspension culture (Sigma, St. Louis, MO, USA, #M-0518) supplemented with 100 units/ml penicillin/streptomycin and 2 mM L-glutamine. For growth in suspension in spinner flasks.
4. Hygromycin.
5. Heat-inactivated fetal bovine serum (FBS)
6. Citric saline: 135 mM KCl, 15 mM sodium citrate; autoclave sterilized.

2.2. Rescue and Amplification of HDAd

1. Monolayers of 116 cells at 70% to 90% confluency in 60-mm and 150-mm dishes.
2. Reagents for calcium phosphate transfection of plasmid DNA into mammalian cells in culture. We use the ProFection® Mammalian Transfection System by Promega, Madison, WI, USA (#E1200).
3. HV and pHDAd.
4. PBS^{++}: PBS supplemented with 0.01 vol 68 mM sterile $MgCl_2$ and 0.01 vol 50 mM sterile $CaCl_2$.
5. Pronase solution: 20 mg/ml pronase in 10 mM Tris–HCl, pH 7.5; preincubate at 56 °C for 15 min followed by 37 °C for 1 h. Aliquot and store at –20 °C.
6. Pronase-SDS solution: 0.5 mg/ml pronase (above) in 0.5% SDS, 10 mM Tris–HCl, pH 7.4, 10 mM EDTA pH 8.0.
7. 40% sucrose solution, filter sterilized.
8. *Optional.* β-gal fix solution: 0.2% glutaraldehyde, 2% paraformaldehyde, 2 mM $MgCl_2$ in PBS (*see* **Subheading 3.2., step 8**).
9. *Optional.* β-gal stain solution: 5 mM $K_4Fe(CN)_6$, 5 mM $K_3Fe(CN)_6$, 2 mM $MgCl_2$ in PBS (*see* **Subheading 3.2., step 8**).
10. *Optional.* X-gal: 20 mg/ml in N,N′-dimethylformamide (*see* **Subheading 3.2., step 8**).

2.3. Large-Scale Production and Purification of HDAd

1. HV.
2. 10 and 100 mM Tris–HCl, pH8.0, autoclave sterilized.
3. 5% sodium deoxycholate, filter sterilize.
4. 2 M $MgCl_2$, autoclave sterilize.
5. DNAase I: 100 mg bovine pancreatic deoxyribosenuclease I in 10 ml of 20 mM Tris–HCl, pH 7.4, 50 mM NaCl, 1 mM dithiothreitol, 0.1 mg/ml bovine serum albumin, 50% glycerol, aliquot and stored at –20 °C.
6. RNAase A (10 mg/ml).
7. CsCl solutions:

Density (g/cc)	CsCl (g)	10 mM Tris–HCl, pH 8.0 (g)
1.35	70.4	129.6
1.25	54.0	146.0

Dissolve CsCl into 10 mM Tris–HCl, pH 8.0 solution in the amounts indicated above to achieve the desired density solution, filter sterilize (*see* **Note 1**).

8. Glycerol, autoclave sterilize.
9. 250 ml (#1965-61002) and 3 l (#1965-61030) spinner flasks (Bellco Glass).
10. Magnetic stirrer (#7785-D200) (Bellco Glass, Vineland, NJ, USA).

11. Beckman SW 40 Ti rotor and ultraclear tubes (#342413).
12. Beckman SW 55 rotor and ultraclear tubes (#344057).
13. Slide-A-Lyzer dialysis cassettes (Pierce, Rockford, IL, USA, #66381).
14. Dialysis buffer (10 mM Tris–HCl pH 8.0)

2.4. Characterization of HDAd

1. Virion lysis buffer: TE supplemented with SDS to 0.1%.
2. Spectrophotometer and cuvettes.
3. Southern blot hybridization materials and reagents.

3. Methods

Production of HDAd can be divided into three parts (*see* **Fig. 2**): *(1)* Rescue, which involves converting pHDAd (the plasmid form) to HDAd (the viral form). *(2)* Amplification, which involves increasing the amount of HDAd by serial coinfections of the producer cells with the HDAd and the HV (each serial coinfection is referred to as a passage). *(3)* Large-scale production to generate large quantities of HDAd.

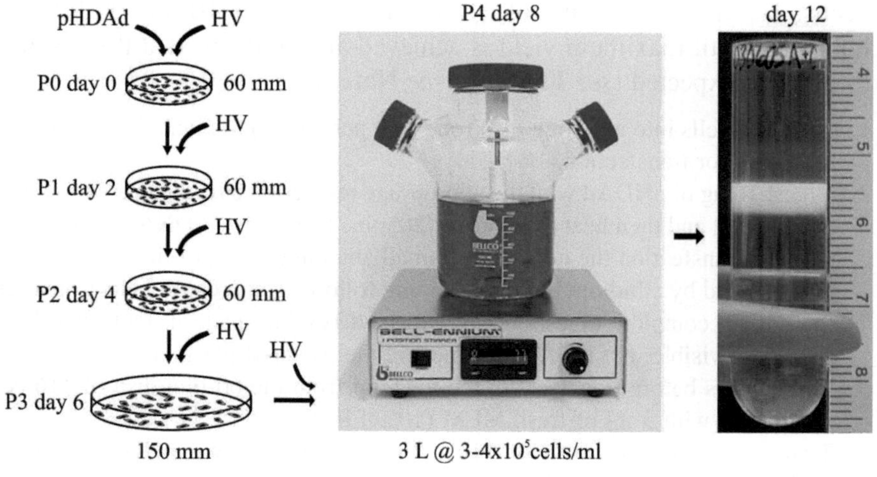

Fig. 2. Overview of rescue (passage 0 or P0), amplification (P1 to P3) and large-scale (P4) production of HDAd. P0 involves transfecting producer cells with pHDAd followed by infection with the HV. Each subsequent serial passage involves coinfecting producer cells with the appropriate amounts of HDAd and HV. The entire process can take as little as 2 weeks with total yields of $>10^{13}$ viral particles (vp) per 3 liter culture and specific yields of >10,000 vp/cell. HV, helper virus; pHDAd, HDAd plasmid.

3.1. Growth of 116 cells in Monolayers

116 cells are maintained in 150-mm dishes in MEM supplemented with FBS (heat inactivated) to 10%, hygromycin to 0.1 mg/ml, penicillin streptomycin to 100 U/ml, and L-glutamine to 2 mM. Split cells 1–2 every 2 days or 1–3 every 3 days when they reach ~90% confluency. Warm media to 37 °C before use.

1. Remove medium by aspiration.
2. Tap the side of the dish to detach cells.
3. Resuspend the cells in an appropriate volume of fresh medium and distribute to new dishes.

In general, 116 cells should be used/infected 2 days after seeding to permit firm attachment of cells. One confluent 150-mm dish of 116 should be split into twenty 60-mm dish for use in 2 days.

3.2. Rescue and Amplification of HDAd

The steps below are for rescue and amplification of one HDAd; scale up accordingly for more than one vector. *Optional:* To ensure that the system is functioning properly, we always include a control HDAd expressing β-galactosidase such as HDΔ28E4LacZ (*see* **Fig. 3B**) *(8)* because its amplification can be easily and quickly monitored by X-gal staining following titration on 293 cells. The titer of this control HDAd should increase 10- to 100-fold with each passage until maximum yield is achieved and confirms that the system is performing as expected (*see* **Fig. 3C**) (*see* **Note 2**).

1. Seed 116 cells into a 60-mm dish (one dish per vector) to reach 70% confluency in 2 days for transfection.
2. Digest 10 μg of pHDAd with the appropriate restriction enzyme in a total volume of 25–50 μl and then heat to 65 °C for 20 min. The digested DNA can be store at 4 °C for transfection the next day. A small amount (0.5 μl) of the digested DNA is visualized by ethidium bromide staining following agarose gel electrophoresis to confirm complete digestion and expected banding pattern. Only two bands should be visible; the HDAd genome and the bacterial plasmid sequences.
3. Three hours before transfection, the medium from the 60-mm dish of 116 cells is replaced with 5 ml of fresh MEM (10% FBS).
4. Transfect the 60-mm dish of 116 cells with the digested pHDAd using the ProFection® Mammalian Transfection System by Promega (#E1200) according to the manufacturer's instructions.
5. The next day, remove the medium from the transfected monolayer and wash twice with 1 ml MEM (10% FBS). Immediately infect the transfected cells with HV at an moi of 5 plaque-forming units (pfu)/cell and PBS^{++} to a total volume of 0.1 ml. Adsorb for 1 h in the incubator rocking the dishes every 10 min.
6. Following adsorption, add 1.5 ml of MEM (5% FBS) to the monolayer. Add another 1 ml of MEM (5% FBS) to the monolayer at 24 hours.

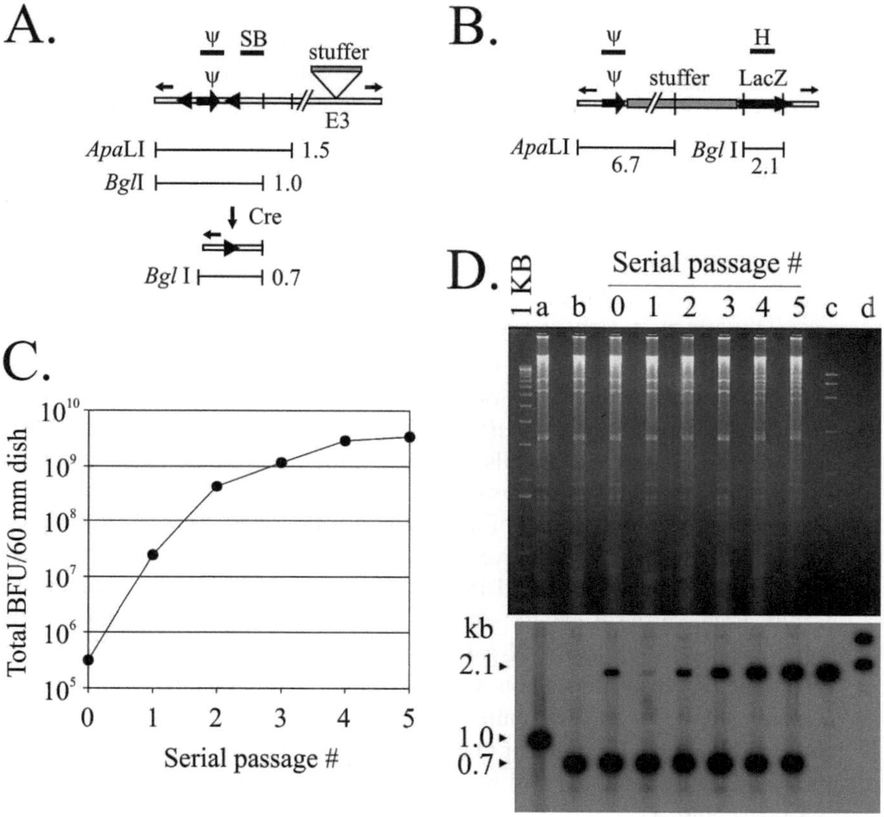

Fig. 3. Analysis of HDAd amplification. (A) The HV AdNG163 is a first generation (E1-deleted) adenovirus that contains a Ψ flanked by loxP sites and a DNA stuffer inserted into the wildtype E3 region to preclude RCA formation *(9)*. The 1 kb BglI fragment is reduced to 0.7 kb following Cre-mediated Ψ excision. (B) The HDAd HDΔ8E4LacZ contains a MCMV-LacZ expression cassette. The sizes of the relevant restriction enzyme fragments and the positions of probe Ψ, HV-specific probe SB and HDAd-specific probe H used for Southern analysis are shown and described in detail elsewhere *(8,18)*. Small triangles represent the loxP sites and the small horizontal arrows represent the viral inverted terminal repeats. (C) Amplification of HDΔ28E4LacZ with 116 cells and AdNG163. The total yield of HDΔ28E4LacZ per 60 mm dish in blue-forming units (BFU) was determined at each serial passage by X-gal staining following titration on 293 cells *(22)*. (D) Analysis of HDΔ28E4LacZ amplification with 116 cells and AdNG163. Total intracellular DNA was extracted from coinfected cells from each serial passage shown in (C) and digested with BglI and hybridized simultaneously with probes SB and H. Lane "a" contains DNA from 293 cells infected with AdNG163. Lane "b" contains DNA from 116 cells infected with AdNG163. Lane "c" contains the HDAd plasmid pΔ28E4LacZ digested with BglI and PmeI. Lane "d" contains a plasmid bearing both probe SB and H fragments digested with AvaII resulting in a 2.3 kb

7. Complete cytopathic effect (CPE) because of HV genome replication and HV gene expression (>90% of the cells rounded up and detached from the dish) should be observed by 48 h post-infection at which time the cells are scraped into the medium (*see* **Note 3**). 0.5 ml of the cell suspension is transferred to a microfuge tube for extraction of total DNA to monitor vector amplification (*see* **Subheading 3.3.**), and the remainder is transferred into a vial and stored at –80 °C after adding 0.1 vol 40% sucrose.
8. For vector amplification, the lysate from **step 7** is subjected to three cycles of freeze (–80 °C) thaw (37 °C). Allow the lysate to equilibrate to 37 °C, mix well and use 0.5 ml to coinfect a 60-mm dish of 90% confluent 116 cells with HV at an moi of 2 pfu/cell. Adsorb for 1 h in the incubator, rocking the dishes every 10 min. 1 ml of MEM (5% FBS) is added to the monolayer 1 h post-coinfection and another 1 ml of MEM (5% FBS) is added to the monolayer 24 h post-coinfection. *Optional:* For the LacZ control vector, infect a 60-mm dish of 90% confluent 293 cells with an appropriate volume (start with 0.1 ml of the transfection/infection lysate and reduce volume accordingly as the titer increases with serial passage) of the lysate (no HV is added) as described above. Twenty-four hours later, remove the medium and fix the monolayer by adding 1 ml of β-gal fix solution and incubating the dish at 37 °C for no more than 5 min. Remove the β-gal fix solution and add 1.5 ml of β-gal stain solution supplement with 0.5 mg/ml X-gal. Incubate the dishes at 37 °C in the dark and determine the vector titer the next day by counting the number of blue cells.
9. Repeat **steps 7** and **8** until maximum HDAd titer is obtained (*see* **Subheading 3.3.**). The number of passages required to achieve maximum HDAd titer may be variable and will be dictated by many variables (*see* **Subheading 3.3.**) (*see* **Note 4**).
10. Infect one 150-mm dish of 90% confluent 116 cells with 0.5 ml of lysate from the passage containing maximum HDAd titer and HV at an moi of 2 pfu/cell. When complete CPE is reached at 48 h post-coinfection, the cells are scraped into the medium, and the cell suspension is centrifuged for 5 min at 750 × *g*. The supernatant is discarded and the cell pellet is resuspended in 1 ml 100 mM Tris–HCl pH 8.0 supplement with glycerol to 10% and frozen at −80 °C. This will be used as inoculum for 3 l culture of 116 cells after three cycles of freeze (–80 °C) thaw (37 °C) (*see* **Subheading 3.6.**).

Fig. 3. fragment encompassing probe SB and a 3.1 kb fragment encompassing probe H. Note that the HDAd-specific bands are visible by ethidium staining in P3 through P5 and that the intensity of the HDAd-specific bands increases with each subsequent passage as determined by Southern analysis. P3 was chosen to make inoculum for 3 l large-scale HDAd production in this example. Southern analysis also reveals that the HV genome bearing a packaging signal (1 kb BglI fragment) is undetectable indicating low-level HV contamination.

3.3. Monitoring HDAd Amplification

To monitor amplification of most vectors that do no contain a reporter such as LacZ, total DNA is extracted from the coinfected cells from each passage and analyzed by ethidium bromide staining following agarose gel electrophoresis (*see* **Fig. 3D**).

1. Spin lysate collected in **step 7** of **Subheading 3.2** in a microcentrifuge for 1 min at 750 × *g* to pellet coinfected cells.
2. Discard the supernatant and add 200 µl of pronase-SDS solution to pellet.
3. Resuspend the cell pellet and incubate overnight at 37 °C to lyse cells.
4. Precipitate DNA by adding 0.5 ml 95% ethanol. Mix by inverting the tube until a visible precipitate is formed.
5. Pellet the DNA by microcentrifugation for 1 min at maximum speed.
6. Wash the DNA twice with 70% ethanol and dry.
7. Resuspend the DNA in an appropriate volume of TE (~40–50 µl). Digest an aliquot with an appropriate restriction enzyme and analyze by ethidium bromide staining following agarose gel electrophoresis. Include relevant controls as shown in **Fig. 3D** (*see* **Note 5**).

The passage(s) with the maximum amount of HDAd is identified as the one in which both the HV- and HDAd-specific bands are visible (*see* **Fig. 3D**) (*see* **Note 6**). If several passages meet this criterion, then the earliest passage should be chosen because the chance of rearrangement of either the vector or helper may increase with higher passage numbers. If no passages meet this criterion, then continue the serial passages until one is obtained (*see* **Note 7**).

3.4. Growth of 116 Cells in Suspension

116 cells are grown in 3-l spinner flasks in Joklik modified MEM supplemented with FBS (heat inactivated) to 5%, hygromycin to 0.1 mg/ml, penicillin streptomycin to 100 U/ml, and L-glutamine to 2 mM. Warm media to 37 °C before use.

1. Transfer spent media from eight confluent dishes of 116 cells into a 3-l spinner flask. Tap the dishes to detach 116 cells and transfer them into the 3-l spinner flask containing the spent media. Add fresh medium to a final volume of 1-l and tighten all lids.
2. Incubate at 37 °C (a humid, CO_2 environment is not necessary) on a magnetic stirrer set at 60 rpm.
3. Add 0.5 l the next day (1.5 l total).
4. Add 0.5 l the next day (2 l total).
5. Add 1 l the next day (3 l total).
6. Count (*see* **Subheading 3.5.**) and coinfect cells next day (*see* **Subheading 3.6.**). The cell density should be 2–4 × 10^5 cells/ml for coinfection.

The cell density should increase each day and can be confirmed by daily counting before addition of medium.

3.5. Counting 116 Cells in Suspension

1. Transfer 2 ml of cells into 15-ml conical tube.
2. Add 2 ml of 2× citric saline (270 mM KCl, 30 mM sodium citrate, autoclave sterilized. Pre-warmed to 37 °C).
3. Vortex for 10 s at maximum setting.
4. Incubate at 37 °C for 5–10 min.
5. Vortex for 10 s at maximum setting.
6. Obtain two independent cell counts using a hemocytometer (*see* **Note 8**).

3.6. Coinfection of 116 Cells in Suspension

1. Remove 0.1 ml of cell suspension from the 3 l culture and place into one well of a 24-well dish containing 1 ml fresh media and incubate at 37 °C. Cells should re-attach and form a monolayer within a few hours, and this serves as a control for the health and appearance of uninfected cells.
2. Harvest cells by centrifugation (750 × *g* for 5 min at room temperature). Save 0.5 l of the spent medium.
3. Resuspend cell pellet in 100 ml of spent medium and transfer to a 250-ml spinner flask.
4. Coinfect cells by adding HV at an moi of 2 pfu/ml and lysate from **Subheading 3.2**, **step 10** containing the HDAd and incubate for 1 h at 37 °C at 60 rpm.
5. Transfer the coinfected cells back into a 3-l spinner flask, add spent medium to 0.5 l and add 1.5 l of fresh MEM (5% FBS) (2 l total volume).
6. Transfer 0.1 ml of coinfected cells into one well of a 24-well dish containing 1 ml fresh MEM (5% FBS). These cells will initially re-attach to the dish but should all round up and detach by 48 h (*see* **Note 9**).
7. Incubate spinner flask for 48 h at 37 °C at 60 rpm.
8. Harvest coinfected cells by centrifugation at 750 × *g* for 5 min at room temperature.
9. Resuspend cells in a total of 15 ml 100 mM Tris–HCl pH 8.0 and transfer to a 50-ml conical tube and proceed with HDAd purification below or can be stored at –80 °C for processing later. For storage at –80 °C, resuspend the cells in 15 ml 100 mM–Tris HCl pH 8.0 supplemented with glycerol to 10%.

3.7. HDAd Purification

1. Add 2.0 ml of 5% sodium deoxycholate to lyse the coinfected cells from **Subheading 3.6.**, **step 9**. The mixture should immediately become thick and gelatinous. Incubate at room temperature for 30 min with frequent mixing.

2. Add 170 μl of 2 M MgCl, 150 μl of RNase A (10 mg/ml), and 150 μl of DNaseI (10 mg/ml) and incubate at 37 °C for 1 h and mix frequently to digest cellular DNA and RNA. The viscosity should be reduced significantly.
3. Remove cell debris by centrifugation at maximum speed in a tabletop centrifuge for 10 min at room temperature.
4. During **step 3**, prepare two CsCl step gradients; add 2 ml 1.35 g/cc CsCl to the bottom of two SW40 ultraclear ultracentrifuge tubes and then careful overlay with 3 ml of 1.25 g/cc CsCl.
5. Carefully overlay supernatant from **step 3** on top of the two CsCl step gradients. If necessary, fill the remainder of the tubes evenly with the 100 mM Tris–HCl pH 8.0.
6. Spin in ultracentrifuge at 151,000 × *g* (35,000 rpm) for 1 h at 4 °C with a SW40 rotor.
7. The HDAd is the lowest band in the step gradient (*see* **Fig. 4A**). Pierce the tube with a 3-cc syringe and 18-G needle below the vector band and slowly retrieve the vector from the tubes.
8. Transfer the vector from both tubes into one SW55 ultraclear tube and fill the tube with 1.35 g/cc CsCl.
9. Spin at 115,000 × *g* (35,000 rpm) overnight at 4 °C with a SW55 rotor.
10. Pierce the tube below the vector band with a 1-cc syringe and 22-G needle and retrieve the vector band from the CsCl gradient and transfer the vector into a new SW55 ultracentrifuge tube. Fill the tube with 1.35g/ml CsCl. Spin again at 115,000 × *g* (35,000 rpm) overnight at 4 °C with a SW55 rotor (*see* **Note 10**).
11. The next day, prepare Slide-A-Lyser dialysis cassette by pre-soaking cassette in 4 °C dialysis buffer according the manufacturer's instructions.
12. Retrieve the virus band from the CsCl gradient with a 1-cc syringe and 22-G needle and transfer into dialysis cassette as per manufacture's instructions. Dialyze overnight at 4 °C with 3 × 500 ml changes of 4 °C dialysis buffer with slow stirring to remove CsCl.
13. Retrieve the virus from the cassette with a 1-cc syringe and 18-G needle. Record the volume of vector in the syringe and transfer into a vial. Add glycerol to a final concentration of 10% and mix well.
14. Remove at least two small aliquots of the vector preparation and transfer them into microfuge tubes for characterization (*see* **Subheading 3.8.**); one for DNA extraction and another for absorbance at 260 nm. The volume of these aliquots will depend on the amount of vector obtained but is typically 10–50 μl. Additional small aliquots can be taken if desired for other characterization (i.e., infectivity, endotoxin, sterility etc., *see* **Subheading 3.8.**).
15. Aliquot vector and store at –80 °C.

Purified HDAd can be used to easily produce more of itself *(8)*. The use of purified HDAd instead of crude lysate greatly simplifies production and improves consistency. In this case, follow the steps outlined in **Subheading 3.6** but use 200 viral particles (vp)/cell of the purified vector instead of the lysate from a 150-mm dish.

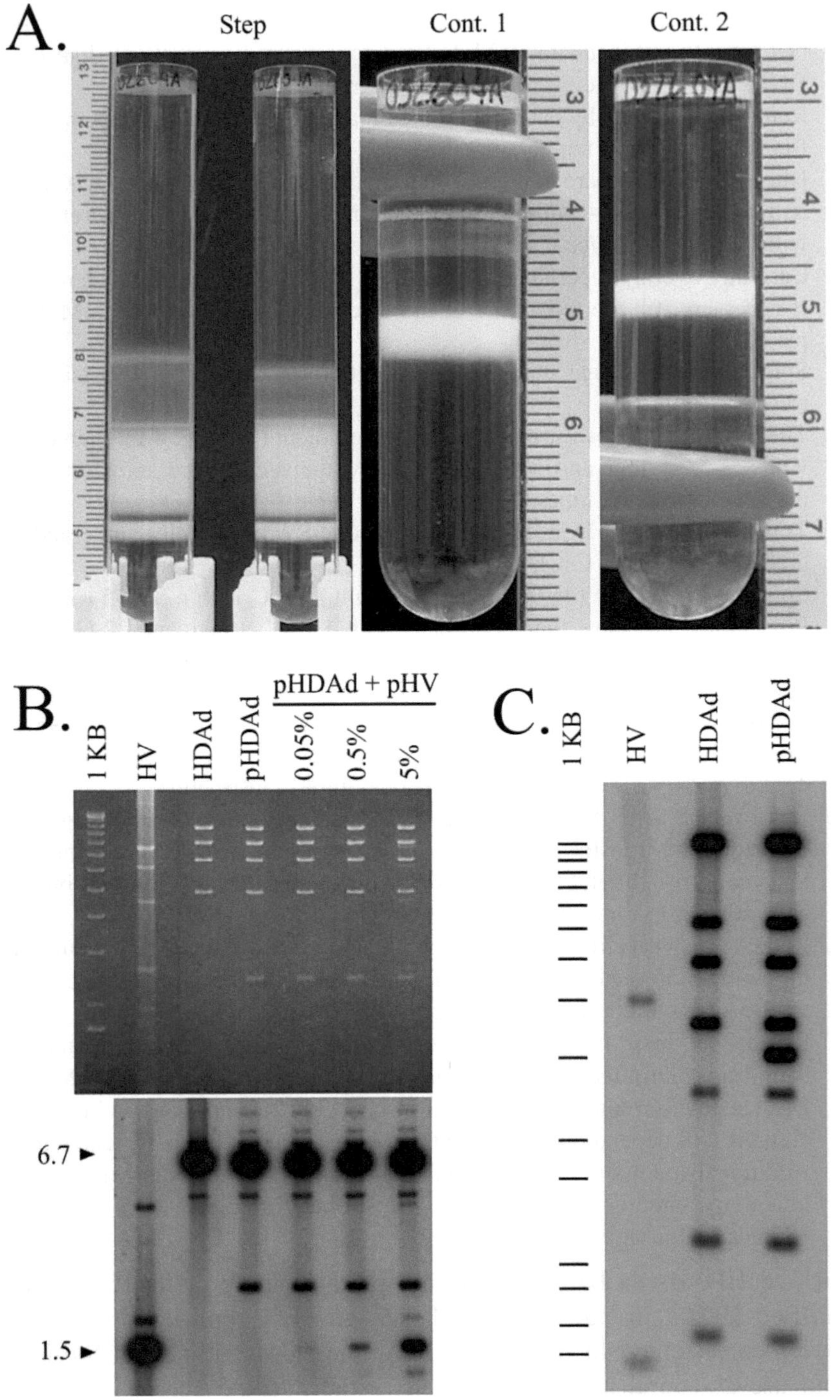
A.
Step
Cont. 1
Cont. 2
B.
1 KB
HV
HDAd
pHDAd
pHDAd + pHV
0.05%
0.5%
5%
6.7 ►
1.5 ►
C.
1 KB
HV
HDAd
pHDAd

Fig. 4.

3.8. Characterization of HDAd

The physical titer is the concentration of total vp in a vector preparation and is obtained by measuring the absorbance at 260 nm following virion lysis ***(21)*** and correction for vector genome size ***(22)***. This method essentially measures the amount of viral DNA in a vector preparation and is expressed as vp/ml.

1. Add virion lysis buffer to the aliquot of vector taken at **step 14** of **Subheading 3.7** to a total volume of 0.5 ml.
2. Prepare a blank sample by adding virion lysis buffer to the same volume of vector vehicle (10 mM Tris–HCl pH 8.0 supplemented with glycerol to 10%).
3. Vortex both samples briefly (avoid bubbles).
4. Incubate samples at 56 °C for 10 min.

Fig. 4. Purification and characterization of HDΔ28E4LacZ produced from 3 l of 116 cells using AdNG163. (A) HDAd is purified by one step and two continuous CsCl gradients (cont. 1 and cont. 2). Note that only a single virus band is visible in all CsCl gradients. The lowest band in the CsCl step gradient is the virions. Total HDAd yield in this example was 2.6×10^{13} vp with a specific yield of 29,126 vp/cell. (B) HV contamination analysis of HDAd. DNA was extracted from the purified virions and digested with ApaLI. The structure of HDΔ28E4LacZ DNA (labeled HDAd) is indistinguishable from the parental pΔ28E4LacZ plasmid (labeled pHDAd), except for the expected absence of the 2.5 kb ApaLI-PmeI fragment bearing the bacterial plasmid sequences present in pΔ28E4LacZ. Southern analysis with probe Ψ revealed the expected 6.7 kb HDΔ28E4LacZ-specific band but no detectable AdNG163-specific 1.5 kb band in all three CsCl gradient fractions (see **Fig. 3** for restriction map and probe locations). The HV lane contains AdNG163 DNA. Control lanes (labeled pHDAd + pHV) consisted of ApaLI and PmeI-digested pΔ28E4LacZ mixed with 10-fold serial dilutions of ApaLI and PacI-digest HV plasmid pNG163 (pHV). Extraneous bands in the HDAd lanes are not rearrangement products because they are also present in the pHDAd lane and thus represent nonspecific hybridization of the probe to the other vector bands. On the basis of comparisons to the controls, the HV contamination of this HDAd preparation was <0.05%. (C) Genomic structure of HDΔ28E4LacZ. DNA was extracted from the purified HDAd presented in (A) and digested with HpaI and analyzed by Southern blotting using pΔ28E4LacZ as the probe. The genomic structure of HDΔ28E4LacZ preparations (labeled HDAd) is indistinguishable from the parental pΔ28E4LacZ (labeled pHDAd) except for the expected absence of the 3.0 kb PmeI fragment bearing the bacterial plasmid sequences in pΔ28E4LacZ indicating an absence of genomic rearrangement. The bands visible in the AdNG163 lane (labeled HV) represent the terminal fragments that hybridize to the homologous Ψ and inverted terminal repeat sequences in the pΔ28E4LacZ probe. 1KB PLUS (Life Technologies) standard shown.

5. Vortex samples for 1 min.
6. Use the blank sample to set reference at 260 and 280 nm, then measure the absorbance of the vector sample at 260 and 280 nm.
7. Calculate vp/ml from absorbance at 260 nm using the formula below. The 260/280 ratio, a measure of relative purity, should be ~1.3.

vp/ml = (absorbance at 260 nm) (dilution factor) (1.1×10^{12}) (36)/(size of vector in kb)

Although beyond the scope of this chapter, HDAd preparations should also undergo additional characterization. Some of these are discussed below along with appropriate citations for more detail. HDAd is invariably contaminated with HV, and accurate determination of this level is critical. Southern blot analysis can be used to determine the HV contamination (*see* **Fig. 4B**). Alternatively, real-time quantitative polymerase chain reaction can also be used for this purpose ***(8)***. Determining HV contamination by plaque assay is less reliable ***(23,24)***. The genomic structure of the HDAd should be verified by ethidium bromide staining or by Southern blot hybridization for greater sensitivity following digestion with the appropriate restriction endonuclease and agarose gel electrophoresis (*see* **Fig. 4B and C**).

In addition to the physical titer (vp/ml), HDAd also possess an infectious titer defined as the proportion of viral particles capable of actually transducing target cells. The infectious titer is expressed as the number of infectious units (IU)/ml. The particle-to-infectious unit ratio (vp : IU) is the infectivity of the vector and represents the proportion of the total particles that are infectious. By definition, the higher the vp : IU ratio, the lower the infectivity. This is an important parameter for meaningful comparisons between different vectors (or between different preparations of the same vector) because erroneous conclusions will be drawn if they have different infectivities. Moreover, because the viral particle itself may mediate dose-dependent acute toxicity ***(16,17,25)***, noninfectious particles (reflected in a high vp : IU ratio) may increase risk and decrease benefit. Therefore, the FDA has recommended that the infectivity of clinical grade Ad vector be less than 30:1 to address this concern ***(26)***. Because HDAds cannot propagate by themselves, methods based on CPE used for early generation Ad vectors are not applicable, and the infectious titer of HDAd without reporter transgenes such as LacZ is difficult to determine. However, a method has recently been developed to determine the infectious titer of HDAd but, because of length constraints, the reader is referred to that publication for details ***(9)***.

If desired and depending on the application, additional characterization (i.e., endotoxin, sterility, adventitious agents, etc.) should be performed according to protocols established for E1-deleted Ad vectors.

3.9. Handling and Storage

Adenoviruses, including HDAd, are relatively stable when handled and stored properly. Vectors are always stored at –70 to –80 °C, and the number of freeze-thaws should be minimized by making multiple aliquots of the vector preparation immediately after purification before the first freeze. For use, the vector should be thawed at room temperature and returned to –70 to –80 °C as soon as possible. Proper handling and storage are critical for preserving vector infectivity. Exposure of vector to dry ice may negatively impact infectivity because of the reduction in pH from the CO_2 *(27)*; therefore, for shipping and transportation of vector in dry ice, special precautions should be taken to ensure proper protection. These precautions also apply to the HV and indeed to all Ad-based vectors.

4. Notes

1. Weigh 1.00 ml to confirm density.
2. The titer of the LacZ control vector typically reaches a peak of 10^8 to 10^9blue-forming units (bfu)/ml. With subsequent passages, the titer may fluctuate but should not drop and remain below 10^7 bfu/ml.
3. CPE is caused by the HV, not the HDAd. If complete CPE is not observed by 48 h post-coinfection, repeat the passage with more HV.
4. Different HDAds may amplify at different rates. Therefore, rate of amplification of the control LacZ vector may not reflect the rate of other vectors.
5. DNA prepared by this method may not dissolve easily. Heat the DNA at 65 °C with occasional vortexing at maximum setting until dissolved.
6. The presence of HV-specific bands in the total cellular DNA is expected and does not represent HV contamination because these HV genomes do not contain a packaging signal; this can be verified by Southern analysis (*see* **Fig. 3D**).
7. A number of factors affect the production efficiency of HDAd. Efficient transfection of the producer cell line with the pHDAd (*see* **Subheading 3.2., steps 1–5**) is critical. The higher the transfection efficiency, the lower the number of passages required to achieve maximum HDAd titer. The amount of HV added to each passage is critical. Ideally, this is the minimum amount required to infect all producer cells to cause complete CPE by 48 h (*see* **Note 3**). However, too much HV may be detrimental for vector propagation by adversely affecting the producer cells *(28)*. The nature of the HDAd also plays a critical role. HDAd with genome sizes outside the optimal packagable range (27.7–37.8 kb) will not amplify efficiently and may undergo rearrangements *(11,12)*. As well, repetitive sequences or unstable elements within the HDAd or homologous sequences between the HDAd and HV may cause rearrangement during vector amplification and should be avoided. Other factors related to the HDAd include the nature of the expression cassette, which may not be compatible with producer

cell growth (i.e., expression of a toxic or otherwise interfering product) as well as the stuffer sequence ***(13)***. Unidentified open reading frames within the HDAd backbone may also contribute in this regard. As well, producer cells should always be maintained in a healthy state as neglect or abuse will negatively affect vector yield. Finally, the protocols described herein are designed specifically for 116 producer cells, AdNG163 or AdNG163R-2 HV and HDAd based on the pΔ28E4 backbone. Results using different reagents may vary.

8. Accurate determination of the cell density at the time of coinfection is critical as it will dictate the amount of HV added and thus the HDAd yield. There should be no large numbers of uncountable cell clumps. If the two independent cell counts are significantly different from each other, then repeat cell count.
9. If a significant number of cells remain attached after 48 h, then the amount of HV added in **step 4** was insufficient.
10. Only a single band is normally visible in all three CsCl gradients (*see* **Fig. 4A**). However, if more than one band is visible, then they should be carefully collected separately for DNA analysis to determine their identity. Two main causes for multiple bands are vector genome rearrangement and HV contamination. If the genome size of the HDAd is outside the optimal size for encapsidation, then rearrangements can occur ***(11,12)***. Alternatively, if the HDAd genome contains repetitive sequences or other unstable elements, then rearrangement may occur. The pHDAd will need to be modified to address these problems. HDAd is invariably contaminated with HV. If HV contamination is high, then two distinct bands will be visible in the continuous CsCl gradient following ultracentrifugation provided that the genome size of the HDAd and HV are sufficiently different. For this reason, HDAd is optimally engineered to have a smaller genome size than the HV (*see* **Note 11**). The chief reason for HV contamination is inefficient excision of the HV packaging signal because of low level Cre expression in the producer cell line ***(8,18)*** and one solution is to use a producer cell line that expresses higher levels of Cre. Other sources of high HV contamination include mutation/rearrangement of the HV that prevents packaging signal excision, but these are generally quite rare ***(18)***. Be sure that every HV stock was generated from a single, well-isolated plaque. Also, bear in mind that the probability of such events increases with increasing number of passages performed.
11. In this case, the HDAd is the upper band, whereas the HV is the lower band. Remove the lower HV band first from the gradient. Then collect the upper HDAd.

Acknowledgments

PN is supported by grants from the National Institutes of Health (P50 HL59314, R01 DK067324, R01 HL083047 and P51 RR13986), the Texas Affiliate of the American Heart Association (0465102Y), and the Cystic Fibrosis Foundation (NG0530 and NG05G0).

References

1. Kim, I.H., Jozkowicz, A., Piedra, P.A., Oka, K., and Chan, L. 2001. Lifetime correction of genetic deficiency in mice with a single injection of helper-dependent adenoviral vector. *Proc. Natl. Acad. Sci. U. S. A.* **98**, 13282–12387.
2. Toietta, G., Mane, V.P., Norona, W.S., Finegold, M.J., Ng, P., McDonagh, A.F., Beaudet, A.L., and Lee, B. 2005. Lifelong elimination of hyperbilirubinemia in the gunn rat with a single injection of helper-dependent adenoviral vector. *Proc. Natl. Acad. Sci. U. S. A.* **102**, 3930–3935.
3. Brunetti-Pierri, N., Nichols, T.C., McCorquodale, S., Merricks, E., Palmer, D.J., Beaudet, A.L., and Ng, P. 2005. Long-term phenotypic correction of canine hemophilia B following systemic administration of helper dependent adenoviral vectors. *Hum. Gene Ther.* **16**, 811–820.
4. Brunetti-Pierri, N., Ng, T., Iannitii, D.A., Palmer, D.J., Beaudet, A.L., Finegold, M.J., Carey, K.D., Cioffi, W.G., and Ng, P. 2006. Improved hepatic transduction, reduced systemic vector dissemination and long-term transgene expression by delivering helper-dependent adenoviral vectors into the surgically isolated liver of nonhuman primates. *Hum. Gene Ther.* **15**, 35–46.
5. Palmer, D.J., and Ng, P. 2005. Helper-dependent adenoviral vectors for gene therapy. *Hum. Gene Ther.* **16**, 1–16.
6. Parks, R.J., Chen, L., Anton, M., Sankar, U., Rudnicki, M.A., and Graham, F.L. 1996. A helper-dependent adenovirus vector system: removal of helper virus by Cre-mediated excision of the viral packaging signal. *Proc. Natl. Acad. Sci. U. S. A.* **93**, 13565–13570.
7. Toietta, G., Pastore, L., Cerullo, V., Finegold, M., Beaudet, A.L., and Lee, B. 2002. Generation of helper-dependent adenoviral vectors by homologous recombination. *Mol. Ther.* **5**, 204–210.
8. Palmer, D.J., and Ng, P. 2003. Improved system for helper-dependent adenoviral vector production. *Mol. Ther.* **8**, 846–852.
9. Palmer, D.J., and Ng, P. 2004. Physical and infectious titers of helper-dependent adenoviral vectors: a method of direct comparison to the adenovirus reference material. *Mol. Ther.* **10**, 792–798.
10. Sandig, V., Youil, R., Bett, A.J., Franlin, L.L. Oshima, M., Maione, D., Wang, F., Metzker, M.L., Savino, R., and Caskey, C.T. (1999). Optimization of the helper-dependent adenovirus system for production and potency in vivo. *Proc. Natl. Acad. Sci. U. S. A.* **97**, 1002–1007.
11. Parks, R.J., and Graham, F.L. 1997. A helper-dependent system for adenovirus vector production helps define a lower limit for efficient DNA packaging. *J. Virol.* **71**, 3293–3298.
12. Bett, A.J., Prevec, L., and Graham, F.L. 1993. Packaging capacity and stability of human adenovirus type 5 vectors. *J. Virol.* **67**, 5911–5921.
13. Parks, R.J., Bramson, J.L., Wan, Y., Addison, C.L., and Graham, F.L. 1999. Effects of stuffer DNA on transgene expression from helper-dependent adenovirus vectors. *J. Virol.* **73**, 8027–8034.

14. Schiedner, G., Hertel, S., Johnston, M., Biermann, V., Dries, V., and Kochanek, S. (2002). Variables affecting in vivo performance of high-capacity adenovirus vectors. *J. Virol.* **76**, 1600–1609.
15. Oka, K., Pastore, L., Kim, I.-H., Merched, A., Nomura, S., Lee, H.-J., Merched-Sauvage, M., Arden-Riley, C., Lee, B., Finegold, M., Beaudet, A., and Chan, L. (2001). Long-term stable correction of low-density lipoprotein receptor-deficient mice with a helper-dependent adenoviral vector expressing the very low-density lipoprotein receptor. *Circulation* **103**, 1274–1281.
16. Brunetti-Pierri, N., Palmer, D.J., Beaudet, A.L., Carey, D., Finegold, M., and Ng, P. 2004. Acute toxicity following high-dose systemic injection of helper-dependent adenoviral vectors into non human primates. *Hum. Gene Ther.* **15**, 35–46.
17. Brunetti-Pierri, N., Palmer, D.J., Mane, V., Finegold, M., Beaudet, A.L., and Ng, P. 2005. Increased hepatic transduction with reduced systemic dissemination and proinflammatory cytokines following hydrodynamic injection of helper-dependent adenoviral vectors. *Mol. Ther.* **12**, 99–106.
18. Ng, P., Evelegh, C., Cummings, C., and Graham, F.L. 2002. Cre levels limit packaging signal excision efficiency in the Cre/loxP helper-dependent adenoviral vector system. *J. Virol.* **76**, 4181–4189.
19. Parks, R.J., Evelegh, C.M., and Graham, F.L. 1999. Use of helper-dependent adenoviral vectors of alternative serotypes permits repeat vector administration. *Gene Ther.* **6**, 1565–1573.
20. Umana, P., Gerdes, C.A., Stone, D., Davis, J.R.E., Ward, D., Castro, M.G., and Lowenstein, P.R. (2001). Efficient FLP recombinase enables scalable production of helper-dependent adenoviral vectors with negligible helper-virus contamination. *Nat. Biotech.* **19**, 582–585.
21. Maizel, J.V., White, D., and Scharff. M.D. 1968. The polypeptides of adenovirus. I. Evidence of multiple protein components in the virion and a comparison of types 2, 7a, and 12. *Virology* **36**, 115–125.
22. Ng, P., Parks, R.J., and Graham, F.L. 2002. Preparation of helper-dependent adenoviral vectors. *Methods Mol. Med.* **69**, 371–388.
23. Reddy, P.S., Sakhuja, K., Ganesh, S., Yang, L., Kayda, D., Brann, T., Pattison, S., Golightly, D., Idamakanti, N., Pinkstaff, A., Kaloss, M., Barjot, C., Chamberlain, J.S. Kaleko, M., and Connelly, S. (2002). Sustained human factor VIII expression in hemophilia A mice following systemic delivery of a gutless adenoviral vector. *Mol. Ther.* **5**, 63–73.
24. Kreppel, F., Biermann, V., Kochanek, S., and Schiedner, G. 2002. A DNA-based method to assay total and infectious particle contents and helper virus contamination in high-capacity adenoviral vector preparations. *Hum. Gene Ther.* **13**, 1151–1156.
25. Muruve, D.A., Cotter, M.J., Zaiss, A.K., White, L.R., Liu, Q., Chan, T., Clark, S.A., Ross, P.J., Meulenbroek, R.A., Maelandsmo, G.M., and Parks, R.J. 2004. Helper-dependent adenovirus vectors elicit intact innate but attenuated adaptive host immune responses in vivo. *J. Virol.* **78**, 5966–5972.

26. Simek, S., Byrnes, A., and Bauer, S. 2002. FDA perspective on the use of the adenovirus reference material. *BioProcessing J.* **1**, 40–42.
27. Nyberg-Hoffman, C., and Aguilar-Cordova, E. 1999. Instability of adenoviral vectors during transport and its implication for clinical studies. *Nat. Med.* **5**, 955–957.
28. Ng, P., Beauchamp, C., Evelegh, C., Parks, R., and Graham, F.L. 2001. Development of a FLP/frt system for generating helper-dependent adenoviral vectors. *Mol. Ther.* **3**, 809–815.

4

Methods for the Production of First Generation Adenoviral Vectors

Donna J. Palmer and Philip Ng

Summary

Adenoviruses (Ads) possess several features that make them attractive mammalian gene transfer vectors. They can efficiently infect a wide variety of quiescent and proliferating cell types from various species to direct high level viral gene expression, their 36 kb double-stranded DNA genome can be manipulated with relative ease by conventional molecular biology techniques, and they can be readily propagated and purified to yield high titer preparations of very stable virus. Consequently, Ads have been extensively used as vectors for recombinant vaccines, for high-level protein production in cultured cells and for gene therapy where high-level, transient transgene expression is desired. This chapter describes in detail methods for the production and characterization of E1-deleted, first generation adenoviral vectors.

Key Words: Adenovirus; Cre; loxP; FLP; frt; suspension; large-scale; spinner; E1; vector; gene therapy.

1. Introduction

1.1. Adenovirus Biology

The adenovirus (Ad) is an icosohedral, non-enveloped capsid containing a linear double-stranded DNA genome of ~36 kb. Of the approximately 50 serotypes of human Ad, serotype 2 (Ad2) and serotype 5 (Ad5) of subgroup C have been the most extensively characterized *(1)*. The 36-kb genome is flanked by inverted terminal repeats (ITRs), which are the only sequences required in *cis* for DNA replication. A *cis*-acting packaging signal, required for encapsidation

From: *Methods in Molecular Biology, vol. 433: Volume 1: Production and In Vivo Applications of Gene Transfer Vectors*
Edited by: J. M. Le Doux © Humana Press, Totowa, NJ

of the genome is located near the ITR at the left end (relative to the conventional map of Ad). The Ad genome can be divided into two regions with respect to the onset of viral DNA replication (*see* **Fig. 1**). The early region genes, E1A, E1B, E2, E3, and E4, are expressed before DNA replication. The E1A transcription unit is the first Ad sequence expressed during viral infection and encodes two major E1A proteins that are involved in transcriptional regulation of the virus and stimulation of the host cell to enter S phase. The two major E1B proteins are necessary for blocking host mRNA transport, stimulating viral mRNA transport and blocking E1A-induced apoptosis. The E2 region encodes proteins required for viral DNA replication and can be divided into two subregions, E2a and E2b. E2a encodes the 72-kD DNA-binding protein and E2b encodes the viral DNA polymerase and terminal protein precursor. The E3 region encodes at least seven proteins, most of which are involved in evasion of the host immune system. This region is dispensable for virus growth

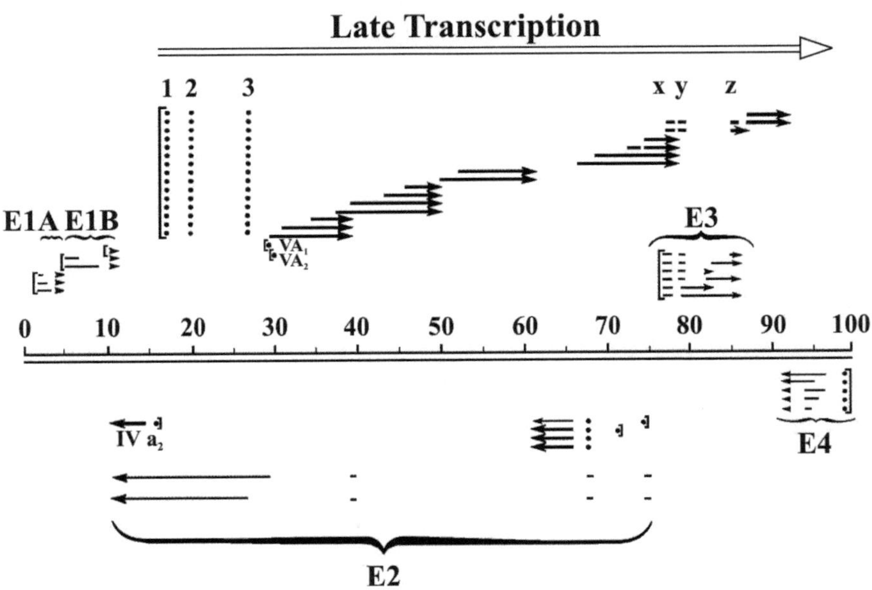

Fig. 1. Transcription map of human adenovirus serotype 5. The 100 map unit (~36 kb) genome is divided into four early region transcription units, E1–E4 and five families of late mRNA, L1–L5, which are alternative splice products of a common late transcript expressed from the major late promoter (MPL) located at 16 mu. Four smaller transcripts are also produced, pIX, IVa, and VA RNAs (I and II). Not shown are the 103 bp inverted terminal repeats located at the termini of the genome involved in viral DNA replication and the packaging signal located from nucleotides 190–380 at the left end of the genome involved in encapsidation.

in cell culture. The E4 region encodes at least six proteins, some of which function in facilitating DNA replication, enhancing late gene expression and decreasing host protein synthesis. The late region genes, L1–L5, are expressed after DNA replication from a common major late promoter and are generated from alternative splicing of a single transcript. Most of the late mRNAs encode virion structural proteins.

Virus infection initiates with the Ad fiber protein binding to specific primary receptors on the cell surface *(2,3)* followed by a secondary interaction between the virion penton base and $\alpha_v\beta_3$ and $\alpha_v\beta_5$ integrins *(4)*. The efficiency with which Ad binds and enters the cell is directly related to the level of primary and secondary receptors present on the cell surface *(5,6)*. Penton–integrin interaction triggers Ad internalization by endocytosis where it escapes from the early endosome into the cytosol before lysosome formation *(7,8)*. The virion is sequentially disassembled during translocation along the microtubule network toward the nucleus where the viral DNA is released into the nucleus *(9)*. Viral DNA replication, beginning 6–8 h post-infection, and assembly of progeny virions occurs in the nucleus. The entire life cycle takes about 24–36 h generating about 10^4 virions per infected cell. In humans, Ads generally cause only relatively mild, self-limiting illness in immunocompetent individuals. Ads have never been implicated as a cause of malignant disease in their natural host. The reader is referred to an excellent review by *(1)* for a more comprehensive discussion of the Ad.

1.2. Adenovirus Vectors

Ads have been extensively used as vectors for recombinant vaccines, for high-level protein production in cultured cells and for gene therapy *(10–14)*. This chapter describes in detail methods for the production of first generation adenoviral vectors. Typically, Ads are modified into vectors by replacing the E1 region with the foreign DNA of interest. This modification serves two important purposes. First, deletion of E1 increases the cloning capacity to ~5 kb. Second, it renders the vectors replication-deficient, which is an important consideration with respect to safety for human gene therapy. These replication-deficient vectors can be propagated in E1-complementing cell lines such as 293 *(15)*. The E3 region can also be deleted from the vector as it is not required for virus propagation in culture. The combination of E1 and E3 deletions results in a cloning capacity of ~8 kb, a size that is more than adequate for most expression cassettes.

The first step in producing an adenoviral vector is construction of the recombinant viral genome. Although a wide variety of methods have been developed for this purpose *(16)*, the first part of this chapter will focus on the two-plasmid rescue system *(17–19)*. The second part of this chapter will focus on the propagation, purification, and characterization of adenoviral vectors, and these protocols are universally applicable.

1.3. Overview of the Two-Plasmid Rescue System

This chapter provides detailed protocols for construction of E1-substituted Ad vectors by site-specific recombination between two plasmids cotransfected into E1-complementing cells such as 293 (*see* **Fig. 2**) *(17–19)*. One of the plasmids used in this method contains the entire Ad5 genome except for certain key deletions. Essential features of this Ad genomic plasmid include an ITR junction, a deletion spanning the E1 region and packaging signal, a target sequence for site-specific recombination and a cassette for expression of the site-specific recombinase. Inclusion of the recombinase expression cassette allows site-specific recombination-mediated vector rescue to be performed in

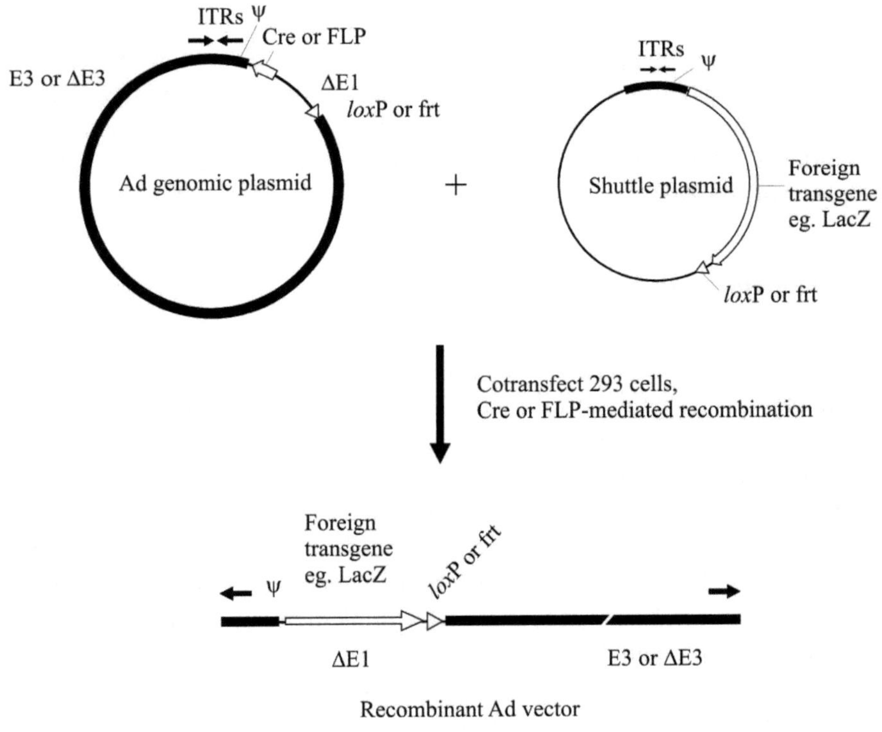

Fig. 2. Construction of Ad vectors by Cre- or FLP-mediated site-specific recombination following cotransfection of 293 cells with an Ad genomic plasmid (*see* **Fig. 3**) and a shuttle plasmid (*see* **Fig. 4**). Ad and bacterial plasmid sequences are represented by thick and thin lines, respectively, and *lox*P or frt sites are represented by "Δ".

any E1-complementing cell. The recombinase expression cassette is not incorporated into the recombinant vector because of its location in the plasmid with respect to the recombinase target site (*see* **Fig. 2**). The second plasmid, called a shuttle, contains an ITR junction, a packaging signal, a polylinker region for insertion of foreign sequences and a target sequence for site-specific recombination. The ITR junctions serve as efficient origins of Ad DNA replication ***(20)***, so that following cotransfection into 293 cells, the plasmids can undergo Ad-mediated DNA replication. However, the deletion of the packaging signal renders the Ad genomic plasmid non-infectious. The presence of the ITR junction in the shuttle plasmid significantly increases the efficiency of vector rescue ***(18)***. Following cotransfection into 293 cells, the two plasmids form a recombinant Ad vector by site-specific recombination. The Ad vector can replicate and be packaged into infectious virions, which cause plaques to form on the monolayer of 293 cells. As neither plasmid alone is capable of producing infectious virus, only the desired recombinant vector is generated following cotransfection. Vector rescue can be achieved by either of two site-specific recombination systems with comparable high efficiencies; the bacteriophage P1 Cre/loxP system ***(17,18)*** or the yeast FLP/frt system ***(19)***.

2. Materials

All cell lines and plasmid DNAs described herein are available from Microbix Biosystems Inc., Toronto, Ontario, Canada (http://www.microbix.com/products/040301.html) under the name AdMax™.

2.1. Ad Genomic Plasmids

Various Ad genomic plasmids are available for construction of vectors (*see* **Fig. 3**). The plasmids pBHGloxE3Cre and pBHGfrtE3FLP are used to generate vectors bearing a wildtype E3 region by Cre-mediated and FLP-mediated recombination, respectively. The maximum foreign DNA insert that can be rescued into an E1-deleted vector with a wildtype E3 region is ~5 kb owing to the size constraints of Ad ***(21)***. The plasmids pBHGloxΔE3(X1)Cre and pBHGfrtΔE3(X1)FLP have a 1864 bp deletion in the E3 region and thus permit rescue into vectors of foreign sequences up to ~7.2 kb by Cre-mediated and FLP-mediated recombination, respectively. The plasmids pBHGloxΔE1,3Cre and pBHGfrtΔE1,3FLP have a 2653 bp deletion in the E3 region and permit rescue into vectors of foreign sequences up to ~8 kb by Cre-mediated and FLP-mediated recombination, respectively. These latter two plasmids offer maximum cloning capacity but vectors bearing this larger E3 deletion may grow slightly slower and result in lower yields (~2-fold) than vectors bearing

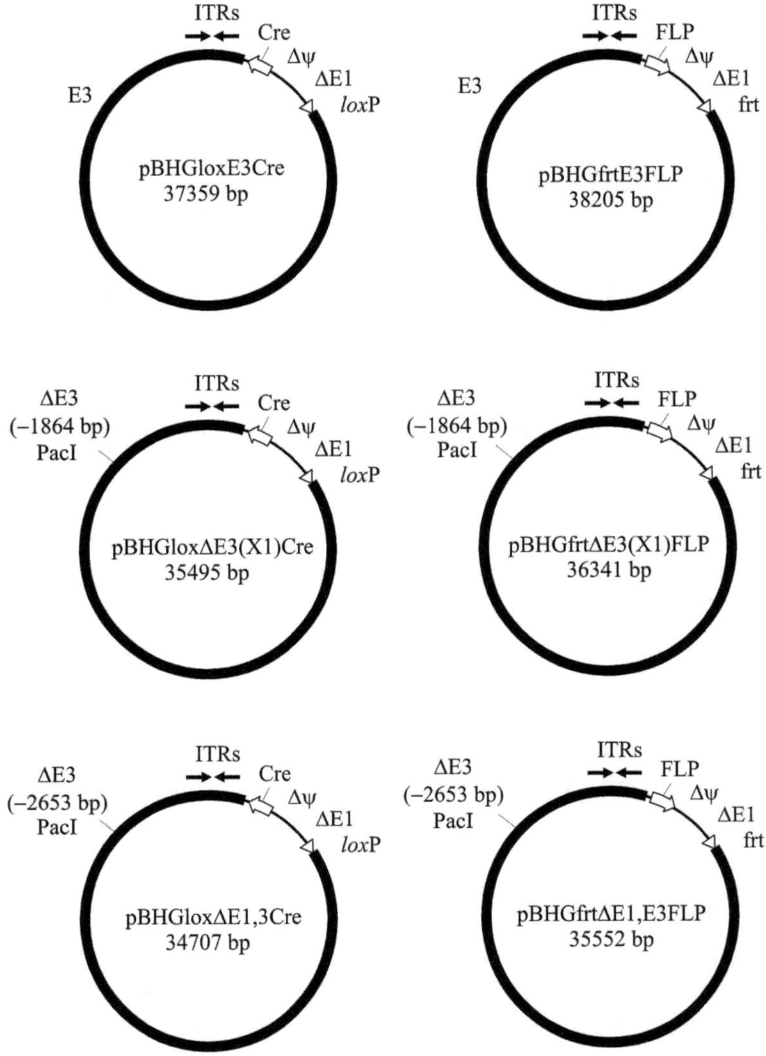

Fig. 3. Ad genomic plasmids used for vector rescue by site-specific recombination in vivo. The plasmids pBHGloxE3Cre, pBHGloxΔE3(X1)Cre, and pBHGloxΔE1,3Cre are used to rescue vectors bearing a wildtype E3 region, a 1864 bp deletion or a 2653 bp deletion of E3 respectively, by Cre-mediated recombination. The plasmids pBHGfrtE3FLP, pBHGfrtΔE3(X1)FLP and pBHGfrtΔE1,3FLP are used to rescue vectors bearing a wildtype E3 region, a 1864 bp deletion or a 2653 bp deletion of E3 respectively, by FLP-mediated recombination. The unique PacI restriction enzyme site in pBHGloxΔE1,3Cre, pBHGloxΔE3(X1)Cre, pBHGfrtΔE3(X1)FLP, and pBHGfrtΔE1,3FLP permit insertion of foreign sequences into the E3 deletion. Ad and bacterial plasmid sequences are represented by thick and thin lines, respectively, and *lox*P or frt sites are presented by a small triangle".

the wildtype or smaller E3 deletion (F. L. Graham; unpublished results). The presence of a unique PacI site in pBHGfrtΔE1,3FLP, pBHGfrtΔE1,3FLP, pBHGloxΔE3(X1)Cre and pBHGfrtΔE3(X1)FLP permits insertion of foreign sequences for rescue into the E3 deletion if desired (*see* **Fig. 3**). The choice of which Ad genomic plasmid to use is dictated by the size of the foreign sequence to be rescued, whether a wildtype or a deleted E3 region is desired in the vector and which site-specific recombination system is preferred/necessitated.

2.2. Shuttle Plasmids

Various small shuttle plasmids are available for insertion of foreign sequences and rescue into Ad vectors by Cre- or FLP-mediated recombination (*see* **Fig. 4**). The shuttle plasmids pDC311, pDC312, pDC511, and pDC512 are designed for the insertion of expression cassettes for rescue into vectors, the first two by Cre-mediated recombination and the last two by FLP-mediated recombination. The shuttle plasmids pDC315, pDC316, pDC515, and pDC516 are designed for the insertion of coding sequences for rescue into vectors, the first two by Cre-mediated recombination and the last two by FLP-mediated recombination. In these plasmids, the site of insertion is flanked by a murine cytomegalovirus (MCMV) immediate early promoter and the SV40 polyadenylation signal, thus permitting high-level transgene expression. The choice of which shuttle plasmid to use is dictated by whether expression from a strong viral promoter is desired, by the orientation of the polylinker and by the site-specific recombination system desired for vector rescue. We have observed that higher expression levels are obtained when the transcription orientation of the transgene is in the same direction as E1 and that the MCMV immediate early promoter is stronger in most cell lines than its more commonly used human counterpart *(22)*.

2.3. Preparation of Plasmid DNA

1. Plasmid DNA (Microbix Biosystems Inc).
2. Sterile LB broth (Lennox) (Difco, Sparks, Maryland, USA) and LB-agar plates supplemented with 50 µg/ml ampicillin.
3. Solution I: 10 mM EDTA, pH 8.0, 50 mM glucose, 25 mM Tris–HCl, pH 8.0, prepared from sterile stock solutions.
4. Solution II: 1% SDS, 0.2N NaOH, freshly prepared.
5. Solution III: 3M potassium acetate, 11.5% glacial acetic acid, autoclave sterilized.
6. 100% Isopropanol.
7. TE: 10 mM Tris–HCl, pH 8.0, 1 mM EDTA pH 8.0, autoclave sterilized.
8. CsCl.
9. 10 mg/ml ethidium bromide.

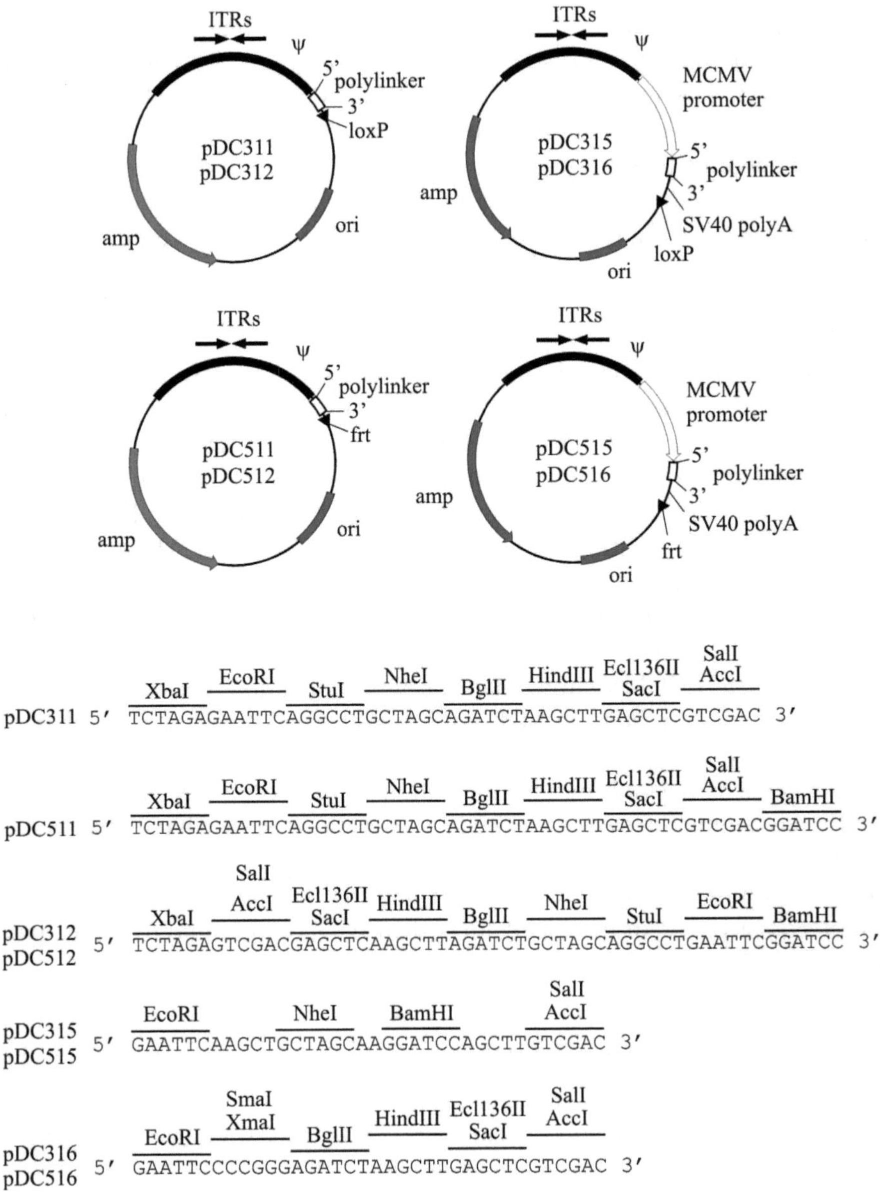

Fig. 4. Shuttle plasmids and their polylinker sequences used for vector rescue by in vivo site-specific recombination. The shuttle plasmids pDC311 (3276 bp), pDC312 (3288 bp), pD315 (3913 bp), and pDC316 (3913 bp) are used to rescue vectors by Cre-mediated recombination. The shuttle plasmids pDC511 (3277 bp), pDC512 (3277 bp), pDC515 (3957 bp), and pDC516 (3957 bp) are used to rescue vectors by FLP-mediated recombination. The plasmids pDC311, pDC312, pDC511, and pDC512 are used for insertion of expression cassettes (inserts with a promoter/enhancer and polyadenylation

2.4. Cell Culture

1. Low passage 293 cells (Microbix Biosystems Inc.).
2. 293N3S cells (Microbix Biosystems Inc.).
3. Complete medium: MEM (Invitrogen, Carlsbad, CA, USA; #61100) containing 10% fetal bovine serum (FBS) (heat inactivated), 100 units/ml penicillin/streptomycin and 2 mM L-glutamine. For adherent growth of 293 and 293N3S cells in monolayers.
4. Joklik's modified-MEM (Sigma, St. Louis, MO, USA; # M-0518) supplemented with 5% FBS (heat inactivated) 100 units/ml penicillin/streptomycin and 2 mM L-glutamine. For suspension growth of 293N3S cells in spinner flasks.
5. Citric Saline: 135 mM KCl, 15 mM sodium citrate, autoclave sterilized.

2.5. Cotransfection

1. Monolayers of low passage 293 cells at 70% confluency in 60-mm dishes.
2. Reagents for calcium phosphate transfection of plasmid DNA into mammalian cells in culture. We use the ProFection® Mammalian Transfection System by Promega, Madison, WI, USA (#E1200).
3. Complete medium (*see* **Subheading 2.4.**).
4. 2× maintenance medium: 2× MEM (Invitrogen; #61100) supplemented with 10% FBS (heat inactivated), 200 units/ml penicillin/streptomycin, 4 mM L-glutamine and 0.2% yeast extract.
5. 1% agarose solution, autoclave sterilized. Store at room temperature and melt in a microwave oven before use.
6. Ad genomic plasmid DNA (*see* **Subheading 2.1.**).
7. Shuttle plasmid DNA with desired foreign sequence inserted (*see* **Subheading 2.2.**).
8. The infectious Ad genomic plasmid pFG140 (Microbix Biosystems Inc.).
9. Phosphate-buffered saline (PBS): 137 mM NaCl, 8.2 mM Na_2HPO_4, 1.5 mM KH_2PO_4, 2.7 mM KCl, autoclave sterilized.
10. PBS^{++}: PBS supplemented with 0.68 mM sterile $MgCl_2$ and 0.5 mM sterile $CaCl_2$.
11. Glycerol, autoclave sterilized.
12. TE (*see* **Subheading 2.3.**).
13. Complete medium (*see* **Subheading 2.4.**).
14. Maintenance medium: MEM (Invitrogen; #61100) containing 5% FBS (heat inactivated), 100 units/ml penicillin/streptomycin and 2 mM L-glutamine.

Fig. 4. signal as well as coding sequence). The plasmids pDC315, pDC316, pDC515, and pDC516 bear a polylinker flanked by the murine cytomegalovirus immediate early promoter/enhancer and SV40 polyadenylation signal and are used for insertion of coding sequences (e.g., cDNAs).

15. Pronase solution: 20 mg/ml pronase in 10 mM Tris–HCl, pH 7.5: preincubate at 56 °C for 15 min followed by 37 °C for 1 h. Aliquot and store at –20 °C.
16. Pronase-SDS solution: 0.5 mg/ml pronase (above) in 0.5% SDS, 10 mM Tris–HCl, pH 7.4, 10 mM EDTA pH 8.0.

2.6. Large-Scale Production of Adenoviral Vectors

2.6.1. Using Monolayers of 293 Cells

1. 150-mm dishes of 80–90% confluent 293 cells.
2. Maintenance medium (*see* **Subheading 2.5.**).

2.6.2. Using Suspension 293N3S Cells

1. Joklik's modified-MEM (Sigma; # M-0518) supplemented with 5% FBS (heat inactivated) 100 units/ml penicillin/streptomycin and 2 mM L-Glutamine.
2. 293N3S cells (Microbix Biosystems Inc.).
3. 250 ml (#1965-61002) and 3 l (#1965-61030) spinner flasks (Bellco Glass, Vineland, NJ, USA).
4. Magnetic stirrer (#7785-D200) (Bellco Glass).

2.7. Purification of Adenovirus by CsCl Banding

1. 10 and 100 mM Tris–HCl, pH8.0, autoclave sterilized.
2. 5% Sodium deoxycholate, filter sterilize.
3. 2 M $MgCl_2$, autoclave sterilize.
4. DNase I: 100 mg bovine pancreatic deoxyribonuclease I in 10 ml of 20 mM Tris–HCl, pH 7.4, 50 mM NaCl, 1 mM dithiothreitol, 0.1 mg/ml bovine serum albumin, 50% glycerol. Aliquot and stored at –20 °C.
5. RNase A (10 mg/ml).
6. CsCl solutions:

Density (g/cc)	CsCl (g)	10 mM Tris–HCl, pH 8.0 (g)
1.35	70.4	129.6
1.25	54.0	146.0

Dissolve CsCl into 10 mM Tris–HCl, pH 8.0 solution in the amounts indicated above to achieve the desired density solution, filter sterilize (*see* **Note 1**). Weigh 1.00 ml to confirm density.

7. Glycerol, autoclave sterilize.
8. Beckman, Fullerton, CA, USA SW 40 Ti rotor and ultraclear tubes (#342413).
9. Beckman SW 55 rotor and ultraclear tubes (#344057).
10. Slide-A-Lyzer dialysis cassettes (Pierce, Rockford, IL, USA).
11. Dialysis buffer: 10mM Tris–HCl pH 8.0.

2.8. Characterization of Adenoviral Vector Preparations

1. All materials listed in **Subheading 2.5**.
2. 10 mM Tris–HCl, pH 8.0.
3. 10% SDS.
4. Pronase-SDS solution (*see* **Subheading 2.5.**).
5. 3 M sodium acetate pH 5.2, autoclave sterilized.
6. 95 and 70% ethanol.
7. Virion lysis buffer: TE supplemented with SDS to 0.1%.
8. Spectrophotometer and cuvettes.

3. Methods

Figure 5 presents an overview of the steps involved in the production of recombinant Ad vectors. First, the gene of interest is inserted into the appropriate shuttle plasmid (*see* **Fig. 4**) by standard molecular biology techniques. Monolayers of 293 cells are then cotransfected with the Ad genomic plasmid and the modified shuttle plasmid. The recombinant vector, generated by site-specific recombination in vivo between the two plasmids, forms a plaque in the cell monolayer. The plaques are isolated, the virus expanded, and the vector DNA is extracted for confirmation by restriction enzyme mapping. Large-scale vector production is then performed using either monolayers of 293 cells or suspension cultures of 293N3S cells and purified by CsCl banding. This chapter provides detailed protocols for each of these steps as well as basic vector characterization.

3.1. Preparation of Plasmid DNA

The foreign DNA of interest is inserted into an appropriate shuttle plasmid and transformed into *Escherichia coli* by conventional molecular biology techniques. The following is a protocol for the preparation of high quality plasmid DNA for cotransfection.

1. Inoculate 5 ml of LB supplemented with 50 μg/ml ampicillin with bacteria bearing the desired plasmid in the morning. Incubate culture at 37 °C with shaking (*see* **Note 2**).
2. Inoculate 500 ml of LB supplemented with 50 μg/ml ampicillin with the above culture in the late afternoon. Incubate overnight at 37 °C with shaking (*see* **Note 3**).
3. Transfer culture to a centrifuge bottle and pellet bacteria by spinning at 6,000 × *g* for 10 min at 4 °C.
4. Resuspend bacterial pellet in 40 ml of cold solution I, so that no cell clumps are visible.

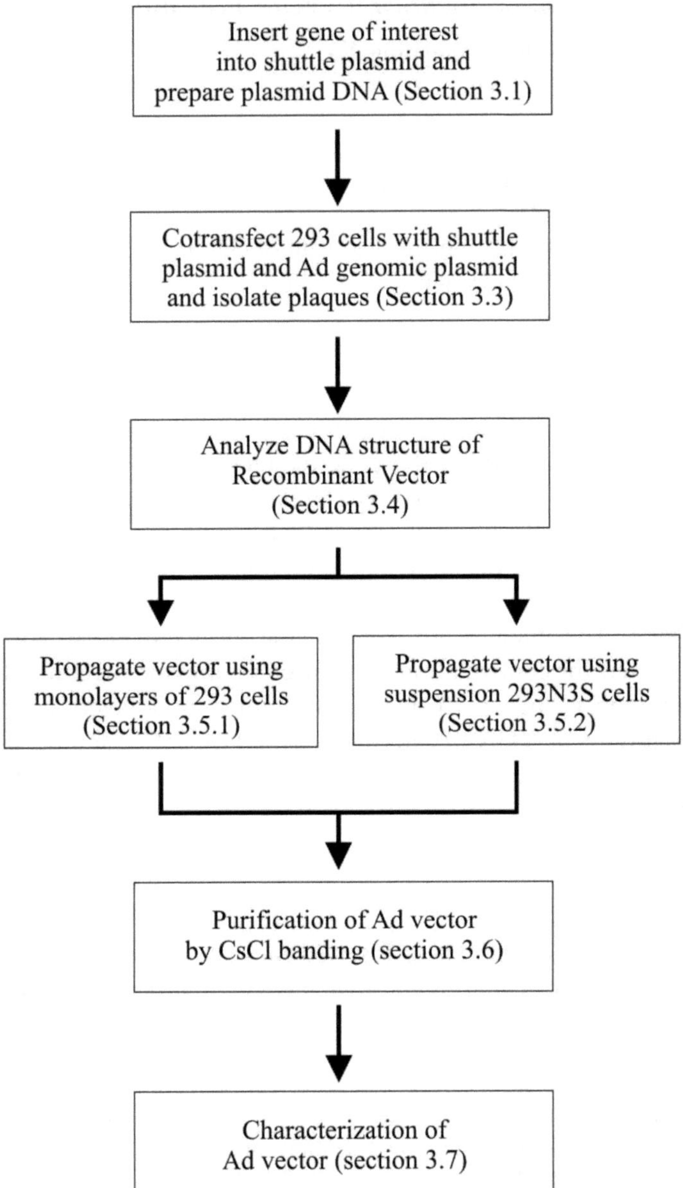

Fig. 5. Overview of the steps involved in rescue, propagation, purification, and characterization of Ad vectors.

5. Add 80 ml of freshly prepared solution II, mix thoroughly but gently by swirling. This should produce a relatively clear, viscous lysate.
6. Add 40 ml of cold solution III, mix thoroughly but gently by swirling and incubate for 20 min on ice. A white precipitate should form and the viscosity should be greatly reduced.
7. Centrifuge at 4 °C for 10 min at 6,000 × *g*.
8. Collect the supernatant by filtering it through 2–3 layers of cheesecloth into a centrifuge bottle.
9. Precipitate plasmid DNA by adding 100 ml of isopropanol mix well and incubate for 30 min at room temperature.
10. Pellet plasmid DNA by centrifuging at 4 °C for 10 min at 6,000 × *g*.
11. Pour off the supernatant and drain at room temperature for 15 min. Keeping the bottle upside down, wipe inside the rim with a clean KimWipe™ to remove all residual isopropanol.
12. Dissolve plasmid DNA pellet in 5 ml TE and transfer to 50-ml conical tube.
13. Add 8.8 g CsCl, mix to dissolve completely, and incubate on ice for 30 min.
14. Centrifuge at maximum speed in a Beckman tabletop centrifuge for 30 min at 5 °C. Slowly collect the supernatant using a 10-cc syringe and 16-Gneedle.
15. Transfer to a Beckman VTi 65.1 ultracentrifuge tube. Add 25 μl of 10 mg/ml ethidium bromide and fill the tube with light paraffin oil. Seal the tube and mix by inversion.
16. Centrifuge in a Beckman VTi 65.1 rotor at 55,000 rpm (288,000 × *g*) for 10–14 h at 14 °C.
17. Remove tube and support it with a stand. The supercoiled plasmid DNA band should be the dominant, if not only, red/pink band in the gradient. Puncture the top of the tube to allow entry of air. Collect plasmid DNA through the side of the tube with a 3-cc syringe and 18-G needle by puncturing the side of tube just below the band (*see* **Note 4**).
18. Transfer plasmid DNA to a 15-ml polypropylene tube containing 5 ml isopropanol, which has been saturated with CsCl in TE, and mix immediately to extract the ethidium bromide into the solvent layer. Allow the phases to separate and remove the ethidium bromide solvent (pink) layer. Repeat extraction until the solvent layer is completely colorless.
19. Add TE to bring the volume up to 4 ml, add 8 ml cold 95% ethanol, and mix by inversion to precipitate the DNA.
20. Centrifuge at top speed for 15 min to pellet DNA. Discard supernatant and wash pellet twice with 10 ml 70% ethanol.
21. Remove as much of the 70% ethanol as possible, allow the pellet to dry and resuspend with an appropriate volume of TE to a concentration of 1–2 μg/μl. Heat at 65 °C with occasional mixing to dissolve the DNA.
22. Determine the plasmid DNA concentration by OD_{260} and digest a sample with appropriate diagnostic restriction enzymes and confirm the structure by agarose gel electrophoresis.

3.2. Cell Culture

3.2.1. 293 Cells

Low passage 293 cells are maintained in 150-mm dishes and are split 1 to 2 or 1 to 3 when they reach ~90% confluency (every 2–3 days). In general, for cotransfections, a confluent 150-mm dish of 293 cells is split into ten to twelve 60-mm dishes for use the next day at ~70% confluency (*see* **Note 5**).

1. Remove medium from 150-mm dish of 293 cells.
2. Rinse monolayer twice with 5 ml citric saline.
3. Remove citric saline from **step 2**, add 0.5 ml citric saline, and leave the dish at room temperature until cells start to round up and detach from the dish (no more than 15 min).
4. Tap the side of the dishes to detach all cells.
5. Resuspend cells with complete medium and distribute into new dishes.

3.2.2. 293N3S Cells

293N3S cells ***(23)*** are suspension adapted 293 cells. 293N3S cells are very useful for large-scale production of Ad as suspension cells are easier to handle in large quantities than adherent cells. Therefore, although vector production can be accomplished using adherent 293 cells, we use suspension 293N3S cells exclusively for large-scale vector production. For routine maintenance, 293N3S cells are grown in 150-mm dishes in complete MEM (*see* **Subheading 2.4.**) as described for 293 cells except that citric saline is not required (**Subheading 3.2.1., step 3** is omitted). For large-scale vector production, 293N3S cells are grown in spinner flasks in Joklik-modified MEM (*see* **Subheading 2.4.**). Pre-warm all cell culture reagents to 37 °C before use.

3.3. Cotransfection

Typically, a large number of plaques are generated by cotransfecting a single 60-mm dish of 293 cells with 2 μg of the shuttle plasmid and 2 μg of the Ad genomic plasmid by site-specific recombination (*see* **Table 1**). However, a number of factors can influence the efficiency of vector rescue including the quality of the DNA, the efficiency of transfection and especially the state of the 293 cells. Another important consideration is that the plaques be well isolated. Therefore, it is recommended that a range of DNA amounts be cotransfected to ensure that the vector is rescued and also that the plaques are well isolated. The infectious Ad genomic plasmid pFG140 ***(20)*** is used as a control for transfection efficiency and under optimal conditions should yield up to ~100 plaques per 0.5 μg. Below is a typical protocol in which four 60-mm dish of 293 cells are cotransfected with 0.5 μg, 2 μg, and 5μg of each plasmid.

Table 1
Vector Rescue Efficiency by In Vivo Site-Specific Recombination[a]

Shuttle plasmid	Ad genomic plasmid	Average plaques/60 mm dish
pCA35loxΔITR	pBHGloxΔE1,3Cre	43
	pBHGloxΔE3(X1)Cre	63
	pBHGloxE3Cre	48
pCA35frtΔITR	pBHGfrtΔE1,3FLP	41
	pBHGfrtΔE3(X1)FLP	27
	pBHGfrtE3FLP	25
	pFG140[b]	103

[a]60-mm dishes of 293 cells were cotransfected with 2 μg of each plasmid and plaques were counted 10 days post-cotransfection.[b]60-mm dishes of 293 cells were transfected with 0.5 μg of pFG140 and plaques were counted 10 days post-cotransfection.

1. Label four 60-mm dishes "A," four dishes "B," four dishes "C" and two dish "D" and seed them with 293 cells to reach ~70% confluency in 1–2 days for cotransfection.
2. In the late afternoon, 1 h before cotransfection, the medium from the 60-mm dishes of 293 cells is replaced with 5 ml of freshly prepared complete medium without washing.
3. We perform calcium phosphate – DNA coprecipitation for plasmid DNA cotransfections into 293 cells using the ProFection® Mammalian Transfection System by Promega (#E1200) according to the manufacturer's instructions. Label four 60-mm dishes "A," four 60-mm dishes "B," four 60-mm dishes "C," and one 60-mm dish "D." For each of the dishes labeled "A," cotransfect 0.5 μg of shuttle plasmid DNA and 0.5 μg of Ad genomic plasmid DNA. For each of the dishes labeled "B," cotransfect 2 μg of shuttle plasmid DNA and 2 μg of Ad genomic plasmid DNA. For each of the dishes labeled "C," cotransfect 5 μg of shuttle plasmid DNA and 5 μg of Ad genomic plasmid DNA. To dish "D" transfect 1 μg of pFG140 DNA. Rock the dishes to distribute the precipitate evenly and return to the incubator.
4. Next day, melt 1% agarose solution in a microwave oven and allow it to equilibrate to 45 °C in a waterbath. Equilibrate 2× maintenance medium to 37 °C. Once the two solutions have equilibrated to the desired temperature, prepare overlay solution by combining 70 ml of melted 1% agarose and 70 ml of 2× maintenance medium.
5. Remove the medium from the cotransfected dishes and add 10 ml of overlay prepared in **step 4** to each dish (*see* **Note 6**).
6. Allow the overlay to solidify at room temperature (10–15 min) and then return the dishes to the incubator. Plaques should begin to appear as early as 5 days post-cotransfection and will continue to appear until about 12–14 days post-cotransfection.

7. Ten days post-cotransfection, pick well isolated plaques from the monolayer by punching out agar plugs with a sterile cotton plugged Pasteur pipet attached to a rubber bulb (*see* **Note 7**).
8. Transfer the agar plugs into 0.5 ml PBS^{++} supplemented with glycerol to 10% in a suitable vial. Vortex briefly and store at –80 °C.

It is strongly recommended that the recombinant vector be isolated from individual plaques following cotransfection especially if large quantities of the vector are to be generated for extensive experimentation. However, plaque purification is not absolutely essential because all infectious virus generated should be the desired recombinant vector ***(17–19)*** and production can be expedited by omitting **steps 4 through 8** in **Subheading 3.3**. In this case, following overnight cotransfection, remove the medium from the monolayers and add 5 ml of maintenance medium. If complete cytopathic effect (CPE) is observed within 7 days post-cotransfection, then proceed from **step 4** in **Subheading 3.4**. If complete CPE is not observed by 7 days post-cotransfection, then proceed as described in **Note 8**.

3.4. Confirmation of Recombinant Vector Genomic Structures from Total Cellular DNA

Once plaques have been isolated, they are expanded for extraction of vector DNA for analysis and to yield a working vector stock as follows:

1. Seed 60-mm dishes of 293 cells (one per plaque) to reach ~90% confluency on the day of use.
2. Thaw virus plaque picks and vortex briefly. Remove medium from 60-mm dishes of 293 cells and add 0.25 ml of the plaque pick. Adsorb for 1 h in the incubator rocking the dishes every 10–15 min.
3. Following adsorption, add 5 ml of maintenance medium and return dishes to incubator until complete CPE is observed (≥90% cells rounded up and detached from dish, usually 3–4 days post-infection) (*see* **Note 8**).
4. Once complete CPE is reached, scrape cells into the medium and transfer 4 ml of the cell suspension into a suitable sized vial and 1 ml into an eppendorf tube. Add 0.4 ml of 40% sucrose to the vial containing the 4 ml of cell suspension and freeze at –80 °C. This crude viral lysate will be used for large-scale vector production. Centrifuge the eppendorf tube containing the 1 ml cell suspension at 750 × g for 1 min. Discard supernatant, add 0.5 ml pronase-SDS solution, and incubate at 37 °C overnight to extract viral DNA.
5. Add 1 ml 95% ethanol to the eppendorf tube above to precipitate DNA by inverting the tube several times.
6. Pellet DNA by spinning in a microcentrifuge (maximum speed for 2 min) and wash pellet twice with 70% ethanol. Let the pellet dry and resuspend in an appropriate volume of TE (50 to 100 μl). Dissolve DNA by heating at 65 °C with occasional vortexing.

7. Digest 5–10 μl of the DNA with an appropriate restriction enzyme and analyze the DNA structure by ethidium bromide staining following agarose gel electrophoresis to verify whether the DNA structure of the recombinant virus is correct.

Once the DNA structure of the vector has been verified, proceed with large-scale vector production.

3.5. Large-Scale Vector Production

Because most of the virus remains associated with the infected cells until very late in infection (i.e., until the cells lyse), high-titer stocks can be prepared easily by concentrating infected 293 cells. The following protocol describes the production of high-titer virus preparations using either monolayers of 293 cells or suspension cultures of 293N3S cells. However, because the greater ease of handling suspension cultures, 293N3S cells are preferable for the production of very large amounts of high-titer viral stocks.

1. Subject the lysate collected in **Subheading 3.4**, **step 4** to three cycles of freeze (–80 °C) thaw (37 °C). Allow the lysate to equilibrate to 37 °C.
2. Remove medium from one 150-mm dish of 90% confluent 293 cells.
3. Mix lysate from **step 1** well and use 0.5 ml to infect the 293 cells in **step 2**. Adsorb for 1 h in the incubator, rocking the dishes every 10 min. 15 ml of MEM (5% FBS) is added to the monolayer 1 h post-coinfection. For large-scale production using adherent 293 cells, proceed to **Subheading 3.5.1**. For large-scale production using suspension 293N3S cells proceed to **Subheading 3.5.2**.

3.5.1. Large-Scale Production from Monolayers of 293 Cells

1. When complete CPE is reached (~48 h post-coinfection, no more than 72 h), the cells (from **Subheading 3.5.**) are scraped into the medium, transferred to a 50-ml conical tube and add 1.5 ml of 40% sucrose. The cell suspension is mixed well and frozen at –80 °C. This viral lysate will be used to infect thirty 150-mm dishes of 90% confluent 293 cells.
2. Subject the viral lysate collated in **step 1** to three cycles of freeze (–80 °C) thaw (37 °C). Allow viral lysate to equilibrate to 37 °C.
3. Remove medium from thirty 150-mm dishes of 293 cells and add 0.5 ml of the viral lysate to each dish.
4. Adsorb for 1 h in the incubator, rocking the dishes every 10–15 min. Following adsorption, add 15 ml maintenance medium and return dishes to the incubator. Examine daily for signs of cytopathic effect.
5. Complete CPE should be reached by 48 h, at which time, the cells are scraped into the medium, and the cell suspension is centrifuged at 800 × *g* for 15 min.
6. Discard the supernatant and resuspend the cell pellet in 15 ml 100 mM Tris–HCl pH 8.0 and proceed to vector purification **Subheading 3.7**. Alternatively, the cells

may be resuspended in 100 mM Tris–HCl pH 8.0 supplemented with glycerol to 10% and frozen for purification at a later time.

3.5.2. Large-Scale Production from Suspension 293N3S Cells

1. Centrifuge the cell suspension from **Subheading 3.5** for 5 min at 750 × *g*. The supernatant is discarded, and the cell pellet is resuspended in 1 ml 100 mM Tris–HCl pH 8.0 supplemented with glycerol to 10% and frozen at –80 °C. This will be used to infect 3 l of 293N3S cells.
2. 293N3S cells are grown in 3-l spinner flasks in Joklik-modified MEM supplemented with FBS (heat inactivated) to 5%, penicillin streptomycin to 100 U/ml and L-glutamine to 2 mM. Warm media to 37 °C before use.
3. Transfer spent media from four confluent dishes of 293N3S cells into a 3-l spinner flask. Tap the dishes to detach 293N3S cells and transfer them into the 3-l spinner flask containing the spent media. Add fresh medium to a final volume of 1-l and tighten all lids.
4. Incubate at 37 °C (a humid, CO_2 environment is not necessary) on a magnetic stirrer set at 60 rpm.
5. Add 0.5 l the next day (1.5 l total).
6. Add 0.5 l the next day (2 l total).
7. Add 1 l the next day (3 l total).
8. Count (*see* **Subheading 3.5.3.**) and coinfect cells next day as described below. The cell density should be 2–4 × 10^5 cells/ml for coinfection. The cell density should increase each day and can be confirmed by daily counting before addition of medium.
9. Remove 0.1 ml of cell suspension from the 3 l culture and place into one well of a 24-well dish containing 1 ml fresh media and incubate at 37 °C. Cells should re-attach and form a monolayer within a few hours and this serves as a control for the health and appearance of uninfected cells.
10. Harvest cells by centrifugation (750 × *g* for 5 min at room temperature). Save 0.5 l of the spent medium.
11. Resuspend cell pellet in 100 ml of spent medium and transfer to a 250-ml spinner flask.
12. Infect cells by adding viral lysate from **step 1** and incubate for 2 h at 37 °C at 60 rpm.
13. Transfer the infected cells back into a 3-l spinner flask, add spent medium to 0.5 l and add 1.5 l of fresh MEM (5% FBS) (2 l total volume).
14. Transfer 0.1 ml of infected cells into one well of a 24-well dish containing 1 ml fresh MEM (5% FBS). These cells will initially re-attach to the dish but should all round up and detach by 48 h.
15. Incubate spinner flask for 48 h at 37 °C at 60 rpm.
16. Harvest coinfected cells by centrifugation at 750 × *g* for 5 min at room temperature.
17. Resuspend cells in a total of 15 ml 100 mM Tris–HCl pH 8.0 and transfer to a 50-ml conical tube and proceed with vector purification (*see* **Subheading 3.6.**)

or can be stored at –80 °C for processing later. For storage, resuspend the cells in 15 ml 100 mM Tris–HCl pH 8.0 supplemented with glycerol to 10%.

3.5.3. Counting 116 Cells in Suspension

1. Transfer 2 ml of cells into 15-ml conical tube.
2. Add 2 ml of 2× citric saline (270 mM KCl, 30 mM sodium citrate, autoclave sterilized. Pre-warmed to 37 °C).
3. Vortex for 10 s at maximum setting.
4. Incubate at 37 °C for 5–10 min.
5. Vortex for 10 s at maximum setting.
6. Obtain two independent cell counts using a hemocytometer.

3.6. Vector Purification by CsCl Ultracentrifugation

1. Add 2.0 ml of 5% sodium deoxycholate to lyse the infected cells from **Subheading 3.5.1** or **3.5.2**. The mixture should immediately become thick and gelatinous. Incubate at room temperature for 30 min with frequent mixing.
2. Add 170 μl of 2 M MgCl, 150 μl of RNase A (10 mg/ml), and 150 μl of DNase I (10 mg/ml) and incubate at 37 °C for 1 h and mix frequently to digest cellular DNA and RNA. The viscosity should be reduced significantly.
3. Remove cell debris by centrifugation at maximum speed in a tabletop centrifuge for 10 min at room temperature.
4. During **step 3**, prepare two CsCl step gradients; add 2 ml 1.35 g/cc CsCl to the bottom of two SW40 ultraclear ultracentrifuge tubes and then careful overlay with 3 ml of 1.25 g/cc CsCl.
5. Carefully overlay supernatant from **step 3** on top of the two CsCl step gradients. If necessary fill the remainder of the tubes evenly with the 100 mM Tris–HCl pH 8.0.
6. Spin in ultracentrifuge at 151,000 × *g* (35,000 rpm) for 1 h at 4 °C with a SW40 rotor.
7. The vector is the lowest band in the step gradient. Pierce the tube with a 3-cc syringe and 18-G needle below the vector band and slowly retrieve the vector from the tubes.
8. Transfer the vector from both tubes into one SW55 ultraclear tube and fill the tube with 1.35 g/cc CsCl.
9. Spin at 115,000 × *g* (35,000 rpm) overnight at 4 °C with a SW55 rotor.
10. The next day, prepare Slide-A-Lyser dialysis cassette by pre-soaking cassette in 4 °C dialysis buffer according the manufacturer's instructions.
11. Retrieve the virus band from the CsCl gradient with a 1-cc syringe and 22-G needle and transfer into dialysis cassette as per manufacture's instructions. Dialyze overnight at 4 °C with 3 × 500 ml changes of 4 °C dialysis buffer with slow stirring to remove CsCl.

12. Retrieve the virus from the cassette with a 1-cc syringe and 22-G needle. Record the volume of vector in the syringe and transfer into a vial. Add glycerol to a final concentration of 10% and mix well.
13. Remove at least two small aliquots of the vector preparation and transfer them into microfuge tubes for characterization (*see* **Subheading 3.7.**); one for DNA extraction and another for absorbance at 260 nm. The volume of these aliquots will depend on the amount of vector obtained but is typically 10–50 µl. Additional small aliquots can be taken if desired for other characterization (i.e., infectivity, endotoxin, sterility, etc.).
14. Aliquot vector and store at –80 °C.

Purified vector can be used to easily produce more of itself. In this case, follow the steps outlined in **Subheading 3.5.1** or **3.5.2** but use 2 pfu/cell of the purified vector instead of the lysate from a 150-mm dish (*see* **Note 9**).

3.7. Characterization of Adenoviral Vectors

Before the recombinant vector is used for experimentation, several basic characterizations should be performed: *(1)* the genomic DNA structure should be verified, *(2)* the physical titer should be determined, and *(3)* the infectious titer should be ascertained. Other characterizations may also be desired such as RCA, endotoxin, sterility, adventious agents, etc.

3.7.1. DNA Structure

The DNA structure of the recombinant vector should be confirmed following large-scale preparation. Virion DNA can be extracted from CsCl banded virus for analysis as follows:

1. An appropriate volume (~25 µl depending on the concentration of the virus) of the purified virus is added to pronase-SDS solution to a final volume of 0.4 ml and incubated overnight at 37 °C to lyse the virions and digest virion proteins.
2. Virion DNA is precipitated by adding 1/10 volume 3 M sodium acetate pH 5.2 and 2.5 volume 95% ethanol and incubating at –20 °C for 15–30 min.
3. Spin in microcentrifuge for 10–15 min at maximum speed.
4. Discard supernatant and wash DNA pellet twice with 70% ethanol.
5. Dry DNA pellet and resuspend in an appropriate volume of TE.

3.7.2. Physical Titer

The physical titer is the concentration of total viral particles (vp) in a vector preparation and is obtained by measuring the absorbance at 260 nm following virion lysis *(24)* and correction for vector genome size. This method essentially measures the amount of viral DNA in a vector preparation and is expressed as vp/ml.

1. Add virion lysis buffer to the aliquot of vector taken at **step 13** of **Subheading 3.6** to a total volume of 0.5 ml.
2. Prepare a blank sample by adding virion lysis buffer to the same volume of vector vehicle (10 mM Tris–HCl pH 8.0 supplemented with glycerol to 10%).
3. Vortex both samples briefly (avoid bubbles).
4. Incubate samples at 56 °C for 10 min.
5. Vortex samples for 1 min.
6. Use the blank sample to set reference at 260 and 280 nm, then measure the absorbance of the vector sample at 260 and 280 nm.
7. Calculate vp/ml from absorbance at 260 nm using the formula below. The 260/280 ratio, a measure of relative purity, should be ~1.3.

$$(OD_{260})(\text{dilution factor})(1.1\times10^{12})(36)/(\text{size in kb})$$

3.7.3. Infectious Titer

The procedure outlined below is used to plaque purify recombinant vectors and is also used to determine the infectious titer of vector stocks (*see* **Note 10**).

1. Seed 60-mm dishes of 293 cells to reach ~70% confluency in 1–2 days for titration.
2. Prepare serial dilutions ($10^{-4}$to 10^{-10}) of the recombinant virus in PBS^{++}.
3. Remove medium from 60-mm dishes of 293 cells and infect with 0.1 ml of the diluted samples. Return dishes to the incubator and adsorb for 1 h, rocking the dishes every 10–15 min.
4. During the adsorption period, melt 1% agarose solution in a microwave oven and equilibrate to 45 °C in a waterbath. Equilibrate 2x maintenance medium to 37 °C.
5. Following 1 h adsorption, combine equal volumes of molten agarose solution with 2x maintenance medium, mix well, and gently overlay dishes with 10 ml (*see* **Note 6**).
6. Allow overlay to solidify for 10–15 min and return dishes to the incubator.
7. Plaques should start to appear about 4 days post-infection and should be counted 12 days post-infection. For isolation of recombinant virus by plaque purification well-isolated plaques should be picked according to **steps 7** and **8** of **Subheading 3.3** around10 days post-infection. The plaque-purified vectors are expanded according to **Subheading 3.4** and used as inoculum for the preparation of high titer viral stocks (*see* **Subheading 3.5**.).
8. Determine the vector concentration in plaque forming units per ml (pfu/ml) as follows:

$$\text{titer} = (\text{number of plaques})\ (\text{dilution factor})/(\text{infection volume})$$

Calculate the titer from dishes bearing 20–70 plaques.

4. Notes

1. Pre-wet filter with a small amount of sterile water.
2. For the large Ad genomic plasmid, bacterial cultures should be started from well-isolated colonies picked from a bacterial plate less than 1-week old.
3. For higher yields of plasmid DNA, richer medium such as Super Both (SB) can be used; LB broth supplemented with 22 g/ml peptone, 15 g/ml yeast extract, 1 g/ml D-glucose and 0.005 N NaOH.
4. Keep tubes covered with foil or otherwise in the dark except when recovering plasmid DNA bands. Do not expose to fluorescent or UV light more than necessary.
5. 293 cells should never be allowed to become overconfluent (neglecting to pass the cells once they have reached 100% confluency), should not be seeded too thinly (<20% confluency), and should have regular medium changes between splits (twice weekly if they are not growing rapidly enough to permit splitting every 2–4 days). It is recommended that a sufficient number of ampoules of the cells be frozen and stored in liquid N_2 to permit initiation of new cultures when the passage number of the lab stocks reaches 40–45 passages or when the cells are no longer behaving well under agar overlays (*see* **Subheadings 3.3.** and **3.7.3.**). Higher passage or poorly adherent cells may be suitable for growth of virus but the properties of the cells are more critical for plaque assays and cotransfections.
6. It is important to perform this step quickly to prevent the overlay solution from solidifying prematurely but gently to prevent disturbing the monolayer.
7. While it is tempting to isolate the plaques as soon as they are visible, it is recommended that they be isolated at about 10 days post-cotransfection. This ensures that the plaques chosen are well isolated and that no other overlapping plaques will form in close proximity.
8. It is important that the DNA be extracted following complete CPE, so that vector DNA bands are clearly visible above the background of cellular DNA. If complete CPE is not reached by 5–6 days post-infection (most likely due to low multiplicity), then scrape the monolayer into the medium and transfer the cell suspension into a suitable vial and supplement with glycerol to 10%. Freeze (–80 °C)-thaw the cell suspension and infect 60-mm dishes of 90% confluent 293 cells with 0.2–0.4 ml as described in **Subheading 3.4**. Complete CPE should be reached within 6 days, and the vector DNA can be extracted.
9. Homologous recombination between the Ad vector and the Ad sequences in 293 cells ***(24)*** results in the emergence of replication competent Ad. The frequency with which Ad vectors recombine with Ad sequences in 293 cells is unknown but in general $E1^+$ RCA replicate faster than $E1^-$ vectors. Consequently the proportion of RCA increase with prolonged propagation of the vectors in 293 cells ***(25)***. To minimize RCA contamination, vectors should not be serially propagated indefinitely. It is recommended that large-scale vector preparations be initiated from a stock prepared following plaque purification (*see* **Subheading 3.7.3.**). If the original plaque purified stock is exhausted, plaque purification can be repeated.

10. To accurately determine vector concentration, titrations should be performed in duplicate or triplicate. The number of plaques should vary in direct proportion to the dilution factor; otherwise, repeat the titration making sure that the samples are thoroughly mixed when setting up the serial dilutions.

Acknowledgments

PN is supported by grants from the National Institutes of Health (P50 HL59314, R01 DK067324, R01 HL083047 and P51 RR13986), the Texas Affiliate of the American Heart Association (0465102Y) and the Cystic Fibrosis Foundation (NG0530 and NG05G0).

References

1. Shenk, T. (1996) Adenoviridae: the viruses and their replication. In *Fields Viology* (Eds., Fields B. N., Knipe D.M., Howely P. M) pp. 2111–2148. Lipponcott-Raven Publishers, Philadelphia, PA.
2. Bergelson, J. M., Cunningham. J. A., Droguett G, et al. (1997) Isolation of a common receptor for Coxsackie B viruses and adenoviruses 2 and 5. *Science* **275**, 1320–1323.
3. Hong, S. S., Karayan, L., Tournier, J., Curiel, D. T., and Boulanger, P. A. (1997) Adenovirus type 5 fiber knob binds to MHC class I alpha2 domain at the surface of human epithelial and B lymphoblastoid cells. *Embo J.* **16**, 2294–2306.
4. Wickham, T. J., Mathias, P., Cheresh, D. A., and Nemerow, G. R.(1993) Integrins alpha v beta 3 and alpha v beta 5 promote adenovirus internalization but not virus attachment. *Cell* **73**, 309–319.
5. Wickham, T. J., Segal, D. M., Roelvink, P. W., et al. (1996) Targeted adenovirus gene transfer to endothelial and smooth muscle cells by using bispecific antibodies. *J Virol.* **70**, 6831–6838.
6. Goldman, M., Su, Q., and Wilson, J. M. (1996) Gradient of RGD-dependent entry of adenoviral vector in nasal and intrapulmonary epithelia: implications for gene therapy of cystic fibrosis. *Gene Ther.* **3**, 811–818.
7. Mellman, I. (1992) The importance of being acidic: the role of acidification in intracellular membrane traffic. *J. Exp. Biol.* **172**, 39–45.
8. Leopold, P. L., Ferris, B., Grinberg, I., Worgall, S., Hackett, N. R., and Crystal, R. G. (1998) Fluorescent virions: dynamic tracking of the pathway of adenoviral gene transfer vectors in living cells. *Hum. Gene Ther.* **9**, 367–378.
9. Greber, U. F., Willetts, M., Webster, P., and Helenius, A. (1993) Stepwise dismantling of adenovirus 2 during entry into cells. *Cell* 75,477–486; Berkner, K. L. (1988) Development of adenovirus vectors for expression of heterologous genes. *Biotechniques* **6**, 616–629.
10. Graham, F. L., and Prevec, L. (1992) Adenovirus-based expression vectors and recombinant vaccines. In *Vaccines: New Approaches to Immunological Problems* (Ed. R. W. Ellis) pp. 363–389. Butterworth-Heinemann, Boston, MA.

11. Hitt, M., Addison, C. L., and Graham, F. L. (1997) Human adenovirus vectors for gene transfer into mammalian cells. *Adv. Pharmacol.* **40**, 137–206.
12. Hitt, M. M., Parks, R. J., and Graham, F. L. (1999) Structure and genetic organization of adenovirus vectors. In *The Development of Human Gene Therapy* (Ed. T. Friedman) pp.61–86. Cold Spring Harbor Laboratory Press, Cold Spring Harbor, NY.
13. Tatsis, N. and Ert, H.C.J. (2005) Adenoviruses as vaccine vectors. *Mol. Ther.* **10**, 616–629.
14. Graham, F. L., Smiley, J., Russell, W. C., and Nairn, R. (1977) Characteristics of a human cell line transformed by DNA from human adenovirus 5. *J. Gen. Viol.* **36**, 59–72.
15. Danthainne, X., and Imperiale, M. J. (2000) Production of first generation adenovirus vectors: a review. *Gene Ther.* **7**, 1707–1714.
16. Ng, P., Parks, R. J., Cummings, D. T., Evelegh, C. M., Sankar, U., and Graham F. L. (1999) A high efficiency Cre/*lox*P based system for construction of adenoviral vectors. *Hum. Gene Ther.* **10**, 2667–2672.
17. Ng, P., Parks, R. J., Cummings, D. T., Evelegh, C. M., and Graham, F. L. (2000) An enhanced system for construction of adenoviral vectors by the two plasmid rescue method. *Hum. Gene Ther.* **11**, 693–699.
18. Ng, P., Cummings, D. T, Evelegh, C. M., and Graham, F. L. (2000) The yeast recombinase flp functions effectively in human cells for construction of adenovirus vectors. *Biotechniqes* **29**, 524–528.
19. Graham, F. L. (1984) Covalently closed circles of human adenovirus DNA are infectious. *EMBO J.* **3**, 2917–2922.
20. Bett, A. J., Prevec, L., and Graham, F. L. (1993) Packaging capacity and stability of human adenovirus type 5 vectors. *J. Virol.* **67**, 5911–5921.
21. Addison, C. L., Hitt, M., Kunsken, D., and Graham, F. L. (1997). Comparison of the human versus murine cytomegalovirus immediate early gene promoters for transgene expression by adenoviral vectors. *J. Gen. Virol.* **78**, 1653–1661.
22. Graham, F. L. (1987) Growth of 293 cells in suspension culture. *J. Gen. Virol.* **68**, 937–940.
23. Maizel, J. V., White, D., and Scharff, M. D. (1968) The polypeptides of adenovirus. I. Evidence of multiple protein components in the virion and a comparison of types 2, 7a, and 12. *Virology* **36**, 115–125.
24. Louis, N., Evelegh, C., and Graham, F. L. (1997) Cloning and sequencing of the cellular/viral junction from the human adenovirus type 5 transformed 293 cell line. *Virology* **233**, 423–429.
25. Lochmuller, H., Jani, A., Haurd, J. Prescott, S., Simoneau, M., Massie, B., Karpati, G., and Acsadi, G. (1994) Emergence of early region 1-containing replication-competent adenovirus in stocks of replication-defective adenovirus recombinants (ΔE1+ΔE3) during multiple passages in 293 cells. *Hum. Gene Ther.* **5**, 1485–1491.

5

Large-Scale Production of Recombinant Adeno-Associated Viral Vectors

Alejandro Negrete and Robert M. Kotin

Summary

Current and future demands of viral vectors for the development of successful preclinical and clinical studies in human gene therapy and possible commercialization of gene therapy products require well-established large-scale production processes. One of the most promising vectors for human gene therapy is recombinant adeno-associated virus vectors (rAAVs). Some of the attractive features of rAAV are broad tissue tropism, low immunogenicity, ability to transduce both mitotic and post-mitotic cells, and long-term gene expression in non-dividing cells. Recently, we developed a novel technology for the production of these vectors exploiting baculovirus expression vectors (BEV) in insect cell cultures. Initially developed in small, shake flask format, this process has been successfully scaled to larger volumes. In an effort to standardize rAAV production in stirred tank bioreactors, we characterized the culture conditions to derive a set of parameters correlated with high rAAV yields. Measuring capacitance and dielectric spectroscopy with a permittivity probe enabled us to determine optimal times of infection and harvest. Consistent yields of rAAV, 2×10^{13} DNase-resistant vector genomes (vg) [1×10^{12} transducing units (tu)] per liter of cell culture were obtained in bioreactors with working volumes ranging from 10 to 40 l. This represents significant progress toward establishing a robust large-scale process at industry level.

Key Words: Adeno-associated virus vectors; gene therapy, large-scale production; recombinant viral vectors; baculovirus-insect cell system, process development.

From: *Methods in Molecular Biology, vol. 433: Volume 1: Production and In Vivo Applications of Gene Transfer Vectors*
Edited by: J. M. Le Doux © Humana Press, Totowa, NJ

1. Introduction

Vectors derived from adeno-associated virus (AAV) transduce various tissues types in vivo including skeletal ***(1,2)***, and cardiac muscles ***(3,4)***, liver ***(5,6)***, central nervous system neurons ***(7,8)***, eye ***(9,10)***, and lung ***(11)*** in human, non-human primate, rodent species, and other animals. In tissue culture models, rAAV has been shown to transduce hematopoietic cells ***(12)*** as well as human cancer cell lines obtained from various tumor types (unpublished observations). Because of the broad tropism of rAAV and ability to transduce dividing and non-dividing cells, a wide range of applications are undergoing pre-clinical development for numerous diseases, arthritis, Alzheimer's disease, solid cancer ***(13–15)*** ocular diseases, alpha-1-antitrypsin deficiency, and Duchenne muscular dystrophy ***(16–18)***. Clinical trials for cystic fibrosis and hemophilia B have been conducted, and a Parkinson's disease trial is currently in progress. Although the trials for cystic fibrosis and hemophilia B did not achieve a therapeutic effect, the vector did not produce any serious adverse events. Therefore, for the doses and routes of administration that were evaluated in the clinical protocols, administering rAAV to humans appears to be safe.

The genus dependovirus of the *Parvoviridae* family consists of the AAV species. The *Parvoviridae* are unique in the biological kingdom for having a linear single-stranded DNA genome. The AAV genome is approximately 5 kilobases (kb) in length and has terminal repetitive sequences (ITR) that are capable of intra-molecular base pairing to form a partially duplex structure at each end of the genome. For AAV DNA replication, two of the four non-structural proteins are required *in trans*, either Rep 78 or Rep 68, and Rep 52 or Rep 40. The three structural proteins VP1, VP2, and VP3 assemble spontaneously to form the non-enveloped, icosahedral capsid. The ITRs are the only AAV genetic elements required in *cis* for replication and encapsidation. In addition, replication and production of both wildtype AAV and rAAV require functions from another, so-called helper viruses such as adenovirus. A recently developed technology for producing rAAVs exploits the baculovirus expression vector (BEV) system. Three BEVs are used to provide the *trans* and *cis* acting components required for rAAV production, and no adenovirus functions are necessary ***(19)***. The baculovirus insect cell system is a powerful technology for overexpression and production of commercially valuable recombinant proteins ***(20)***. Some advantages of this baculovirus-insect cell system include growth in serum-free culture medium, high tolerance to osmolality, linear scalability in suspension cultures, and high protein expression levels. Although this insect cell-based production system has the potential to alleviate the major bottleneck in the manufacturing of rAAVs, the practical aspects of large-scale production and processing of vector remain unresolved.

Online monitoring of insect cell cultures used in the production of recombinant proteins with the BEV system provides a valuable tool for the optimization, operation, and control of the production process *(21–25)*. Determination of capacitance or its normalized measure, ratio frequency dielectric permittivity, provides a means to estimate viable biomass in insect cell cultures *(26)*. This measurement has become very important because cell density at the time of infection has a very strong influence on the final volumetric yield of product *(20,27)*. Spectroscopy analysis of the relative permittivity at different frequencies determines the critical frequency where the cell is polarized. This value facilitates the determination of the precise time of infection in the bioreactor and hence the optimal time for harvesting. Detailed characterization of the conditions and kinetics of rAAVs production is essential for process optimization and bioprocessing scale-up.

2. Materials

2.1. Insect Cell Culture and Cell Count

1. Sf9 cells, serum-free medium adapted (GIBCO, Carlsbad, CA, USA).
2. HyQ®SFX-Insect serum-free media (HyClone, Logan, UT, USA).
3. Baculovirus Bac-Rep, Bac-Cap, and Bac-GFP *(19)*.
4. Stericup™ Presterilized vacuum-driven disposable filtration system with 0.22 μm polyethersulfone membrane (Millipore, Bellerica, MA, USA).
5. Erlenmeyer cell culture flasks 125 ml and 1 l with vented cap (Corning, Corning, NY, USA).
6. BioFlo 110 Modular Benchtop Fermentor. Jacketed glass vessel, capacity of 10 l working volume, pitch blade impeller, four gases mixer, primary control unit (New Brunswick Scientific, Edison, NJ, USA) (*see* **Note 1**).
7. Bioreactor 40 l. Jacketed stainless steal vessel, capacity of 40 l working volume, pitch blade impeller, TRYTON controller (Pierre Guerin Technologies, Mauze, France) (*see* **Note 2**).
8. Gentamicin, 50mg/ml (Quality Biologicals Inc., Gaithersburg, MD, USA).
9. Guava®Viacount®Flex (Guava Technologies Inc., Hayward, CA, USA).
10. Pluronic® F-68, 10% (GIBCO).
11. AcroVent®Device with 0.2μm PTFE membrane (Pall Life Sciences, East Hills, NY, USA).
12. Aluminum foil 6 × 6" (VWR, West Chester, PA, USA).
13. Steam indicator tape (3M, St. Paul, MN, USA).
14. Avery Denninson™ cable ties (Avery Denninson, Fitchburg, MA, USA).
15. Screw compressor clamp (VWR).
16. OAKTON® Standard Buffer Solutions (pH 4.0, pH 7.0) (OAKTON, Vernon Hills, IL, USA).
17. Fogale online biomass monitoring system (Fogale Nanotech, Nimes, France).

18. Autoclavable 20 l carboy with spigot (Nalge Nunc International, Rochester, NY, USA).
19. Phosphate Buffer Solution (PBS), pH 7.4 (Biosource, Carlsbad, CA, USA).
20. SciPure™ Automated TFF System (SciLog, Middleton, WI, USA) (*see* **Note 3**).
21. Polyethersulfone SARTOCON Cassette, 100KDa cut off (Sartorious, Goettingen, Germany).

2.2. Poly-Ethylene Glycol (PEG) Precipitation and Gradient Ultracentrifugation

1. PEG, WT 8000 (Sigma-Aldrich, St. Louis, MO, USA).
2. CsCl ultrapure (Invitrogen, Carlsbad, CA, USA).
3. Slide-A-Lyzer®Mini Dialysis Units, 10,000 MWCO (PIERCE, Rockford, IL, USA).
4. PBS, pH 7.4 (Biosource).
5. 2M $MgCl_2$ (Quality Biologicals Inc.).

2.3. SDS–Polyacrylamide Gel Electrophoresis (SDS–PAGE)

1. Precast NuPAGE® 4–12% Novex Bis-Tris gel (Invitrogen; *see* **Note 4**).
2. NuPAGE® LDS Sample Buffer, 4× (Invitrogen).
3. NuPAGE® Reducing Agent, 10× (Invitrogen).
4. NuPAGE® MOPS SDS running buffer (Invitrogen).
5. NuPAGE® Antioxidant (Invitrogen).
6. SeeBlue® Plus pre-stained standard (Invitrogen).
7. Xcell Surelock™ mini cell.
8. Simply Blue™Safe Stain (Invitrogen).

2.4. Western Blotting

1. NuPAGE® Transfer buffer, 20× (Invitrogen).
2. Methyl alcohol, anhydrous (Mackrodt Chemicals, Phillipsburg, NJ, USA).
3. Magic Mark™ standards to antibodies (Invitrogen).
4. Nitrocellulose membrane paper sandwich, 0.45 μm pore size (Invitrogen).
5. Blotting pads (Invitrogen).
6. PBS with 0.05% Tween 20, pH 7.4 (Sigma-Aldrich).
7. 5% instant non-fat dry milk (*see* **Note 5**) dissolved in PBS with 0.05% Tween 20, pH 7.4 (Sigma-Aldrich).
8. Western Lightning™ Chemiluminescence reagent plus (PerkinElmer Life Sciences Inc, Boston, MA, USA).
9. Glad® Cling Wrap (Oakland, CA, USA).
10. Kodak scientific imaging film, Biomax MS (Kodak, Rochester, NY, USA).
11. Hypercassettte™ (Amersham Biosciences, Piscataway, NJ, USA).
12. For CAP VP: Polyclonal rabbit anti-AAV5 capsid antibody (American Research Products, Inc., Belmont, MA, USA).

13. Anti-rabbit immunoglobulin G (IgG) (whole molecule) antibody (Sigma-Aldrich).
14. For GFP: Goat anti-GFP HPP (Fitzgerald Industries International Inc., Concord, MA, USA).
15. For REP: Mouse anti-AAV REP protein (Research Diagnostics Inc., Flanders, NJ, USA).
16. ECL™ anti-mouse IgG, Horseradish peroxidase linked whole antibody-from sheep (GE Healthcare UK, Buckinghamshire, UK).

2.5. Transduction Assay

1. HEK 293 cells (GIBCO).
2. Dulbecco's Modified Eagle Medium Nutrient Mixture F12 (D-MEM/F-12) (GIBCO).
3. Nunclon™ Surface 80 cm^2 vented tissue culture flask (Nunc, Rosklide, Denmark).
4. Fetal bovine serum (FBS), (HyClone).
5. PBS, pH 7.4 (Biosource).
6. Trypsin–EDTA (GIBCO).
7. Costar® 24-well cell culture cluster (Corning Incorporated).
8. Falcon 5-ml polystyrene round-bottom tube with cell-strainer cap (Becton Dickinson, Franklin Lakes, NJ, USA).
9. FACSCalibur flow cytometer (Becton Dickinson).

2.6. Baculovirus Titration

1. PBS, pH 7.4 (Biosource).
2. Bovine albumin (Sigma-Aldrich).
3. Anti-baculovirus envelope gp64 protein anybody (eBioscience, San Diego, CA, USA).
4. Guava-Express™ goat $F(a\,b')_2$-anti mouse IgG-PE anybody (Guava Technologies Inc.).

2.7. qPCR

1. Quanti Tect™ SYBR® green (Qiagen, Valencia, CA, USA)
2. Primer CMV-up: 5′-AACGCCAATAGGGACTTTCC-3′ (Integrated DNA Technologies, Coralville, IA, USA).
3. Primer CMV-down: 5′-GGGCGTACTTGGCATATGAT-3′ (Integrated DNA Technologies).
4. Plasmid pFBGR (made in-house).
5. TE buffer, pH 8.0 (Quality Biological Inc., Gaithersburg, MD, USA).
6. PCR Qualified Water (Quality Biological Inc.).
7. iCycler iQ™ PCR plates, 96-well (Bio-Rad Laboratories, Hercules, CA, USA).
8. iCycler iQ™ optical tape (Bio-Rad Laboratories).

3. Methods

3.1. Insect Cell Culture and AAV Production in Flasks

1. Rapidly thaw the cells contained in the cryogenic vial using a 37°C water bath.
2. Asceptically transfer the content to a 125-ml Erlenmeyer cell culture flask containing 27 ml of HyQ®SFX insect serum-free medium pre-warmed at 27°C.
3. Incubate at 27 ± 0.5°C in an orbital shaker platform rotating at 120–140 rpm.
4. Cells are sub-cultured until cell viability is higher than 90% (*see* **Note 6**).
5. For cell count, mix gently in an eppendorf tube 2 μl of Guava®Viacount Flex®, 20 μl of cell culture and 378 μl of HyQSFX® insect serum-free medium. Load two wells of a 96-well plate. Read samples using Guava EasyCyte® Mini (*see* **Note 7**).
6. In a 1-l Erlenmeyer cell culture flask, inoculate 200 ml of HyQSFX® insect serum-free medium with Sf9 cells at 1×10^6 cells/ml final cell density (*see* **Note 8**).
7. Incubate at 27 ± 0.5°C in an orbital shaker platform rotating at 120–140 rpm. Determine cell density 24 h after inoculation. Sf9 cells at cell density of 2×10^6 cells/ml and viability higher than 90% can be infected with three different baculoviruses at multiplicity of infection (MOI) of 3 (*see* **Note 9**).
8. Incubate at 27 ± 0.5°C in an orbital shaker platform rotating at 120–140 rpm for 72 h (*see* **Note 10**).

3.2. AAV Production in Bioreactor

3.2.1. For 10-l Bioreactor

1. Calibrate pH probe with standard buffer solutions to pH 4.0 and pH 7.0.
2. Set pH, dissolved oxygen (DO) and biomass probes in place.
3. Fill bioreactor with PBS, pH 7.4 submerging probes at least 5 cm.
4. Assemble the bioreactor and tighten all fittings except one.
5. Place a filter device at the exhaust of the condenser, the end of the sampling port, and the gas inlet. Secure connections with cable ties.
6. Clamp tubing from vent, inoculation, harvest, and sample ports.
7. Cover tubing and filter end with aluminum foil and tape them with steam indicator tape.
8. Autoclave bioreactor ≥20 min, 121°C, 15 psi (*see* **Note 11**).
9. Re-circulate cold water through the jacket. Set temperature control to 27°C and agitation speed to 200 rpm. Start sparging air at flow rate of 0.1-l/min.
10. When temperature and DO readings are stable calibrate DO probe to 100%.
11. Disconnect DO probe from the main unit and calibrate to zero.
12. Stop agitation and sparging and switch off temperature control.
13. Replace buffer with HyQ®SFX-Insect serum-free media (*see* **Note 12**).
14. Add Gentamicin at 25 μg/ml final concentration using a syringe through the port with septum (*see* **Note 13**).
15. Restart agitation at 100 rpm and temperature control at 27°C.

16. Set DO control to 30%. Once temperature is stabilized, calibrate biomass probe to zero. Bioreactor is ready for inoculation.
17. Inoculate bioreactor with Sf9 cells through inoculation port to get a final cell density of 1×10^6 cells/ml (*see* **Note 14**).
18. Start recording permittivity.
19. Monitor cell growth using Guava®Viacount®Flex. Cell density must be 2×10^6 cells/ml for infection.
20. Sterile filter baculovirus for infection using a presterilized vacuum-driven disposable filtration unit. Pool the three types of baculovirus (Rep, VP, and GFP) required for an MOI of 3 and pump them into the bioreactor through inoculation port.
21. Monitor and record temperature, agitation speed, pH, DO, and biomass during the entire process.
22. Take a 200 ml sample for comparative analysis and monitor cell density every 24 h. Keep unused samples at –80°C for future analysis.
23. At 72 hpi stop agitation, temperature control, sparge, and biomass recording. Harvest cell culture through harvest port. An example of the parameters monitored is shown in **Fig. 1**.

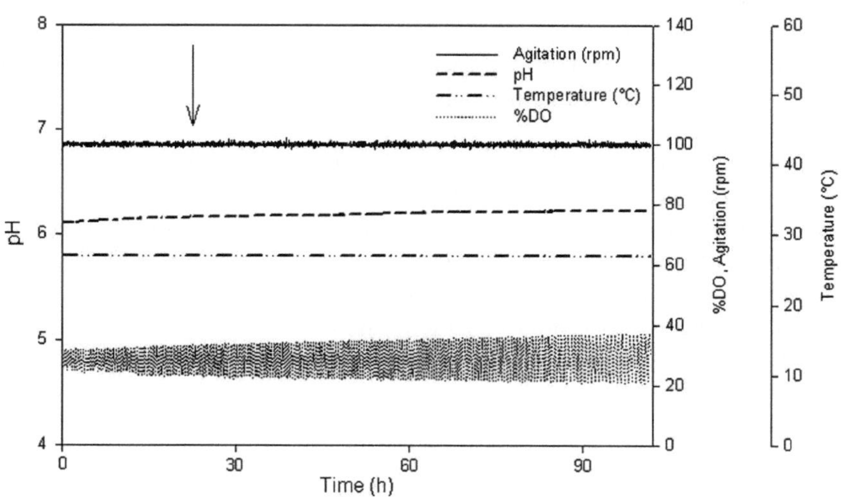

Fig. 1. Parameters commonly controlled during the production of rAAVs. Insect cells (Sf9) were grown in suspension in HyQSFX serum-free media in a 10 l stirred tank bioreactor. The set point for the parameters monitored included agitation 100 rpm, temperature 27°C and DO 30%. In this system, pH does not change during cell growth. Hence, pH was only monitored. Cells at a cell density of 2×10^6 cells/ml in exponential phase were infected with three different baculoviruses (Rep, GFP, and VP) at a MOI of 3 (time of infection denoted by the arrow). After infection cells require more oxygen. For this reason, DO fluctuates ±10 units.

24. Concentrate cell culture 10-fold by tangential flow filtration using a Sartocon® cassette 100 kDa MW cut-off.
25. Freeze concentrated material at –80°C for future processing.

3.2.2. For 40-l Bioreactor

1. Calibrate pH probe with standard buffer solutions to pH 4.0 and pH 7.0.
2. With all probes in place fill bioreactor with PBS pH 7.4 covering at least 5 cm above probes.
3. Sterilize bioreactor (*see* **Note 15**).
4. When temperature reaches 121°C calibrate DO probe to zero.
5. After sterilization process is over, sparge 2 l/min of air and set agitation control to 200 rpm. Once readings of DO and temperature (27°C) are stable, calibrate DO probe at 100% (*see* **Note 16**).
6. Switch off airflow and set agitation control back to 100 rpm.
7. Replace PBS with HyQ®SFX-Insect serum-free media under sterile conditions and add Gentamicin at 25 μg/ml final concentration and Pluronic® F-68 at 0.1% final concentration.
8. After temperature is stabilized at 27°C calibrate biomass probe to zero.
9. Inoculate bioreactor with Sf9 cells to get a final cell density of 1×10^6 cells/ml.
10. Start recording permittivity.
11. Sparge air only if DO value decreases below the set point (*see* **Note 17**).
12. Infect cells with a MOI of 3 of each of the three types of baculovirus (Bac-Rep, Bac-Cap, and Bac-GFP) when cell density is 2×10^6 cells/ml in a 40 l volume (*see* **Note 18**).
13. Take a 200-ml sample for analysis and monitor cell density every 24 h. Keep the samples at –80°C for future analysis.
14. At 72 hpi stop agitation, temperature control, sparge, and permittivity recording.
15. Harvest cell culture through harvest port and collect it in previously sterilized carboy. An example of the biomass monitoring is shown in **Fig. 2**.
16. Cell culture is concentrated 10-fold by tangential flow filtration using a Sartocon® cassette, 100 KDa cut off. Concentrated material can be freeze at –80°C for future use.

3.3. Recovery and Purification of AAV by PEG Precipitation and CsCl Gradient Ultracentrifugation

1. Freeze and thaw the 200-ml harvested culture three times (*see* **Note 19**).
2. Centrifuge at 2000 × *g* for 10 min and discard the cell pellet.
3. Add 50% PEG stock solution (w/v) to supernatant (final concentration 2%). Shake gently overnight at 4°C.
4. Centrifuge at 5000 × *g* for 30 min and discard the supernatant.
5. Resuspend the pellet with 12 ml of CsCl stock solution (refractive index 1.372; corresponding to a density of 1.4 g/cm^3).

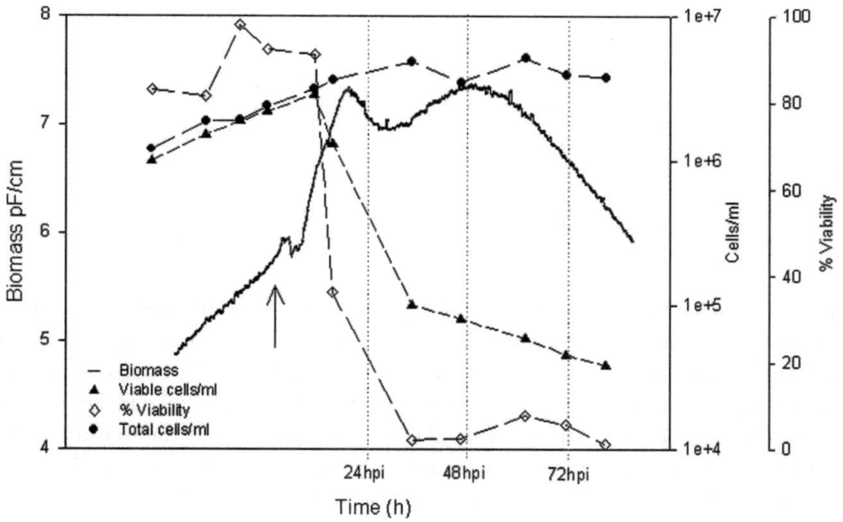

Fig. 2. Biomass monitoring during the production of rAAVs. Biomass was measured online as value of permittivity during the process. Cell growth is directly proportional to permittivity in non-infected cells. Other parameters such as total cell density, viable cell density, and viability were monitored offline and determined by Guava cell count. Cells in exponential phase and a cell density of 2 × 10^6 were infected with three different baculovirus (Rep, VP, and GFP) at a MOI of 3 (time of infection denoted by the arrow). After infection cells stop growing. At this point, permittivity correlates with the increase in cell diameter. Synchronized infection was determined around 18 hpi from the permittivity profile. After infection, cell density does not show significant increase, whereas viable cell density and thus viability decrease. The optimum harvest time is 72 h from the synchronized infection *(28)*.

6. Ultracentrifuge at 38,000 rpm (178,000 × *g* average) SW41 rotor (Beckman Coulter) at 20°C for 72 h.
7. Harvest 1 ml fractions and measure the refractive index. The fraction with refractive index corresponding to a density of 1.372 g/cm^3 contains the rAAV capsids with vector DNA.
8. Dialyze an aliquot of the fraction containing rAAV with 1 l of PBS + 2 mM $MgCl_2$ for ≥1 h using Slide-A-Lyzer®Mini Dialysis Units for further analysis (*see* **Note 20**).

3.4. SDS–PAGE Analysis

1. Samples are prepared as follows: 5 μl NuPAGE® LDS Sample Buffer, 4×; 2 μl NuPAGE® Reducing Agent, 10×; and 13 μl of dialyzed sample.
2. Heat samples at 70°C for 10 min.

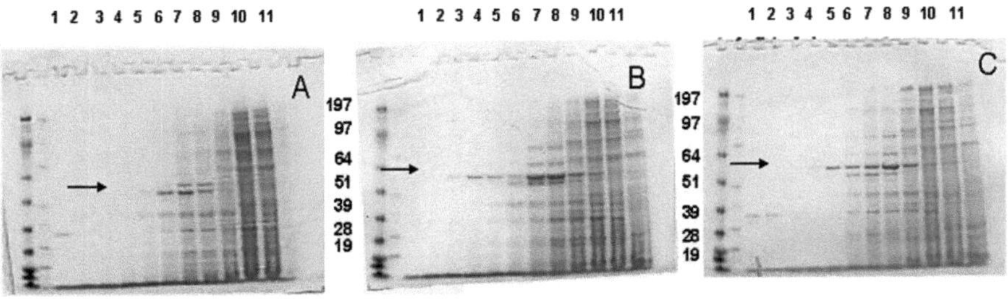

Fig. 3. Protein distribution across CsCl gradient at different times post-infection. Protein expression was analyzed by SDS–PAGE at 24 hpi (**A**); 48 hpi (**B**), and 72 hpi (**C**). Gels were stained with Simply Blue™ safe stain. The full capsids of rAAVs have a density of 1.4085 g/cm^3. The major capsid protein; VP3 can be visualized in the gel (denoted by the arrow) from 24 hpi (**A**) but in a lower density than the rAAV full capsid. This corresponds to empty particles. At 48 hpi, the VP3 protein is detected in the density range of the full capsids. At the time of harvest (72 hpi) (**C**), VP3 packaged in full capsids reaches the highest concentration. Densities (g/cm^3) of the CsCl fractions for the 24 hpi blots (**A**) correspond to line number 1, 1.4973; 2, 1.4625; 3, 1.4318; 4,1.4075; 5,1.3832; 6, 1.3631; 7, 1.3525; 8, 1.3271; 9, 1.3091; 10, 1.2933; and 11, 1.2859. For 48 hpi blots (**B**) densities are 1, 1.4952; 2, 1.4614; 3, 1.4329; 4, 1.4128; 5, 1.3821; 6, 1.3578; 7, 1.3409; 8, 1.3313; 9, 1.3165; 10, 1.3049; and 11, 1.2964. For 72 hpi (**D**) densities are 1, 1.5502; 2, 1.5101; 3, 1.4656; 4, 1.4329; 5, 1.4032; 6, 1.3747; 7, 1.3493; 8, 1.3313; 9, 1.3113; 10, 1.2975; and 11, 1.2838 *(28)*.

3. Fill the upper chamber of the Xcell Surelock™ mini cell with NuPAGE® MOPS SDS running buffer.
4. Load on the gel 10 μl SeeBlue® Plus pre-stained standard. Load 20 μl of prepared sample (*see* **Note 21**).
5. Fill the lower chamber of the Xcell Surelock™ mini cell with NuPAGE® MOPS SDS running buffer and add 500 μl NuPAGE® Antioxidant to the upper chamber.
6. Run electrophoresis at 190 V for 50 min.
7. Stain the gel with 20 ml Simply Blue™ Safe Stain for 1 h with gentle shaking.
8. Replace the Simply Blue™ Safe Stain with fresh distilled water to distain for 1 h. An example of the results produced is shown in **Fig. 3**.

3.5. Western Blot Analysis

1. Prepare samples for SDS–PAGE.
2. Load 10 μl SeeBlue® Plus pre-stained standard and 3.5 μl Magic Mark™ standard on the gel.
3. Running conditions for electrophoresis are the same than those for SDS–PAGE.
4. Submerge blotting pads, filter paper, and nitrocellulose membrane in transfer buffer (*see* **Note 22**).

5. Place two blotting pads, filter paper, gel, membrane, filter paper, and two blotting pads between the two electrodes of the blot module avoiding the formation of bubbles.
6. Run transfer electrophoresis in transfer buffer at 28 V for 1.5 h.
7. Block the membrane with 10 ml of 5% non-fat milk for 30 min with gentle shaking.
8. Replace the blocking milk with the first antibody and incubate at room temperature with gentle shaking for 1 h (*see* **Note 23**).
9. Wash membrane with 10 ml PBS with Tween 20, pH 7.4 for 5 min with gentle shaking. Repeat three times.
10. Replace washing buffer with secondary antibody and incubate at room temperature with gentle shaking for 1 h (*see* **Note 24**).
11. Wash membrane with 10 ml PBS with Tween 20, pH 7.4 for 5 min with gentle shaking. Repeat three times.
12. Replace washing buffer with 5 ml of Western Lightning™ Chemiluminescence reagent plus and incubate at room temperature with gentle shaking for 1 min (*see* **Note 25**).
13. Drain the chemiluminescence reagent and blot excess reagent by gently touching edge of membrane to paper towel or blotting paper. Wrap the membrane in cellophane cling wrap.
14. Fix the membrane into the Hypercassettte™ and expose the Kodak scientific imaging film for 30 s in a dark room (*see* **Note 26**). An example of the results produced is shown in **Fig. 4**.

3.6. Transduction Assay

1. Rapidly thaw the HEK 293 cells contained in the cryogenic vial using a 37°C water bath.
2. Transfer the content to a Nunclon™ Surface tissue culture flask containing 27 ml of D-MEM/F-12 10% FBS pre-warmed at 37°C.
3. Incubate at 37 ± 0.5°C in a 5% CO_2 environment.
4. Cells are sub-cultured until cells are 80% confluent. Discard media and wash cells with 5 ml of PBS.
5. Discard PBS and add 3 ml of Trypsin–EDTA. Incubate at 37 ± 0.5°C in a 5% CO_2 environment for 5 min.
6. Add 7 ml of D-MEM/F-12 10% FBS. Transfer 3 ml of this suspension to a Nunclon™ Surface tissue culture flask containing 20 ml of D-MEM/F-12 10% FBS pre-warmed at 37°C.
7. For cell count, mix gently in an eppendorf tube 2 µl of Guava®Viacount®Flex, 20 µl of cell culture and 378 µl of D-MEM/F-12 10% FBS. Load two wells of a 96-well plate. Read samples using Guava EasyCyte Mini.
8. Seed 0.5 ml of HEK 293 cells in a Costar® 24-well cell culture cluster to a cell density of 2×10^4 cells/well.
9. Incubate at 37 ± 0.5°C in a 5% CO_2 humidified environment for 24 h.

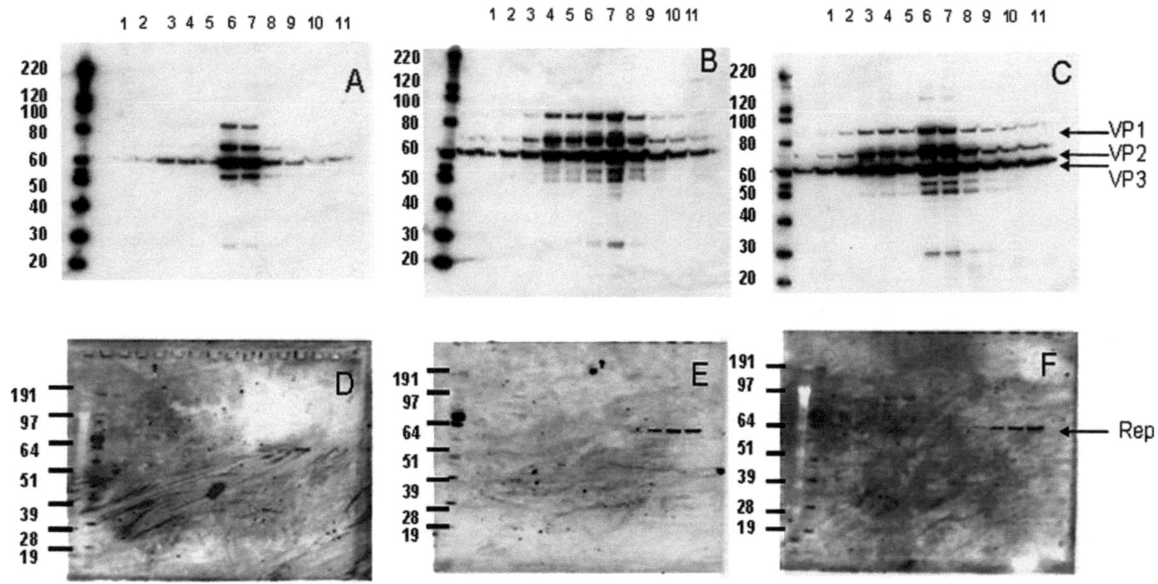

Fig. 4. Western blot of rAAVs distribution across CsCl gradient at different times post-infection. A better visualization of the components of the rAAV capsid (VP1, VP2, and VP3) was obtained using anti-capsid antibodies. Western blot analysis shows the separation of full and empty particles according to their density across the gradient. At early stages of infection (24 hpi) (**A**), there is expression of the three rAAVs constituent proteins. However, only at later stages (48 hpi) (**B**) and (72hpi) (**C**), it is possible to detect the maximum concentration of capsid components, with the highest production of full capsids (density 1.4085 g/cm^3) at 72 hpi (**C**). The bands observed using anti-Rep antibodies at 24, 48, and 72 hpi (**D**, **E**, and **F**, respectively) correspond to Rep protein from baculovirus. This protein is not packaged into rAAV and is not present in the fractions containing full particles. Densities (g/cm^3) of the CsCl fractions for the 24 hpi blots (**A**, **D**) correspond to line number 1, 1.4973; 2, 1.4625; 3, 1.4318; 4,1.4075; 5,1.3832; 6, 1.3631; 7, 1.3525; 8, 1.3271; 9, 1.3091; 10, 1.2933; and 11, 1.2859. For 48 hpi blots (**B**, **E**) densities are 1, 1.4952; 2, 1.4614; 3, 1.4329; 4, 1.4128; 5, 1.3821; 6, 1.3578; 7, 1.3409; 8, 1.3313; 9, 1.3165; 10, 1.3049; and 11, 1.2964. For 72hpi (**C**, **F**) densities are 1, 1.5502; 2, 1.5101; 3, 1.4656; 4, 1.4329; 5, 1.4032; 6, 1.3747; 7, 1.3493; 8, 1.3313; 9, 1.3113; 10, 1.2975; and 11, 1.2838 ***(28)***.

10. Make three 10-fold serial dilutions of samples in D-MEM/F-12 10% FBS.
11. Inoculate wells by duplicate using 1 μl of each dilution. Leave two wells as control.
12. Incubate at 37 ± 0.5°C in a 5% CO_2 environment for 48 h.
13. Discard medium. Add 200 μl Trypsin–EDTA per well.
14. Incubate at 37 ± 0.5°C in a 5% CO_2 environment for 5 min.
15. Add 300 μl D-MEM/F-12 and resuspend cells with the pipette.
16. Filter suspension using Falcon tube with cell-strainer cap.
17. Read GFP expression in FACSCalibur flow cytometer (*see* **Note 27**).

3.7. qPCR

1. Dilute sample 1:100 in PCR qualified water.
2. Make serial dilutions of plasmid pFBGR from 10e-4 to 10e-8 in TE-buffer to use as standards.
3. Prepare a master mix that contains per sample: 25 μl Sybr Green, 0.3 μl CMV upper primer (1 mM final concentration), 0.3 μl CMV lower primer (1 mM final concentration), and 23 μl water (*see* **Note 28**).
4. Add 49 μl of master mix per well into PCR plate.
5. Add 1 μl of standards or diluted sample. Mix well and seal with optical tape.
6. Run program (95°C, 3 min; 95°C, 45 s; 40 cycles of 95°C, 30 s; 55°C, 30 s; 72°C, 30 s; 4°C hold).
7. Calculate sample concentration from standard curve (vg/ml = copy number from PCR × 1000 × 100).

4. Notes

1. Glass bioreactors with working volumes ranging from 1 to 15 l are available from several sources. Regardless of configuration, probes are required for monitoring and regulating pH, DO, and temperature. Agitation, e.g. impeller speed has to be accurately controlled. Recording the parameters during the process is advised. As these glass bioreactors are sterilized in autoclaves, the interior dimensions of the autoclave have to be considered.
2. These types of bioreactors are available in different working volumes. These have to be equipped with pH, DO, temperature, and agitation controllers and a recording unit. The sterilization process in this case is in situ utilizing house steam or a clean steam generator.
3. Smaller configurations of filtration units, for processing up to 10 l, are available from different sources.
4. The NuPAGE® system from Invitrogen was used in this study. However, this can be replaced by any electrophoresis system.
5. The instant non-fat dry milk can be obtained in any supermarket.
6. Cell count can be performed using any other automatic or manual method that determines cell density and cell viability (e.g., Trypan blue).

7. Cell density must be at least 4 × 10^6 cells/ml to split 1:2. Cultured cells may require from 2 to 3 passages before cell viability reaches at least 90%.
8. Cell density should not be lower than 1 × 10^6 and higher than 6 × 10^6 cells/ml. In 1-l flasks the recommended cell culture volume is lower than 400 ml to ensure good oxygenation.
9. Sf9 cells must be in exponential phase for infection. It is recommended the addition of all baculovirus simultaneously.
10. After infection cell growth is stopped. Viability will start decreasing after 24 hpi and GFP expression becomes visible under UV microscopy. The highest expression of GFP is observed after 48 hpi.
11. Before taking the bioreactor out of the autoclave, tightly close the loose fitting and unclamp the vent filter. After this point, all steps involving connections to the bioreactor are made under sterile conditions.
12. The volume of HyQ®SFX-Insect serum-free media to be used depends on the volume of inoculum and the volume of baculovirus stock needed for infection.
13. To maintain aseptic conditions during injection through the septum, fill the septum port with 70% ethanol.
14. Introduce cells into the bioreactor using a pump for 2–3 s and then continue siphoning or gravity feed, thereby preventing cell damage. Sparging air is necessary until DO decreases below to the set point. This normally occurs after 24 h post inoculation or at even longer times. For better control of DO, use the four gases mix mode with air, oxygen, and nitrogen only, carbon dioxide. During cell growth and infection, pH does not show significant differences. Hence, pH control is not necessary.
15. If clean steam is unavailable, bioreactors can be sterilized by batching the vessel with dH_2O and heating the jacket to pressurize the vessel to 15 psi and 121°C for 15 min.
16. After this point, all steps involving connections to the bioreactor are made under sterile conditions.
17. Sparging air normally is not necessary until cells duplicate and reach a density of 2 × 10^6 cells/ml.
18. To prevent foaming, after infection it might be necessary to sparge pure oxygen instead of air to maintain the DO set point of 30%.
19. Freeze at –80°C and thaw in water bath at 37°C in the same Erlenmeyer flask.
20. If various samples are to be dialyzed simultaneously, it is recommended to exchange the dialysis buffer for fresh buffer at least once. Dialyzed sample will be used for SDS–PAGE and Western blot. Because of the large dilutions, it is not necessary to dialyze the sample for transduction assay and qPCR.
21. The amount of sample loaded on the gel is based on a 10-well precast gel.
22. The composition of transfer buffer is 50 ml NuPAGE® Transfer buffer 20×, 100 ml methyl alcohol (anhydrous), and 850 ml distilled water.
23. The first antibody is diluted in 5% milk. The following examples are for specific antibody stocks: *(1)* polyclonal rabbit anti-AAV5 capsid antibody, 1:5000, *(2)* goat anti-GFP HPP, 1:5000, *(3)* mouse anti-AAV Rep protein, 1:1000.

24. Secondary antibody for GFP detection is not necessary. For anti-rabbit IgG (whole molecule) antibody use 1:5000 dilution, for ECL™ anti-mouse IgG, horseradish peroxidase linked whole antibody – from sheep – use 1:5000 dilution.
25. Prepare the Western Lightning™ Chemiluminescence reagent plus fresh as required by mixing 1:2 oxidizing reagent and enhanced luminol reagent.
26. Exposure time may vary from 10 s to 5 min depending on the intensity of the signal.
27. To determine the AAV titer consider only the dilutions where between 1 and 25% of the cells are GFP positive. In this range, the percentage of positive cells is directly proportional to the number of virus in the sample assuming that one cell is infected by a single virus.
28. It is recommended to prepare master mixes for all reactions to analyze, including triplicates for each sample.

Acknowledgments

The authors thank Joseph Shiloach and Loc Trinh for providing the facilities for 40 l production and Geoffrey Esteban for the spectroscopy analysis.

References

1. Herzog, R. W. (2004) AAV-mediated gene transfer to skeletal muscle. *Methods Mol. Biol.* **246**, 179–194.
2. Kessler, P. D., Dubielzig, R., Grimm, D., and Kleinschmidt, J. A. (1996) Gene delivery to skeletal muscle results in sustained expression and systemic delivery of a therapeutic protein. *Proc. Natl. Acad. Sci. U. S. A.* **93**, 14082–14087.
3. Su, H., Huang, Y., Takagwa, J., Barcena, A., Rakawa-Hoyt, J., Grossman, W., and Kan, Y. W. (2006) AAV serotype-1 mediates early onset of gene expression in mouse hearts and results in better therapeutic effect. *Gene Ther.* [Advanced online publication].
4. Muller, O. J., Leuchs, B., Pleger, S. T., Grimm, D., Franz, W. M., Katus, H. A., and Kleinschmidt, J. A. (2006) Improved cardiac gene transfer by transcriptional and transductional targeting of adeno-associated viral vectors. *Cardivasc. Res.* **1**, 70–78.
5. Snyder, R. O., Miao, C. H., Patijn, G. A., Spratt, S. K., Danos, O., Nagy, D., Gown, A. M., Winther, B. B., Meuse, L., Cohen, L. K., Thompson, A. R., and Kay, M. A. (1997) Persistent and therapeutic concentrations of human factor IX in mice after hepatic gene transfer of recombinant AAV vectors. *Nat. Genet.* **16**, 270–276.
6. Manno, C. S., Arruda, V. R., Pierce, G. F., Glader, B., Ragni, M., Rasko, J., Ozelo, M. C., Hoots, K., Blatt, P., Konkle, B., Dake, M., Kaye, R., Razavi, M., Zajko, A., Zehnder, J., Nakai, H., Chew, A., Leonard, D., Wright, J. F., Lessard, R. R., Sommer, J. M., Tigges, M., Sabatino, D., Luk, A., Jiang, H.,

Mingozzi, F., Couto, L., Ertl, H. C., High, K. A., and Kay, M. A. (2006) Successful transduction of liver in hemophilia by AAV-Factor IX and limitations imposed by the host immune response. *Nat. Med.* **12**, 342–347.

7. Kaplitt, M. G., Leone, P., Samulski, R. J., Xiao, X., Pfaff, D. W., O'Malley, K. L., and During, M. J. (1994) Long-term gene expression and phenotypic correction using adeno-associated virus vectors in the mammalian brain. *Nat. Genet.* **8**, 148–154.
8. Xia, H., Mao, Q., Eliason, S. L., Harper, S. Q., Martins, I. H., Orr, H. T., Paulson, H. L., Yang, L., Kotin, R. M., and Davidson, B. L. (2004) RNAi suppresses polyglutamine-induced neurodegeneration in a model of spinocerebellar ataxia. *Nat. Med.* **10**, 816–820.
9. Flannery, J. G., Zolotukhin, S., Vaquero, M. I., LaVail, M. M., Muzyczka, N., and Hauswirth, W. W. (1997) Efficient photoreceptor-targeted gene expression in vivo by recombinant adeno-associated virus. *Proc. Natl. Acad. Sci. U. S. A.* **94**, 6916–6921.
10. Auricchio, A., and Rolling, F. (2005) Adeno-associated viral vectors for retinal gene transfer and treatment of retinal diseases. *Curr. Gene Ther.* **5**, 339–348.
11. Halbert, C. L., Allen, J. M., and Miller, A. D. (2002) Efficient mouse airway transduction following recombination between AAV vectors carrying parts of a larger gene. *Nat. Biotechnol.* **20**, 697–701.
12. Handa, A., Muramatsu, S., Qiu, J., Mizukami, H., and Brown, K. E. (2000) Adeno-associated virus (AAV)-3-based vectors transduce haematopoietic cells not susceptible to transduction with AAV-2-based vectors. *J. Gen. Virol.* **81**, 2077–2084.
13. Carter, B. J. (2005) Adeno-associated virus vectors in clinical trials. *Hum. Gene Ther.* **16**, 541–550.
14. Fukuchi, K. I., Tahara, K., Kim. H., D., Maxwell, J. A., Lewis, T. L., Ccavitti-Loper, M. A., Kim, H., Ponnazhagan, S., and Lalonde, R. (2006) Anti-Abeta single chain antibody delivery via adeno-associated virus for treatment of Alzheimer's disease. *Neurobiol Dis.* [Advanced online publication].
15. Lucas, A., Kremer, E. J., Hemmi, S., Luis, J., Vignon, F., and Lazennec, G. (2003) Comparative transductions of breast cancer cells by three DNA viruses. *Biochem. Biophys. Res. Commun.* **4**, 1011–1016.
16. De, B. P., Heguy, A., Hackett, N. R., Ferris, B., Leopold, P. R., Lee, J., Pierre, L., Gao, G., Wilson, J. M., and Crystal, R. G. (2006) High levels of persistent expression of alpha 1-antitrypsin mediated by the nonhuman primate serotype rh.10 adeno-associated virus despite preexisting immunity to common human adeno-associated viruses. *Mol. Ther.* **1**, 67–76.
17. Grieger, J. C. and Samulski, R. J. (2005) Adeno-associated virus as a gene therapy vector: development, production and clinical applications. *Adv. Biochem. Eng. Biotechnol.* **99**, 119–145.
18. Athanasopoulos, T. Graham, I. R., Foster, H., and Dickson, G. (2004) Recombinant adeno-associated viral (rAAV) vectors as therapeutic tools for Duchenne muscular dystrophy (DMD). *Gene Ther.* **1S**, 109–121.

19. Urabe, M., Ding, C., and Kotin, R. M. (2002) Insect cells as a factory to produce adeno-associated virus type 2 vectors. *Hum. Gene Ther.* **13**, 1935–1943.
20. Power, J., Greenfield, P. F., Nielsen, L., and Reid, S. (1992) Modelling the growth and protein production by insect cells following infection by a recombinant baculovirus in suspension culture. *Cytotechnology* **9**, 149–155.
21. Bédard, C., Jolicoeur, M., Jardin, B., Tom, R., Perret, S., and Kamen, A. (1994) Insect cell density in bioreactor cultures can be estimated from on-line measurements of optical density. *Biotechnol Tech.* **8**, 605–610.
22. Akhnoukh, R., Gerlinde K., and Schugerl, K. (1996) On-line monitoring and control of the cultivation of *Spodoptera frugiperda* Sf9 insect cells and [beta]-galactosidase production by *Autographa californica* virus vector. *Enzyme Microb. Tech.* **18**, 220–228.
23. Kamen, A. A., Bédard, C., Tom, R., Perret, S., and Jardin, B. (1996) On-line monitoring of respiration in recombinant-baculovirus-infected and uninfected insect cell bioreactor cultures. *Biotechnol. Bioeng.* **50**, 36–48.
24. Palomares, L. A., and Ramirez, O. T. (1996) The effect of dissolved oxygen tension and the utility of oxygen uptake rate in insect cell culture. *Cytotechnology* **22**, 225–237.
25. Zeiser, A., Elias, C. B., Voyer, R., Jardin, B., and Kamen, A. A. (2000) On-line monitoring of physiological parameters of insect cell cultures during the growth and infection process *Biotechnol. Prog.* **16**, 803–808.
26. Zeiser, A., Bedard, C., Voyer, R., Jardin, B., Tom, R., and Kamen, A. A. (1999) On-line monitoring of the progress of infection in Sf-9 insect cell cultures using relative permittivity measurements. *Biotechnol. Bioeng.* **63**, 122–126.
27. Licari, P., and Bailey, J. E. (1992) Modeling the population dynamics of baculovirus-infected insect cells: Optimizing infection strategies for enhanced recombinant protein yields. *Biotechnol. Bioeng.* **39**, 432–441.
28. Negrete, A., Esteban, G., and Kotin, R. M. (2007) Dielectric Spectroscopy monitoring for optimum harvest time determination during large-scale production of recombinant adeno-associated vectors. *Appl. microbial. Biotechnol. In press.* doi: 10.1007/s00253-007-1030-9.

6

Construction and Production of Recombinant Herpes Simplex Virus Vectors

William F. Goins, David M. Krisky, James B. Wechuck, Shaohua Huang, and Joseph C. Glorioso

Summary

Virus vectors have been employed as gene transfer vehicles for various pre-clinical and clinical gene therapy applications. Replication-competent herpes simplex virus (HSV) vectors that replicate specifically in actively dividing glial tumor cells have been used in Phase I–II human trials in patients with glioblastoma multiforme (GBM), a fatal form of brain cancer. Research during the last decade on the development of HSV vectors has resulted in the engineering of recombinant vectors that are totally replication defective, non-toxic, and capable of long-term transgene expression. This chapter describes methods for the construction of recombinant genomic HSV vectors based on the HSV-1 replication-defective vector backbones, steps in their purification, and their small-scale production for use in cell culture experiments as well as studies in animals.

Key Words: Herpes simplex virus; gene therapy; gene transfer; virus vectors; virus purification; virus production.

1. Introduction

Herpes simplex virus (HSV) is one of eight members of the human herpesviruses including HSV-1 and HSV-2 serotypes, varicella zoster virus (VZV), cytomegalovirus (CMV), Epstein-Barr virus (EBV), human herpesvirus 6 (HHV-6), HHV-7, and Kaposis-sarcoma herpes virus (KSHV), all of which cause some form of disease in the human host and are capable

From: *Methods in Molecular Biology, vol. 433: Volume 1: Production and In Vivo Applications of Gene Transfer Vectors*
Edited by: J. M. Le Doux © Humana Press, Totowa, NJ

of persisting for life within that host. HSV-1 is a large neurotropic virus containing 152 Kb of linear double-stranded DNA encoding approximately 83 gene products *(1)*. The HSV genome shown in **Fig. 1A** is composed of two segments, the unique long (U_L) and unique short (U_S), each of which is flanked by inverted repeats that contain important viral transcriptional regulatory genes. With few exceptions, the viral genes are present as contiguous transcription units, which make their genetic manipulation relatively straightforward for the construction of recombinant vectors. The HSV particle depicted in **Fig. 1B** is composed of over 34 proteins consists of a nucleocapsid composed of structural capsid proteins surrounded by an envelope containing glycoproteins essential for virus attachment and penetration into cells. Between the capsid and the envelope is a protein matrix known as the tegument. The tegument contains a number of structural proteins, foremost of which is VP16 *(2)* that acts in concert with cellular transcription factors to activate HSV immediate early (IE) gene promoters, which in turn enable the expression of the IE transcriptional regulatory genes that activate the remainder of the lytic life cycle cascade of gene expression that ultimately results in the production of progeny virus

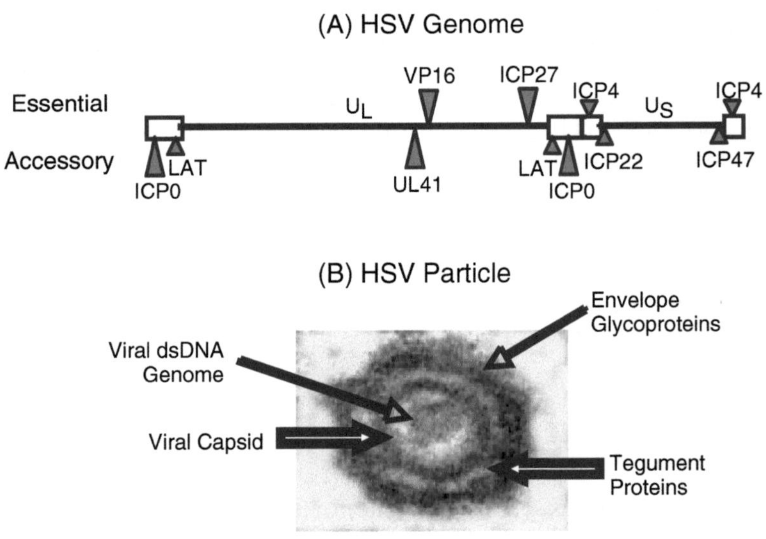

Fig. 1. HSV-1 genome organization and virion structure. (**A**) Schematic representation of the HSV-1 genome, showing the unique long (U_L) and unique short (U_S) segments, each bounded by inverted repeat elements. The location of several important essential genes (VP16, ICP27, and ICP4) which are required for viral replication in vitro, and accessory genes (ICP0, LAT, UL41, ICP22, and ICP47), which may be deleted without dramatically affecting replication in tissue culture, are indicated. (**B**) Electron microscopic depiction of the HSV virion denoting the capsid containing the 152 Kb viral genome, the tegument, and glycoprotein-containing envelope.

particles and the lysis of the infected cell. In addition, the tegument also contains the UL41 (virion host shutoff, vhs) gene product that shuts off host protein synthesis *(3)*.

During natural infection in the human host, the virus initially replicates in epithelial cells of the skin or mucosa. Progeny virions are then taken up by sensory nerve termini of the peripheral nervous system (PNS) and carried via retrograde axonal transport to peripheral nerve cell bodies. At this point, the viral DNA genome is injected through a modified capsid penton portal into the nucleus, after which two alternative forms of the viral life cycle may ensue. The virus may enter the lytic form of the replication cycle, in which expression of viral IE genes serves to transactivate expression of early (E) genes whose products are the principal components of the DNA replication complex. Of note, IE gene expression does not require the de novo expression of viral gene products but is augmented by the HSV tegument protein VP16. Expression of the early genes enables viral DNA synthesis that leads to the production of progeny viral genome concatemers. Following viral DNA synthesis, in conjunction with IE gene products, the late (L) genes are then transcribed. These genes encode primarily the structural proteins such as the capsid, tegument, and viral glycoproteins present within the virion envelope. These late genes are required for viral particle assembly within the nucleus, the budding of the particle through a modified portion of the nuclear membrane, transport of that particle to the cell surface, and egress from the cells with release of fully infectious progeny virus particles. Alternatively, the virus may enter a latent state, in which over 80 viral gene products characteristic of lytic infection are either not transcribed or are transcriptionally silenced over time by methods that are not completely understood. The ability of the virus to enter either the lytic or the latent form of the virus life cycle hold true for replication-competent (oncolytic) vectors, however replication-defective vectors, except for replication-defective vectors that have been rendered deficient through the deletion of one or more essential HSV gene products, usually IE transcriptional regulatory factors such as infected cell polypeptide (ICP) 4 and 27, fail to replicate and thus directly enter a latent-like state.

HSV possesses numerous biological features (*see* **Fig. 2**) that make it attractive as a gene delivery vehicle for gene transfer to the nervous system and other tissues *(4–6)*. The virus possesses a very wide host-range and is able to infect both non-dividing cells such as neurons and dividing cells at extremely high efficiencies *(6–9)*. The virus is capable of establishing a latent infection in neurons as part of the natural biology of the virus, a state in which viral genomes become transcriptionally silenced, and the genomes themselves persist as intranuclear episomal elements in the absence of integration into host chromosomes. Completely replication-defective viruses can be constructed, which retain the ability to establish persistent quiescent genomes in neurons,

(A) Advantages

- Episomal Persistence (dsDNA circular)
- Broad Host Range
- Infects Non-dividing and Dividing Cells
- Large Genome Size (152 Kb)
- High Titers Produced (10^8-10^{10} particles/ml)
- High Transduction Efficiency (1-3 particles/cell)
- Redosing Possible Even in immune Host
- Long-Term Transgene Expression in Neurons

(B) Disadvantages

- Transient Transgene Expression
- Episomal Persistence
- Residual Cytotoxicity

Fig. 2. HSV-1 vector properties. (**A**) The advantages of HSV-1 replication-defective vectors for various gene transfer and therapeutic applications are detailed. HSV-1 is highly infectious with a broad host range and can infect both dividing and non-dividing cells at very low MOIs. Replication-defective vector genomes remain as circular episomes and do not integrate into the host genome and are able to express transgenes long-term from the HSV latency promoter system. HSV vectors can also be readily propagated to high titers and can be re-administered even in a pre-immune host. (**B**) The limited number of HSV-1 vector disadvantages that are listed make HSV more ideal for short-term applications in dividing cells while long-term applications in non-dividing cells such as neurons are still achievable.

but which are unable to replicate (or reactivate) in contrast to wild-type virus that may be reactivated from latency. These persistent genomes are devoid of lytic gene expression but retain the ability to express transgenes long-term using the latency viral promoter ***(10–12)***. The large viral genome (152 Kb), and the fact that many viral genes can be removed as contiguous segments without dramatically affecting virus titers, has enabled the incorporation of large ***(13)*** or multiple ***(14)*** transgenes. Finally, as the HSV genes are expressed in a sequential, interdependent order during lytic infection ***(15)***, the simple removal of the essential immediate early (IE) gene ICP4 blocks expression of later genes in the gene expression cascade ***(16)***. This results in a first generation replication-defective vector that is incapable of producing virus particles or expressing early or late gene products. However, these vectors are still toxic to cells in culture ***(17)*** because the expression of the remaining IE genes. Second and third generation vectors deleted for combinations of these multiple IE genes are less cytotoxic than first-generation vectors ***(18–20)***. A third-generation vector deleted for ICP4, ICP27, and ICP22 (TOZ.1) containing an ICP0 promoter-*lac*Z expression cassette exhibited reduced toxicity in neurons in culture ***(18)***.

We have developed methods to systematically introduce foreign genes into the HSV-1 genome by homologous recombination ***(21)***, specifically using the TOZ.1 vector backbone. This vector can only be propagated using the ICP4/ICP27 complementing cell line (7b) that has been engineered to avoid overlap of these sequences with the deletions present within the virus to

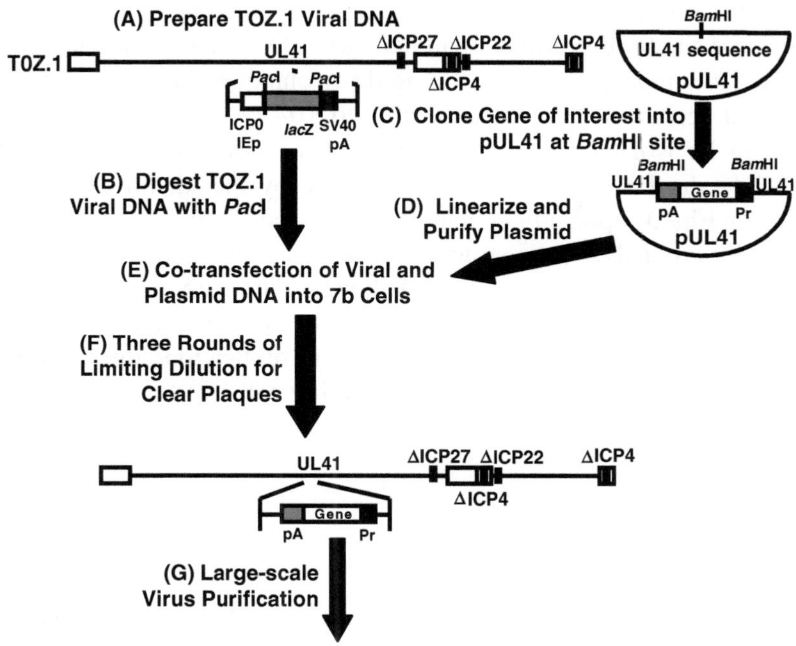

Fig. 3. Construction and production of a replication-defective recombinant HSV-1 vector. (**A**) Purified DNA from the replication-defective HSV-1 vector TOZ.1, deleted for ICP4, ICP27, and ICP22 contains an ICP0 promoter driven *lac*Z expression cassette that possesses unique *Pac*I restriction sites flanking the reporter gene. (**B**) The purified TOZ.1 viral DNA is digested with *Pac*I. (**C**) The therapeutic gene of interest, driven by the desired promoter (Pr), in an expression cassette containing a polyadenylation signal (pA) is cloned into the unique *Bam*HI site within the pUL41 plasmid. (**D**) The gene cassette with the UL41 flanking sequences is then linearized and purified. To achieve homologous recombination of the therapeutic gene of interest into TOZ.1, (**E**) transfect purified *Pac*I digested TOZ.1 DNA into 7b cells along with the UL41 transfer plasmid containing the gene of interest. If homologous recombination has occurred, the gene of interest will have replaced *lac*Z and (**F**) positive recombinants can be identified by the production of a clear plaque phenotype compared to the blue plaque phenotype produced by the parental vector in limiting dilution assays. (**G**) Following purification, high-titer virus stocks can be readily produced.

eliminate the chance of homologous recombination and rescue of the mutant viruses during propagation in the complementing line *(22)*. One can readily construct new vectors using this TOZ.1 vector backbone by homologous recombination of a plasmid containing a cassette expressing the gene of interest inserted into the *UL41* gene sequence *(18,21)*. The TOZ.1 vector (*see* **Fig. 3**) was designed to *(1)* express a reporter gene (*lac*Z) in the UL41 locus, so that recombination of the transgenic cassette into the UL41 locus results in loss of reporter gene activity, which is readily assayed, and *(2)* contain a unique *Pac*I endonuclease site for digestion of parental virus DNA that substantially reduces the non-recombinant background. Using this technique, the proportion of viral plaques that represent recombinants rises to 10–50% *(21)*. The new recombinants can be isolated following homologous recombination of the shuttle plasmid into the *Pac*I-digested TOZ.1 genome by the identification of clear plaques (*see* **Fig. 3**) upon a blue plaque background. Following three rounds of limiting dilution analysis, the structure of the recombinants is then confirmed by Southern blot or PCR analysis. Although the methods detailed in this chapter concentrate on the generation and use of replication-defective HSV vectors, these techniques can also be applied to replication-competent vectors except that they do not require a complementing cell line for their growth and propagation.

2. Materials

2.1. Cell Culture

1. Dulbecco's Eagles modified essential medium (DMEM) supplemented with nonessential amino acids, 100 U/ml penicillin G, 100 µg/ml streptomycin sulfate, 2 mM glutamax, and 10% fetal bovine serum (FBS) [DMEM-10%FBS: all reagents from Invitrogen (Gibco), Bethesda, MD, USA]. Store at 4°C.
2. Methylcellulose overlay (1.0%): Add 25 g methylcellulose (Aldrich, Milwaukee, WI, USA) to 100 ml phosphate-buffered saline (PBS) pH 7.5 in a 500-ml sterile bottle containing a stir bar. Autoclave the bottle on liquids cycle for 45 min. After the solution cools, add 350 ml of DMEM supplemented with nonessential amino acids, 100 U/ml penicillin G, 100 µg/ml streptomycin sulfate, and 2 mM glutamax, mix well and place the bottle on a stir plate at 4°C overnight. Once the methylcellulose has entered solution, add 50 ml of FBS. Store at 4°C (*see* **Note 1**).

2.2. Cells

1. Vero (African green monkey kidney; ATCC#CCL81, Rockville, MD, USA) cells or 7b complementing cells *(22)* are required to propagate HSV-1 replication-competent or replication-defective viruses.

2.3. Buffers and Solutions

1. Tris-buffered saline (TBS) pH 7.5: 50 mM Tris–HCl, pH 7.5, 150 mM NaCl, 1 mM ethylenediamine tetraacetic acid (EDTA). Store at room temperature.
2. Cell lysis buffer: 10 mM Tris–HCl, pH 8.0, 10 mM EDTA. Store at room temperature.
3. Tris-EDTA (TE): 10 mM Tris–HCl, pH 8.0, 1 mM EDTA. Store at room temperature.
4. TE equilibrated phenol : chloroform : isoamyl alcohol (25:24:1 v/v). Store at 4°C (*see* **Note 2**).
5. Isopropanol (Sigma, St. Louis, MO, USA). Store at –20°C.
6. X-gal staining solution that contains the chromogenic substrate X-gal at a final concentration of 300 μg/ml in TBS containing 14 mM $K_4Fe(CN)_6$, 14 mM $K_3Fe(CN)_6$. The staining solution without X-gal can be made, aliquoted, and stored at –20°C until use. X-gal stock (25 mg/ml; Research Products International Corp., Mt. Prospect, IL, USA) is highly insoluble in PBS or other buffers and must be dissolved in either dimethyl formamide (DMF, Sigma) or dimethyl sulfoxide (DMSO, Sigma) (*see* **Note 3**).
7. PBS (1×) pH 7.5: 135 mM NaCl, 2.5 mM KCl, 1.5 mM KH_2PO_4, 8.0 mM Na_2HPO_4, pH 7.5. Store at room temperature.
8. Glycerol (Sigma). Store at room temperature.
9. 70% ethanol. Store at –20°C.
10. 60% iodixanol (OptiPrep, Invitrogen). Store at 4°C.
11. Proteinase K (Boehringer Mannheim, Indianapolis, IN, USA). Store at –20°C.
12. *Pac*I restriction endonuclease (New England Biolabs, Beverly, MA, USA). Store at –20°C.
13. Lipofectamine 2000 (Invitrogen). Store at 4°C.
14. 1 M HEPES, pH 7.35 (Invitrogen). Store at 4°C.

2.4. Nucleic Acids

1. Transfer plasmid p41 ***(21)*** for recombination into the UL41 locus. Other transfer plasmids such as pBS or pZ1 can also be employed ***(21)*** that will enable the transfer of the expression cassette into the thymidine kinase (tk) locus or ICP47 loci of the vector, respectively.
2. Plasmid containing gene of interest with promoter and polyA site.
3. TOZ.1 virus ***(18,21)***.

2.5. Equipment

1. 850-cm^2roller bottles (Falcon/Becton-Dickinson, San Diego, CA, USA).
2. 6-well, 12-well, and 96-well flat-bottomed plates (Falcon/Becton-Dickinson).
3. T-75-cm^2 flasks (Falcon/Becton-Dickinson).
4. Cell scrapers (Falcon/Becton-Dickinson).
5. 15-ml conical polypropylene tubes (Falcon/Becton-Dickinson).

6. Phase Lock Gel^{TM} Heavy Tube (Eppendorf, Westbury, NY, USA).
7. 13-ml Beckman 17 mm path-length seal tubes (Beckman, Fullerton, CA, USA).
8. 3-ml syringe, needle (Falcon/Becton-Dickinson).
9. Wide-bore pipette tips (BioRad, Hercules, CA, USA).
10. Cup-horn sonicator (Virtis, Gardiner, NY, USA).
11. Preparative and table-top centrifuges and Ultracentrifuge (Beckman).
12. Nutator rocker platform (Clay Adams, Becton-Dickinson).
13. CO_2 incubator (Forma/Thermo, Waltham, MA, USA).
14. 500-ml polypropylene centrifuge bottle (Beckman).
15. Multi-channel pipetor (Eppendorf).
16. Mini-Prep kit (Qiagen, Valencia, CA, USA).
17. Parafilm.

3. Methods

The protocols contained here describe the methods necessary to construct and purify recombinant genomic HSV vectors. Although the chapter details the procedures for constructing and producing a replication-defective HSV vector, these same methods can be applied to replication-competent genomic HSV vectors like the oncolytic vectors employed in the glioblastoma multiformae (GBM) clinical trials ***(23,24)***. We have also provided methods for the production and purification of high-titer vector stocks once an isolate is identified and purified through three rounds of limiting dilution analysis. We focus here on the construction of replication-defective vectors, but the same series of steps could be employed for the engineering of replication-competent genomic HSV vectors. The only major difference between the two is that the replication-defective vectors require the use of a cell line that expresses HSV gene products that are deleted from the genome of the vector *in trans* to complement the missing essential genes.

3.1. Isolation of Viral DNA for Transfection

The quality of viral DNA preparation is an important factor as it determines the frequency of generating the desired recombinant virus. Although restriction digests of viral DNA and Southern blotting can be used as diagnostics, we have found that these techniques do not accurately predict whether the viral DNA is intact and infectious once transfected into cells to produce virus. Thus, to evaluate whether the viral DNA is infectious and capable of producing an optimal number of plaques, we routinely transfect 1 μg of viral DNA into the complementing cells and examine the cell monolayers for plaque formation. Moreover, we have optimized the protocol for the production of highly infectious viral DNA and routinely obtain preps in which 1 μg of purified viral DNA yields 100–1000 plaques. The DNA produced by this procedure (*see* **Fig. 3**) is not composed exclusively of viral DNA, but also contains cellular DNA that acts as carrier when precipitating the DNA during the isolation procedure, thereby increasing the yield of DNA and obviating the need for addition of a carrier. The presence of the cellular DNA also increases the overall efficiency of forming a precipitate, thereby increasing the chance of obtaining the desired recombinant (*see* **Note 4**).

1. Add 2.4×10^7 plaque forming units (PFU) of TOZ.1 virus to 8×10^6 complementing 7b cells [multiplicity of infection, i.e., number of virus particles per cell (MOI = 3)] and incubate for 1 h at 37°C with rocking in a 15-ml conical tube.
2. Pipet the infected cells into a T-75 flask and grow in a 37°C 5% CO_2 incubator for approximately 18–24 h. All cells should be infected with virus showing cytopathic effects (CPE; i.e., rounding of the cells) and remain adherent to the flask (*see* **Note 5**).
3. Dislodge the cells from the flask with a cell scraper and transfer the cells in the media into a 15-ml conical polypropylene tube.
4. Centrifuge for 10 min at 2060 × *g* at 4°C and remove supernatant to pellet the cell mass.
5. Resuspend in 1 ml lysis buffer containing 0.1 mg/ml proteinase K (*see* **Note 6**).
6. Incubate the tube with continuous agitation at 37°C overnight on a Nutator rocker platform.
7. Transfer the slurry solution into a Phase Lock Gel™ Heavy Tube (Eppendorf).
8. Add 1 ml phenol : chloroform : isoamyl alcohol (25:24:1) and mix gently for about 1–2 min (*see* **Note 7**).
9. Centrifuge the tube for 5 min at 3020 × *g* in a tabletop centrifuge to separate the phases.
10. Remove the aqueous phase on top of the phase-lock and transfer to a new tube (*see* **Note 8**).
12. Add 2 volumes of isopropanol and mix thoroughly by inversion.
13. Spool the precipitated DNA (white slurry) onto a glass Pasteur pipette (*see* **Note 9**).
14. Transfer spooled DNA to a new tube and rise once with 70% ethanol.
15. Allow the DNA to briefly air dry. Do not dry completely.
16. Add 0.5 ml TE buffer to the dry pellet and incubate at 25°C overnight. Resuspend the DNA by gently pipetting the solution using a wide-bore pipette tip. The use of wide-boar pipetman tips (Biorad, Hercules, CA, USA) will help prevent shearing of the viral DNA, thereby increasing the infectivity of the viral DNA prep.
17. Measure DNA concentration spectrophotometrically ($O.D._{260nm}$). DNA concentration ($O.D._{260nm}$ × Dilution factor × 50 (*see* **Note 10**).

3.2. Construction of Recombinant Virus

To engineer the desired recombinant virus, the gene of interest to be inserted into the virus must first be cloned into the transfer plasmid (p41, pBS, or pZ1) that contains at least 500–1000 bp of flanking HSV-1 sequences (*see* **Fig. 3C**). In the example delineated in this chapter, we will employ the p41 plasmid that contains HSV flanking sequences that enable recombination of the gene expression cassette into the UL41 gene locus of the vector. The addition of 500–1000 bp of flanking sequence is needed to achieve a higher frequency of producing and isolating the recombinant. Flanking sequences as small as 100–200 bp will produce recombinants, but at a very reduced frequency. The p41 plasmid contains a unique *Bam*HI restriction site for cloning of the expression cassette into the transfer plasmid. The expression cassette should consist of the cDNA of interest driven by the promoter of interest as well as a polyadenylation signal. As mentioned previously, other transfer plasmids such as pBS or pZ1 can be employed containing HSV-1 flanking sequences for recombination into either the tk or the ICP47 loci ***(21)***. Once the transfer plasmid containing the gene of interest expression cassette has been constructed, it is then possible to recombine these sequences into the virus at the UL41 locus. This is accomplished by transfecting both linearized plasmid and *Pac*I restriction enzyme digested purified TOZ.1 viral DNA into 7b permissive cells (*see* **Fig. 3E**). Digestion of the TOZ.1 viral DNA with *Pac*I results (*see* **Fig. 3B**) in viral DNA that is cut specifically at the UL41 locus. Transfection of *Pac*I digested TOZ.1 viral DNA alone yields few if any viral plaques. By incorporating this step into the entire process, the number of background plaques that occur following transfection of both the digested viral DNA and the plasmid DNA is dramatically reduced. This same schema can be used to insert gene cassettes into wildtype virus, except one does not digest the viral DNA with *Pac*I, and thus will have a higher background frequency of non-recombinants, and the process does not require complementing cells and thus standard Vero cells can be employed. To readily detect the recombinants containing the desired therapeutic gene of interest, the parental virus backbone should possess a reporter gene (eGFP, *lac*Z) cassette (*lac*Z for TOZ.1) at the desired site of recombination, in this case UL41. Positive recombinants obtained from the transfection will produce clear plaques because of recombination of the gene cassette into the UL41 locus compared with the blue plaque phenotype of the parental virus enabling rapid identification (*see* **Fig.3F**).

1. Clone your gene of interest into the pUL41 shuttle plasmid at the *Bam*HI site (*see* **Note 11**).
2. One day before transfection, seed 5 × 10^5 7b cells in a 6-well tissue culture plate in DMEM/10%FBS. This will ensure that cells are nearly confluent the next day.

3. Digest viral DNA with *Pac*I at 37°C for 6 h according to the manufacturer's protocol (*see* **Note 12**).
4. Transfect the cells with the viral + plasmid DNA mix using Lipofectamine-2000 (Invitrogen), following the manufacturer's instructions (*see* **Note 13**). Use molar equivalents of plasmid and virus (i.e., viral DNA is 152 Kb in length) at 2–3 different ratios (5× virus/1× plasmid; 1/1; and 1/5). The amount of viral DNA should be based on the number of plaques that undigested viral DNA produces when transduced in the absence of plasmid (*see* **Note 14**). It is important to linearize the plasmid construct before transfection to increase the recombination frequency compared with that obtained with uncut supercoiled plasmid. Digestion of the plasmid to release the insert followed by purification of the restriction fragment does not increase the recombination frequency. Although the frequency is the same, use of purified fragment is superior because no chance exists for the insertion of plasmid vector sequences into the virus by semi-homologous recombination with any complementary sequences such as promoters or polyadenylation sites.
5. After incubation steps, add fresh DMEM/10%FBS and incubate at 37°C.
6. It usually takes 3–5 days for plaques to develop depending on the virus and the cell line. With wildtype virus and Vero cells, plaques will be visible in 2–3 days.
7. Once plaques have formed, harvest media and cells using a cell scraper.
8. Subject cells/media to 3 cycles of freeze/thaw and sonicate the cells 3 times for 15 s each on setting 5 using a cup-horn sonicator (Virtis, Gardiner, NY, USA).
9. Centrifuge at 2060 × *g* for 5 min at 4°C to remove cell debris.
10. Store supernatant at –80°C for use as a stock.

3.2.1. Determine the Titer of the Stock of Recombinant Virus

1. Prepare a series of 10-fold dilutions (10^{-2} to 10^{-10}) of the virus stock in serum-free media.
2. Add 100 μl of each dilution to a well of a 12-well tissue culture plate containing 4×10^5 7b cells/well (*see* **Note 15**).
3. Incubate the plates at 37°C in a CO_2 incubator for 1 h, then add 1 ml DMEM/10% FBS and place in the incubator overnight.
4. Within the next 24 h, remove the media and overlay the monolayer with 1 ml of 1% methylcellulose/10% FBS solution to limit virus spread and produce readily visible plaques.
5. Incubate the plates for 3–5 days until well-defined plaques appear, depending on the virus and the cell line used.
6. Aspirate the methylcellulose overlay and stain with 1% crystal violet solution (in 50:50 methanol : dH_2O v/v) for 5 min. Remove stain, rinse with water, and air dry.
7. Count plaques and calculate number of PFU per 1 ml of original stock.

3.2.2. Limiting Dilution Analysis to Isolate and Purify Recombinants

1. Add 30 PFU of titered original stock virus to 1 ml containing 2×10^6 cells in suspension (DMEM/10%FBS) in a 15-ml conical tube and place the tube on a Nutator rocker platform at 37°C for 1.5 h. Cover the cap with parafilm to prevent leaking and contamination.
2. Add 9 ml of fresh media, mix, and plate 100 µl in each well of a 96-well flat-bottomed tissue culture plate using the multichannel pipettor.
3. Incubate the plates at 37°C in a CO_2 incubator for a period of 3–5 days until plaques appear, depending on the virus and cell line employed. Score the wells for the number of plaques. Theoretically, there should be approximately 30 individual plaque wells/plate. Most wells should lack plaques, while some may have two or more plaques.
4. If recombination between the transgene cassette with UL41 flanking sequences and the virus has occurred, the gene of interest will have replaced the *lac*Z reporter gene, X-gal staining can now be employed to determine whether each well contains plaques derived from the desired recombinant virus.
5. Transfer the media from wells of the 96-well plate to a new 96-well plate, wrap the plate with parafilm and store that plate at –80°C for use as a stock for the next round of limiting dilution.
6. Mix X-gal dissolved in DMF or DMSO with $K_4Fe(CN)_6/K_3Fe(CN)_6$ solution (*see* **Note 16**).
7. Pipet 100 µl of X-gal staining solution in each well of the 96-well plate of infected cells.
8. Incubate at 37°C for between 1 and 18 h until blue color appears.
9. Score wells that have clear plaques.
10. Select a well containing only single clear plaques, as these were formed from virus recombinants in which the gene of interest has replaced *lac*Z in TOZ.1.
11. Carry out two additional rounds of limiting dilution/plaque isolation using the stock of virus stored at –80°C, as in **steps 1–10** above. At this point, the virus stock can be used to produce a midi-stock for the eventual preparation of a high-titer stock for general experimental use. At the same time, this stock is used to produce viral DNA to confirm the presence of the insert as well as the absence of the deleted sequences by Southern blot analysis.

3.3. Virus Stock Preparation and Purification

The following procedure entails the preparation of a virus stock from two 850-cm^2 roller bottles worth of cells (*see* **Fig. 3G**) that can be scaled up or down depending on specific needs. Additional roller bottles can be used or 10-layer Nunc flasks or Celligen bio-reactors. This protocol calls for the use of iodixanol (OptiPrep) gradients; however, other methods employing filtration systems can also be employed. The goal is to purify virus particles away from cellular and extracellular debris. Additional downstream purification steps may be added to further eliminate contaminating cellular DNA and protein.

1. Seed two 850-cm^2 roller bottles with 2 × 10^7 complementing cells/bottle in 100 ml DMEM/10% FBS per bottle and incubate at 37°C. Add 1.75 ml 1 M HEPES (pH 7.35) per bottle to buffer the medium if the roller bottles are not incubated in a CO_2 incubator.
2. Infect with virus when the cell monolayer is approximately 75% confluent (~2–3 days). Aspirate the medium, add virus (MOI = 0.01–0.05) in 10 ml fresh serum-free DMEM and incubate at 37°C for 1.5 h (*see* **Note 17**).
3. At the end of the adsorption period, add 90 ml of DMEM lacking FBS plus 1 ml of 1 M HEPES and incubate at 37°C for 2–3 days. The bottles are observed by light microscopy to determine the optimal time for harvesting virus. This time has been shown to vary dramatically depending on the particular vector backbone and the cell line. Generally, viral CPE should be obvious with the majority of the cells rounded up yet not detached from the roller bottle surface.
4. Use a cell scraper to detach the cells into the media and harvest the cell suspension.
5. Centrifuge at 2060 × *g* at 4°C for 10 min.
6. Decant supernatant into a 500-ml polypropylene centrifuge bottle and store on ice.
7. The cell pellet is resuspended in 2–5 ml of serum-free DMEM, subjected to three rounds of sonication to disrupt the cell pellet and release cell-associated virus.
8. Centrifuge at 2060 × *g* for 10 min at 4°C.
9. Decant supernatant, add to the supernatant from **step 6** above, and store on ice.
10. To harvest virus from the supernatant, centrifuge the combined supernatants for 1 h at 18,600 × *g* in a Beckmann preparative centrifuge JLA10.5 rotor.
11. Discard the supernatant. The virus pellet is then resuspended in 500 μl of PBS per roller bottle of virus-infected cells by thoroughly vortexing and pipeting (*see* **Note 18**).
12. Transfer the suspension to 13-ml Beckman 17-mm path-length seal tubes. Make the total volume up to 13 ml with 20% Iodixanol (dilute 60% Iodixanol to 20% with 1× PBS).
13. Seal and centrifuge for 4.5 h at 342,000 × *g* in a NVT65 rotor in a Beckmann XL-90 ultracentrifuge.
14. Three bands usually appear in the gradient. The upper band contains cellular DNA and protein contaminants. Insert a needle, and using a 3-ml syringe, aspirate 1–1.5 ml of the viral band with approximately 60% of the virus being present in this second band from the bottom of the gradient. Occasionally, with wildtype, a higher band of defective particles is visible. If there are three lower bands in the gradient, take the middle band. Alternatively, the tubes can be fractionated and the titer of each fraction determined.
15. Aliquot and store viral stocks at –80°C. Iodixanol (OptiPrep) serves as a cryoprotectant. Titer an aliquot of virus as described in **Subheading 3.2.1** above. If needed, the virus can be further concentrated by centrifugation in a microfuge at 2060 × *g* for 10 min at 4°C. Following centrifugation, the viral pellet is

resuspended in a minimal volume of PBS (1–2 ml) containing glycerol. It is necessary to add glycerol to a final concentration of 10% to the virus stock to properly cryopreserve the virus because this resuspended stock lacks OptiPrep. Virus stocks should be maintained at a low passage. Use one vial of a newly prepared stock as a stock for preparing all future stocks. To reduce the chance of rescuing wildtype virus during the propagation of viruses carrying deletions of essential gene(s), stocks should be routinely prepared from single plaque isolates.

4. Notes

1. Stir methylcellulose overlay media before each use as methylcellulose tends to settle at bottom of bottle.
2. Stable for 6–9 months, check pH routinely before use. If pH is outside normal range (i.e., too acidic or basic), discard and remake.
3. X-gal solution in DMSO or DMF must be made fresh each time before addition to the staining solution. Failure to do so may result in poor staining or no signal.
4. If necessary, pure viral DNA can be prepared from virus harvested solely from the supernatant of infected cells or from virus particles that have been gradient purified. The yield of DNA obtained in this instance is significantly reduced and carrier DNA should be added during precipitation steps.
5. The MOI is chosen to insure that all cells are infected at roughly the same time, so that all cells should show the same level of CPE at approximately18–24 post infection. It is important to wait until CPE occur, as this is an indication that viral DNA replication has occurred, which will result in optimal levels of viral DNA. It is also crucial to make sure that the infection has not proceeded too far because this equates with viral genome being packaged into newly synthesized virus particles. Thus, if most or all cells have come off the flask, this indicates that the infection is too far along for harvesting high levels of good quality viral DNA.
6. Insure that the cell pellet is resuspended. The lysis buffer should rapidly dissolve the cell pellet. If some cell debris remains, it will resuspend once the tube is rocked overnight.
7. It is important to rock the tubes gently. DO NOT VORTEX as this will shear the large viral DNA fragments! However, it is crucial to rock vigorously enough, so that the aqueous and organic phases mix sufficiently.
8. It is important to remove the liquid closest to the phase-lock surface as this is the region where the viral DNA tends to be found. Always use larger pipettes (5 ml or greater) or wide-bore pipette tips to keep from shearing the viral DNA.
9. Take a Pasteur pipette and using a Bunsen burner, bend the end of the pipette into a closed hook. This when cooled, can be used to spool the viral DNA. If no clear precipitate forms which is clearly visible as a white slurry or clump, incubate the tube overnight at –20°C to allow the DNA to precipitate. On the following day, one can simply spin down the precipitate for 1 h at 18,600 × *g*

in a Beckmann preparative centrifuge at 4°C. Wash the pellet once with 70% ethanol, dry briefly, and resuspend as in **step 16**.

10. The A_{260}/A_{280} ratio should be 1.70 or greater, thus relatively free of contaminating protein.
11. The transfer plasmid DNA can be prepared by various methods. Large-scale plasmid preparations are not necessary as plasmid DNA prepared using Mini-Prep kits such as the Qiagen Mini-kit (Qiagen) is of sufficient purity to deliver high transduction efficiencies.
12. Do not digest the viral DNA overnight to avoid degrading the DNA.
13. The choice of specific transfection reagent is crucial, as some transfection reagents (Lipofectamine Plus, Fugene, etc.) work poorly or not at all when transfecting large DNAs such as the 152 Kb HSV genome, while Lipofectamine, Lipofectamine 2000, and the standard calcium phosphate method work very efficiently.
14. Approximately 1 µg of viral DNA should yield 100–1000 plaques when transfected alone into complementing cells.
15. To obtain a more accurate titer, the titration should be performed in duplicate or triplicate.
16. It is important to filter the X-gal/$K_4Fe(CN)_6/K_3Fe(CN)_6$ solution as crystals develop that would interfere with the observation of plaques.
17. It is important to employ low MOIs to generate the virus stock as high MOIs result in the introduction of unwanted mutations throughout the viral genome.
18. It is important to thoroughly resuspend the pellet to get an even suspension of particles, but over vortexing is also not good as it can damage the particles and render them non-infectious.

References

1. Roizman, B. and Knipe, D.M. (2001) Herpes simplex viruses and their replication, in *Fields Virology*, 4th edition (Knipe, D.M. and Howley, P.M., eds.), Lippincott Williams and Wilkins, Philadelphia, PA, pp. 2399–2459.
2. Mackem, S. and Roizman, B. (1982) Structural features of the herpes simplex virus alpha gene 4, 0, and 27 promoter-regulatory sequences which confer alpha regulation on chimeric thymidine kinase. *J. Virol.* **44**, 939–949.
3. Oroskar, A. and Read, G. (1989) Control of mRNA stability by the virion host shutoff function of herpes simplex virus. *J. Virol.* **63**, 1897–1906.
4. Burton, E.A., Wechuck, J.B., Wendell, S.K., Goins, W.F., Fink, D.J., and Glorioso, J.C. (2001) Multiple applications for replication-defective herpes simplex virus vectors. *Stem Cells* **19**, 358–377.
5. Goins, W.F., Wolfe, D., Krisky, D.M., Bai, Q., Burton, E.A., Fink, D.J., and Glorioso, J.C. (2004) Delivery using herpes simplex virus: an overview. *Methods Mol. Biol.* **246**, 257–299.

6. Wolfe, D., Goins, W.F., Yamada, M., Moriuchi, S., Krisky, D.M., Oligino, T.J., Marconi, P.C., Fink, D.J., and Glorioso, J.C. (1999) Engineering herpes simplex virus vectors for CNS applications. *Exp. Neurol.* **159**, 34–46.
7. Glorioso, J., Goins, W., Meaney, C., Fink, D., and DeLuca, N. (1994) Gene transfer to brain using herpes simplex virus vectors. *Ann. Neurol.* **35**, S28–34.
8. Haarr, L., Shukla, D., Rodahl, E., Dal Canto, M., and Spear, P. (2001) Transcription from the gene encoding the herpesvirus entry receptor nectin-1 (HveC) in nervous tissue of adult mouse. *Virology* **287**, 301–309.
9. Mata, M., Zhang, M., Hu, X., and Fink, D. (2001) HveC (nectin-1) is expressed at high levels in sensory neurons, but not in motor neurons of the rat peripheral nervous system. *J. NeuroVirol.* **7**, 1–5.
10. Goins, W.F., Lee, K.A., Cavalcoli, J.D., O'Malley, M.E., DeKosky, S.T., Fink, D.J. and Glorioso, J.C. (1999) Herpes simplex virus type 1 vector-mediated expression of nerve growth factor protects dorsal root ganglia neurons from peroxide toxicity. *J. Virol.* **73**, 519–532.
11. Goins, W.F., Sternberg, L.R., Croen, K.D., Krause, P.R., Hendricks, R.L., Fink, D.J., Straus, S.E., Levine, M. and Glorioso, J.C. (1994) A novel latency-active promoter is contained within the herpes simplex virus type 1 U_L flanking repeats. *J. Virol.* **68**, 2239–2252.
12. Goins, W.F., Yoshimura, N., Ozawa, H., Yokoyama, T., Phelan, M., Bennet, N., deGroat, W.C., Glorioso, J.C. and Chancellor, M.B. (2000) Herpes simplex virus vector-mediated nerve growth factor expression in bladder and afferent neurons: potential treatment for diabetic bladder dysfunction. *J. Urol.* **165**, 1748–1754.
13. Akkaraju, G.R., Huard, J., Hoffman, E.P., Goins, W.F., Pruchnic, R., Watkins, S.C., Cohen, J.B. and Glorioso, J.C. (1999) Herpes simplex virus vector-mediated dystrophin gene transfer and expression in MDX mouse skeletal muscle. *J. Gene Med.* **1**, 280–289.
14. Krisky, D.M., Marconi, P.C., Oligino, T.J., Rouse, R.J., Fink, D.J., Cohen, J.B., Watkins, S.C., and Glorioso, J.C. (1998a) Development of herpes simplex virus replication-defective multigene vectors for combination gene therapy applications. *Gene Ther.* **5**, 1517–1530.
15. Honess, R. and Roizman, B. (1974) Regulation of herpes simplex virus macro-molecular synthesis. I. Cascade regulation of the synthesis of three groups of viral proteins. *J. Virol.* **14**, 8–19.
16. DeLuca, N.A., McCarthy, A.M., and Schaffer, P.A. (1985) Isolation and characterization of deletion mutants of herpes simplex virus type 1 in the gene encoding immediate-early regulatory protein ICP4. *J. Virol.* **56**, 558–570.
17. Johnson, P., Miyanohara, A., Levine, F., Cahill, T., and Friedmann, T. (1992) Cytotoxicity of a replication-defective mutant herpes simplex virus type 1. *J. Virol.* **66**, 2952–2965.
18. Krisky, D.M., Wolfe, D., Goins, W.F., Marconi, P.C., Ramakrishnan, R., Mata, M., Rouse, R.J., Fink, D.J., and Glorioso, J.C. (1998b) Deletion of multiple immediate-early genes from herpes simplex virus reduces cytotoxicity and permits long-term gene expression in neurons. *Gene Ther.* **5**, 1593–1603.

19. Samaniego, L., Webb, A., and DeLuca, N. (1995) Functional interaction between herpes simplex virus immediate-early proteins during infection: gene expression as a consequence of ICP27 and different domains of ICP4. *J. Virol.* **69**, 5705–5715.
20. Wu, N., Watkins, S.C., Schaffer, P.A., and DeLuca, N.A. (1996) Prolonged gene expression and cell survival after infection by a herpes simplex virus mutant defective in the immediate-early genes encoding ICP4, ICP27, and ICP22. *J. Virol.* **70**, 6358–6368.
21. Krisky, D., Marconi, P., Oligino, T., Rouse, R., Fink, D., and Glorioso, J. (1997) Rapid method for construction of recombinant HSV gene transfer vectors. *Gene Ther.* **4**, 1120–1125.
22. Marconi, P., Krisky, D., Oligino, T., Poliani, P.L., Ramakrishnan, R., Goins, W.F., Fink, D. J., and Glorioso, J.C. (1996) Replication-defective HSV vectors for gene transfer in vivo. *PNAS* **93**, 11319–11320.
23. Markert, J., Medlock, M., Rabkin, S., Gillespie, G., Todo,T., Hunter, W., Palmer, C., Feigenbaum, F., Tornatore, C., Tufaro, F., and Martuza, R. (2000). Conditionally replicating herpes simplex virus mutant, G207 for the treatment of malignant glioma: results of a phase I trial. *Gene Ther.* **7**, 867–874.
24. Rampling, R., Cruickshank, G., Papanastassiou, V., Nicoll, J., Hadley, D., Brennan, D., Petty, R., MacLean, A., Harland, J., McKie, E., Mabbs, R., and Brown, M. (2000) Toxicity evaluation of replication-competent herpes simplex virus (ICP 34.5 null mutant 1716) in patients with recurrent malignant glioma. *Gene Ther.* **7**, 859–866.

7

Plasmid-Based Gene Transfer in Mouse Skeletal Muscle by Electroporation

Jonathan D. Schertzer and Gordon S. Lynch

Summary

Advances in molecular biology and genetics have provided an abundance of descriptive biological data. Ascribing physiologic roles to these data is an important task to translate this information into improvements in human health. Non-viral transgene delivery of a transgene is a promising approach for defining the physiologic role of a specific protein in vivo. This methodology also has therapeutic potential for the treatment of various diseases. Plasmid DNA injection followed by controlled electric pulses across the injection site (i.e., electroporation or electrotransfer) can dramatically increase transgene expression in skeletal muscle, but in some methods can result in local tissue damage. Muscle damage and subsequent repair processes may confound the measurement of physiological, biochemical, and molecular properties for a given experiment. In addition, it is essential to eliminate damage to existing muscle fibers if such procedures are to be used for the treatment of myopathies and muscle wasting conditions. The measurement of muscle contractile parameters (i.e., function) is an important end-point for assessing the role of ectopically delivered proteins in vivo. As such, methodologies that allow the delivery of a transgene to levels that have a physiological effect and do not compromise muscle function will aid in the characterization of the role of specific proteins in vivo.

Key Words: Non-viral; injury; transfection efficiency; electroporation; skeletal muscle; gene therapy.

From: *Methods in Molecular Biology, vol. 433: Volume 1: Production and In Vivo Applications of Gene Transfer Vectors*
Edited by: J. M. Le Doux © Humana Press, Totowa, NJ

1. Introduction

The sequencing of various mammalian and non-mammalian genomes and simultaneous measurement of many transcripts using microarray technology in response to diverse stimuli has provided a plethora of descriptive biological information. Ascribing physiologic roles to much of these data through meticulous functional characterization of the genes and gene products is a daunting task. Integrating molecular, cellular, and physiologic information is necessary for optimizing the translation of research to improvements in human health. The postnatal delivery of a transgene is a promising approach for defining the physiologic role of a specific protein in vivo. This methodology also has therapeutic potential for the treatment of various diseases. Compared to viral gene delivery techniques, non-viral methods can generally accommodate larger transgenes, are relatively inexpensive, are associated with a lower cytotoxicity and immunogenic responses, and limit possibility of insertional mutagenesis ***(1,2)***. However, it has been reported that plasmid-based gene transfer by direct injection into skeletal muscle has a very low transfection efficiency of approximately 1% of fibers ***(3,4)***, which is one of the factors that limits its experimental and potential therapeutic use for in vivo application.

Various factors affect the efficiency of gene transfer by direct plasmid DNA injection, such as the design of the plasmid construct, age of the host, and vehicle used for injection ***(5–7)***. To increase transfection efficiency of non-viral gene delivery, several strategies have been developed, including the use of different DNA solutes that are designed to inhibit endogenous enzymes that degrade foreign DNA (DNases), manipulation of specific injection techniques such as the speed and volume of injection, and passing a controlled electric field across the injection site ***(6–10)***.

Plasmid DNA injection followed by controlled electric pulses across the injection site (i.e., electroporation or electrotransfer) can dramatically increase transgene expression and reduce intra-individual variability compared with injection of plasmid DNA alone ***(9)***. Despite many efforts to optimize electroporation parameters, variability in transgene expression and the amount of tissue damage has been reported ***(11–13)***. Although currently employed electroporation protocols may be sufficient for certain objectives, the damage caused by high voltage electroporation is often a hindrance in skeletal muscle. It is essential to eliminate damage to existing muscle fibers if such procedures are to be used for the treatment of myopathies and muscle wasting conditions. In addition, local tissue damage and subsequent repair processes may confound the measurement of physiological, biochemical, and molecular properties for a given experiment. As such, there is merit in recent methods that have optimized electroporation parameters to limit (or eliminate) deleterious effects on skeletal muscle function ***(14)***. Several factors appear to be important in optimizing such protocols, including the voltage applied, the duration of electric pulses, the number of electric pulses, pre-treatment of the muscle, and the DNA solute used for injection ***(11,14,15)***.

This chapter describes an electroporation protocol designed to optimize plasmid-based gene transfer in mouse skeletal muscle whilst minimizing muscle fiber damage. It describes how in vivo transfection efficiency after electroporation-assisted gene transfer can be confirmed by the measurement of various reporter constructs [i.e., green fluorescent protein (GFP) or beta-galactosidase (*lac*Z)] in transverse muscle sections. Alternatively, changes in the level of a given protein can be assessed by immunoblotting or enzyme-linked immunosorbent assay (ELISA). Methods are described for the measurement of muscle damage from histological assessment of transverse muscle sections that reveal general muscle architecture. Sections stained with hematoxylin and eosin stain (H&E) can reveal changes in muscle structure that are indicative of muscle fiber damage. Importantly, this chapter also describes methods for the measurement of the contractile properties of skeletal muscle that provide a sensitive functional assessment and verification of the totality of muscle damage. Functional measurements of the electroporated muscle allow for the determination of a principal physiological outcome after in vivo protein manipulation. As such, the characterization of a given protein in vivo can be achieved if the method of delivery of a transgene does not compromise the reliable measurement of muscle function. The methods describe how transfection efficiency can be optimized to induce a quantifiable physiologic outcome.

2. Materials

2.1. Plasmid DNA Preparation

1. A suitable mammalian expression vector (i.e., plasmid) (*see* **Note 1**).
2. DH5α strain of *Escherichia coli* (*see* **Note 2**).
3. SOC medium: 2% tryptone, 0.5% yeast extract, 0.5% sodium chloride.
4. LB: 10 g/l tryptone, 5 g/l yeast extract, and 5 g/l sodium chloride.
5. LB-agar: 10 g/l tryptone, 5 g/l yeast extract, 5 g/l sodium chloride, and 15 g/l of bacto-agar.
6. The transformation of bacteria by electroporation requires a suitable electrical stimulator and cuvette such as the Precision Pulse Electro Cell Manipulator (ECM 630) with the 630B safety cuvette stand (BTX, Genetronics, San Diego, CA, USA). Alternatively, transformation can be achieved in chemically competent bacteria.
7. The Qiagen EndoFree Plasmid Giga Kit (Qiagen, Valencia, CA, USA) and other endotoxin-free plasmid DNA kits are commercially available.
8. Electro-conductive gel (Medical Equipment Services, Richmond, VIC, Australia) (*see* **Note 3**).
9. Sodium pentobarbitone (Nembutal, Rhone Merieux, Pinkenba, QLD, Australia).

2.2. Plasmid DNA Injection and Electroporation

1. Small caliber (29–30 G) needles and syringes are required to inject plasmid DNA into the tibialis anterior (TA) muscles of mice.
2. Calipers with embedded platinum plate electrodes are required for electroporation. The calipers should allow for fine manipulation of the embedded electrodes and provide an accurate measurement of the distance between the electrodes when placed across the TA muscle. The electrodes should be designed to be the approximate size of a typical mouse TA muscle (i.e., each approximately 27–30 mm^2; *see* **Fig. 1**).
3. Electroporation requires a constant-current variable-voltage Grass stimulator (Grass S88, Quincy, MA, USA) (or equivalent) attached to the electrodes and calipers.
4. Prepare 1 ml aliquots of Type IV-S hyaluronidase (0.5 U/μl) at store at –20°C. Aseptically processed hyaluronidase supplied by Sigma-Aldrich (St. Louis, MO, USA) generally contains 750–1500 U/mg.

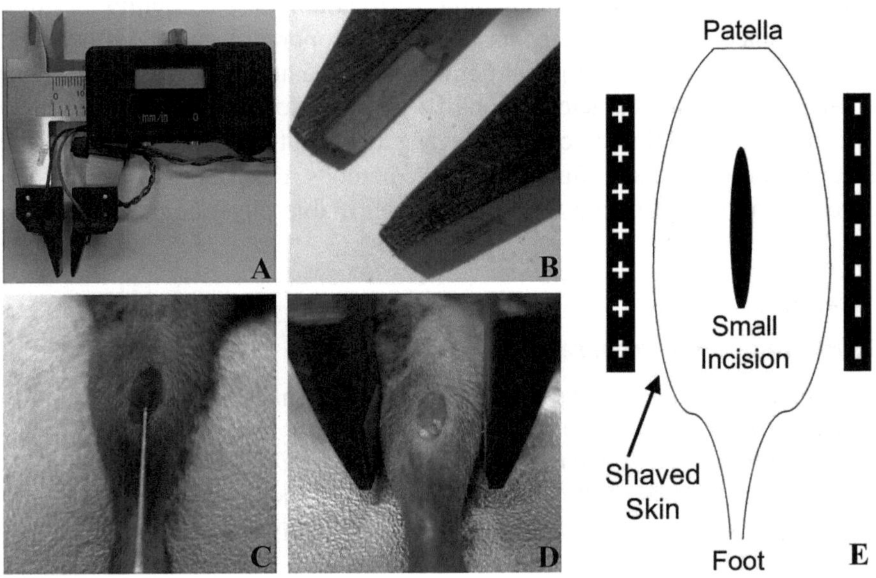

Fig. 1. A simplified representation of plasmid injection and electroporation of mouse tibialis anterior (TA) muscle. Typical digital calipers that have been modified to be used for electroporation (**A**). Platinum plate electrodes have been embedded in the jaws of the digital calipers (**B**). A small incision in the skin on the anterior aspect of the TA allows for reliable and accurate intramuscular injection of plasmid DNA (**C**). After intramuscular injection of plasmid DNA, the electrodes are covered in conductive gel and placed on the closely shaved skin on both sides of the TA muscle. The electrodes deliver transcutaneous pulses and do not contact the exposed muscle (**D**). A schematic representation of electroporation of mouse TA muscle (**E**).

2.3. Assessment of Muscle Function

1. The measurement of TA muscle function in situ requires a suitable platform (which is usually Plexiglas) to secure the anesthetized mouse and for appropriate positioning of the servomotor. The platform is jacketed to allow for circulation of heated water to maintain the optimal temperature of the preparation.
2. It requires a dual mode servomotor such as Model 305B-LR from Aurora Scientific (Richmond Hill, ON, Canada) with appropriate sensitivity and a lever arm suitable for attaching the tendon of the TA muscle. The servomotor controller allows for displacement of the level arm (to either shorten or lengthen the muscle or to keep it at a fixed length) and for the simultaneous measurement of force production.
3. It requires appropriate computer applications such as LabView software (National Instruments, Austin,TX, USA) driving a personal computer with an onboard controller (i.e., PCI-MIO-16XE-10, National Instruments) interfaced with the transducer-servomotorcontrol/feedback hardware. Suitable systems are commercially available (Aurora Scientific, Aurora, ON, Canada).

3. Methods

3.1. Plasmid DNA Preparation

Before producing large amounts of suitable plasmid DNA for use in gene transfer, the plasmid DNA must be transduced into an appropriate bacteria (i.e., DH5α strain of *E. coli*), which allows replication of the plasmid during growth. If the plasmid is supplied in transduced bacteria, omit **steps 1–6**.

1. Mix 40 μl of the DH5α strain of *E. coli* cell suspension with 3 μl of plasmid DNA (0.1–1.0 μg) in an ice-cold polypropylene tube and equilibrate for 1 min.
2. Transfer the mixture to an ice-cold 0.2 cm gap electroporation cuvette and electroporate once at 25 μF, 200 Ω, and 2.5 kV (*see* **Note 4**).
3. Immediately add 1 ml of SOC medium to the cuvette and resuspended the cells with a Pasteur pipette.
4. Transfer the cell suspension into a polypropylene tube and incubate at 37°C for 1 h with vigorous shaking (>300 rpm).
5. Spread the cell suspension evenly on a 10 cm tissue culture dish coated with solidified selective medium (i.e., LB-agar containing 100 μg/ml of ampicillin or 50 μg/ml of kanamycin) and incubate overnight at 37°C.
6. Pick a single colony with a sterile pipette tip and place it into a polypropylene tube containing 1–5 ml of LB containing the appropriate antibiotic. Incubate for 6–8 h at 37°C with vigorous shaking (>300 rpm).
7. Prepare glycerol stocks by mixing equal parts of the cell suspension and glycerol. Aliquot the suspension and store at –80°C.
8. Prepare endotoxin-free plasmid DNA using the Qiagen EndoFree Plasmid Giga Kit (Qiagen) (or equivalent) according to the manufacturer's instructions.

9. During the final step of plasmid DNA preparation, resuspend the DNA in a suitable volume of 50% (vol : vol) saline : water such that it can later be diluted to 2 μg/μl. For example, if a typical yield is 9 mg of plasmid DNA during the preparation, resuspend the DNA in a volume less than 4.5 ml. Determine the concentration of plasmid DNA in solution and dilute with an appropriate amount of 50% (vol : vol) saline : water to the final concentration of 2 μg/μl.
10. Store 40–50 μl aliquots of the plasmid DNA (2 μg/μl) at –80°C.

3.2. Plasmid DNA Injection and Electroporation

1. Anesthetize a mouse by intraperitoneal (i.p.) injection of sodium pentobarbitone (Nembutal) at a dose of 60 mg/kg body mass. For example, prepare Nembutal at a concentration of 6 mg/ml in isotonic saline. For a mouse with a body mass of 20 g, inject 200 μl of Nembutal (6 mg/ml). The mouse should be at a suitable depth of anesthesia ~10 min after injection. Maintain an adequate depth of anesthesia with subsequent i.p. injections of ~100 μl of Nembutal (6 mg/ml) to keep the mouse unresponsive to tactile stimulation. In general, this is done every 45 min, but there can be significant individual variability in the response to anesthetics.
2. After the mice are anesthetized deeply, shave the entire hindlimb closely, and surgically expose a small (approximately 3 mm) portion of the anterior aspect of the TA muscle by cutting the overlying skin without damaging the muscle (*see* **Note 5**; *see* **Fig. 1**).
3. Inject the TA muscle with 30 μl (0.5 U/μl) hyaluronidase and monitor the anesthetized mouse for 2 h on a heat pad. Keep the exposed TA muscle moist with warmed (37°C) isotonic saline.
4. Before plasmid DNA injection, completely cover the plate electrodes with conductive gel.
5. Inject plasmid DNA (40 μl, 2 μg/μl) into the TA muscle a single motion (*see* **Note 6**; *see* **Fig. 1**).
6. Immediately after plasmid DNA injection, place the electrodes against the closely shaved skin on either side of the small surgical incision. Do not allow the electrodes to contact the exposed portion of the muscle (*see* **Fig. 1**).
7. Measure the distance between the electrodes and adjust the stimulator to deliver a voltage equivalent to 75–100 V/cm.
8. Deliver three transcutaneous pulses (each 20 ms in duration) across the TA muscle at a voltage of 75–100 V/cm and a frequency of 1 Hz.
9. Reverse the polarity (i.e., invert the electrodes) and deliver three more pulses.
10. Close the small incision with surgical clips (i.e., Michel clips) or surgical suture and place the mouse on heat pad to allow for recovery from the anesthesia.
11. A number of days later (i.e., 3–7 days or as required), measure transfection efficiency using an appropriate method, such as the number of fibers expressing a reporter gene (*see* **Fig. 2**).

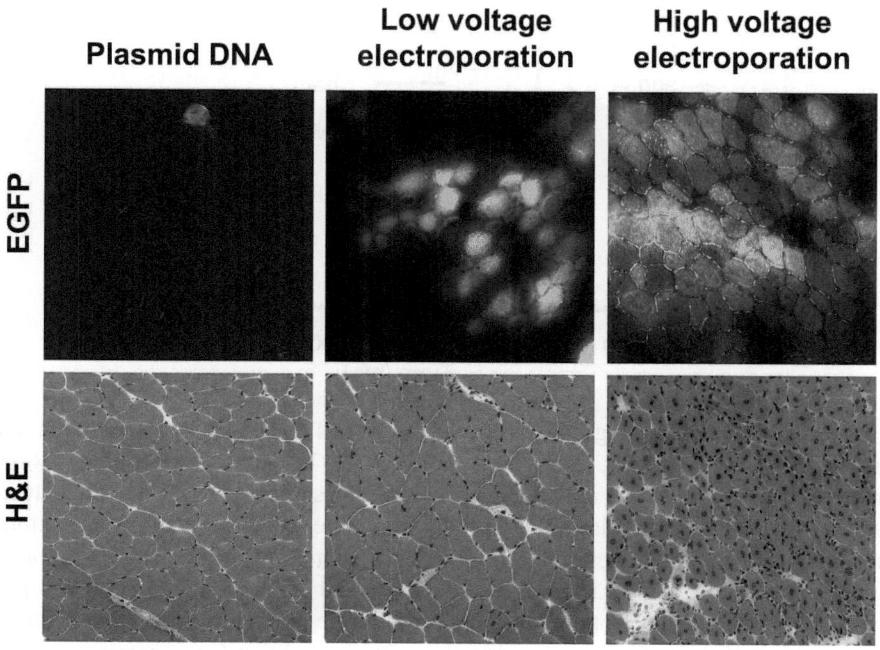

Fig. 2. Transfection efficiency and tissue damage after plasmid DNA injection or electroporation at low and high voltages. Transfection efficiency (top) is measured by the using a reporter construct encoding enhanced green fluorescent protein (EGFP) in transverse sections of tibialis anterior (TA) muscles. The extent of muscle damage (bottom) is assessed from the morphological characteristics of muscle fibers after staining nuclei with hematoxylin (blue; dark) and the cytosol with eosin (pink; light) (H&E). Transfection efficiency is very low after intramuscular injection of plasmid DNA, but electroporation at higher voltages (i.e., 200 V/cm) dramatically increases transfection efficiency as evident by a large percentage of EGFP-positive myofibers. However, this procedure also results in tissue damage as evident from the presence of regenerating myofibers, which contain a centrally located nucleus (or nuclei). Lower voltage electroporation (i.e., 75–100 V/cm) increases transfection efficiency compared with intramuscular injection of plasmid DNA, but does not result in significant tissue damage.

3.3. Assessment of Muscle Function

1. Once mice are adequately anesthetized, shave the right (or left) hindlimb and expose the entire TA muscle.
2. Throughout the procedure, keep the exposed tendon, muscle, and nerve moist with warmed isotonic saline.
3. Cut the exposed tendon of the TA muscle several millimeters distal to the myotendinous junction and tie secure knots on the tendon with 5-0 surgical suture.

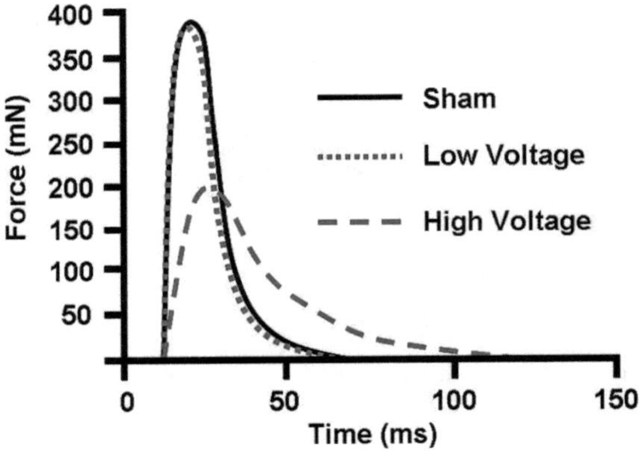

Fig. 3. Sample recordings of force characteristics in TA muscles of mice after electroporation at high (i.e., 200 V/cm) or low (i.e., 75–100 V/cm) voltages. Representative force tracings are shown during in situ measurement of the maximal twitch response (single brief stimulation) at optimal muscle length. Note the lower maximum force and prolongation in the time course of twitch contraction after electroporation at a high voltage, effects that are indicative of damaged and regenerating skeletal muscle. In contrast, low voltage electroporation does not alter the maximum force or time course of contraction of the twitch compared to the control condition (i.e., sham surgery).

4. Immobilize the knee by passing a pin behind the patellar tendon without damaging the surrounding tissue. The pin should affix to the base of the platform.
5. Using the suture from **step 3**, tie the tendon of the muscle to the lever arm of the dual-mode servomotor.
6. Immerse the entire TA muscle in warmed paraffin or mineral oil. The oil is used to maintain the muscle preparation at the desired temperature (i.e., 36–37°C).
7. Expose a small portion of the femoral nerve by making an incision in the thigh muscles.
8. Place two wire (simulating) electrodes on the femoral nerve and stimulate muscle contraction by a supramaximal voltage (i.e., 10 V) with square wave pulses of 300 ms in duration.
9. All stimulation parameters and contractile responses can be controlledand measured using appropriate computer applications.
10. Appropriate contractile measurements are dependent on a given experiment (*see* **Note 7**; *see* **Fig. 3**).

4. Notes

1. Plasmid-based gene transfer in mammalian tissue requires the use of an appropriate expression vector encoding the protein of interest. The plasmid must contain an appropriate antibiotic resistance insertion (i.e., ampicillin or kanamycin) allowing for selective growth when transduced into bacteria. The plasmid

sequence must also contain a strong promoter to maximize transgene expression [i.e., cyclomegalovirus (CMV) or myosin light chain (MLC)], the full-length cDNA sequence encoding the protein of interest, a polyadenylation sequence to increase mRNA stability [i.e., simian vacuolating virus 40 (SV40) polyA] and may contain various enhancer sequences. Many suitable mammalian expression vectors that encode a specific protein are commercially available. Conversely, the desired cDNA can be inserted into an expression vector using standard subcloning techniques.

2. Any suitable electrically competent bacteria can be used for propagation (i.e., growing) plasmid DNA. Alternatively, chemically competent bacteria can be transformed with plasmid DNA, a technique that negates the procedure described in **step 2** of plasmid DNA preparation. Chemically competent bacteria are routinely transformed with DNA by incubating the DNA with the bacteria on ice for 10 min, then inducing heat-shock for 60 min by incubating at 37°C.
3. Any high-quality conductive gel suitable for electrocardiography or ultrasonic diagnosis should be suitable for electroporation. The gel should be balanced to a neutral pH.
4. There is significant flexibility to change the electrical parameters for electroporation to transduce bacteria. In addition, the parameters are dependent on the constituents of the suspension such as calcium and sodium. A reduction in the current or voltage may result in a lower number of transformants. The most important factor in this procedure is to ensure that electroporation does not "arc," where the bacteria are destroyed due to excessive current or voltage. If this occurs, reduce the voltage and/or current or modify the constituents of the buffer (usually by removing residual sodium contamination).
5. In our experience, surgical exposure of a small portion of the anterior aspect TA muscle is the best way to ensure accurate and reproducible intramuscular injections. Care must be taken not to damage the TA muscle or surrounding connective tissue. Damage to the muscle will result in a reduction in the holding capacity of the muscle because the solution will seep out during injection. In addition, laceration injury severely compromises muscle structure and function.
6. In our experience, intramuscular injection of 40 μl of a solution fills a mouse TA muscle to holding capacity and allows for distribution throughout the muscle. This injection technique maximizes the penetration of the transgene along the entire length (i.e., longitudinal axis) of the muscle ***(16)***. Multiple intramuscular injections were not used in the mouse TA because it leads to escape of the plasmid DNA by seeping out of the previous injection site. In addition, an excessive number of injections may compromise the integrity of the muscle and cause damage.
7. It is common to determine optimal muscle length by progressively increasing the length in small increments until maximum twitch force is obtained. Typical values for in situ measurements of an undamaged/uninjured TA muscle in a 12–16-week-old C57BL/6 mouse are maximum (tetanic) force of ~1500–1700 mN occurs at a stimulation frequency of 200–300 Hz; maximum force corrected

for the cross-sectional area of the muscle (i.e., specific force) is ~260–300 kN m^{-2}; maximum twitch force is slightly more variable and is ~350–500 mN. A lower value for specific force generally indicates muscle damage.

References

1. Hacein-Bey-Abina S., Von Kalle C., Schmidt M., McCormack M.P., Wulffraat N., Leboulch P., Lim A., Osborne C.S., Pawliuk R., Morillon E., Sorensen R., Forster A., Fraser P., Cohen J.I., de Saint Basile G., Alexander I., Wintergerst U., Frebourg T., Aurias A., Stoppa-Lyonnet D., Romana S., Radford-Weiss I., Gross F., Valensi F., Delabesse E., Macintyre E., Sigaux F., Soulier J., Leiva L.E., Wissler M., Prinz C., Rabbitts T.H., Le Deist F., Fischer A. and Cavazzana-Calvo M. (2003) LMO2-associated clonal T cell proliferation in two patients after gene therapy for SCID-X1. *Science* **302**, 415–419.
2. Wells D.J. (2006) Viral and non-viral methods for gene transfer into skeletal muscle. *Curr. Opin. Drug Discov. Devel.* **9**, 163–168.
3. McMahon J.M. and Wells D.J. (2004) Electroporation for gene transfer to skeletal muscles: current status. *BioDrugs* **18**, 155–165.
4. Wolff J.A., Malone R.W., Williams P., Chong W., Acsadi G., Jani A. and Felgner P.L. (1990) Direct gene transfer into mouse muscle in vivo. *Science* **247**, 1465–1468.
5. Danko I., Williams P., Herweijer H., Zhang G., Latendresse J.S., Bock I. and Wolff J.A. (1997) High expression of naked plasmid DNA in muscles of young rodents. *Hum. Mol. Genet.* **6**, 1435–1443.
6. Hartikka J., Bozoukova V., Jones D., Mahajan R., Wloch M.K., Sawdey M., Buchner C., Sukhu L., Barnhart K.M., Abai A.M., Meek J., Shen N. and Manthorpe M. (2000) Sodium phosphate enhances plasmid DNA expression in vivo. *Gene Ther.* **7**, 1171–1182.
7. Blomberg P., Eskandarpour M., Xia S., Sylven C. and Islam K.B. (2002) Electroporation in combination with a plasmid vector containing SV40 enhancer elements results in increased and persistent gene expression in mouse muscle. *Biochem. Biophys. Res. Commun.* **298**, 505–510.
8. Andre F.M., Cournil-Henrionnet C., Vernerey D., Opolon P. and Mir L.M. (2006) Variability of naked DNA expression after direct local injection: the influence of the injection speed. *Gene Ther.* **13**, 1619–1627.
9. Mir L.M., Bureau M.F., Gehl J., Rangara R., Rouy D., Caillaud J.M., Delaere P., Branellec D., Schwartz B. and Scherman D. (1999) High-efficiency gene transfer into skeletal muscle mediated by electric pulses. *Proc. Natl. Acad. Sci. U. S. A.* **96**, 4262–4267.
10. Niedzinski E.J., Chen Y.J., Olson D.C., Parker E.A., Park H., Udove J.A., Scollay R., McMahon B.M. and Bennett M.J. (2003) Enhanced systemic transgene expression after nonviral salivary gland transfection using a novel endonuclease inhibitor/DNA formulation. *Gene Ther.* **10**, 2133–2138.

11. Lee M.J., Cho S.S., Jang H.S., Lim Y.S., You J.R., Park J., Suh H., Kim J.A., Park J.S. and Kim D.K. (2002) Optimal salt concentration of vehicle for plasmid DNA enhances gene transfer mediated by electroporation. *Exp. Mol. Med.* **34**, 265–272.
12. Lefesvre P., Attema J. and van Bekkum D. (2002) A comparison of efficacy and toxicity between electroporation and adenoviral gene transfer. *B.M.C. Mol. Biol.* **3**, 12.
13. Taylor J., Babbs C.F., Alzghoul M.B., Olsen A, Latour M., Pond A.L., and Hannon K. (2004) Optimization of ectopic gene expression in skeletal muscle through DNA transfer by electroporation. *B.M.C. Biotechnol.* **4**, 11.
14. Schertzer J.D., Plant D.R. and Lynch G.S. (2006) Optimizing plasmid-based gene transfer for investigating skeletal muscle structure and function. *Mol. Ther.* **13**, 795–803.
15. McMahon J.M., Signori E., Wells K.E., Fazio V.M. and Wells D.J. (2001) Optimisation of electrotransfer of plasmid into skeletal muscle by pretreatment with hyaluronidase – increased expression with reduced muscle damage. *Gene Ther.* **8**, 1264–1270.
16. Schertzer J.D. and Lynch G.S. (2006) Comparative evaluation of IGF-I gene transfer and IGF-I protein administration for enhancing skeletal muscle regeneration after injury. *Gene Ther.* **13**, 1657–1664.

8

Chitosan Nanoparticle-Mediated Gene Transfer

Dongwon Lee and Shyam S. Mohapatra

Summary

Recent advances in genomics and proteomics have led to the evolution of DNA-based therapeutics and the use of gene therapy for treatment of a wide range of human diseases. However, poor cellular uptake and rapid in vivo degradation of DNA-based therapeutics are the major drawbacks of gene therapy. Viral and nonviral gene transfer vectors have been developed to facilitate the cellular internalization and preserve their activity until the successful transgene expression. Chitosan, a natural polysaccharide, is biocompatible and biodegradable. Chitosan has been reported to form nanocomplexes with DNA to protect them against DNase degradation and transfer DNA effectively and safely into cells in culture cell as well as in vivo.

Key Words: Plasmid DNA; gene expression; transfection; chitosan; liposomes; western blotting; flow cytometry.

1. Introduction

The basic concept of gene therapy is to deliver foreign genetic material into the somatic cells of a patient to correct or supplement defective genes responsible for inborn and acquired diseases *(1)*. The ideal gene delivery system must have well-defined physicochemical characteristics and the following properties: ease of assembly with protection of DNA before and after cell uptake, and capability of bypassing or escaping endocytic pathways *(2)*. Successful exogenous DNA delivery to cells can be accomplished using vectors, which can be classified as viral vector-mediated systems and nonviral vector-mediated systems *(3)*. Viral systems are by far the most effective means of DNA delivery, but their applications have been limited by toxicity, immunogenicity, limited

From: *Methods in Molecular Biology, vol. 433: Volume 1: Production and In Vivo Applications of Gene Transfer Vectors*
Edited by: J. M. Le Doux © Humana Press, Totowa, NJ

DNA carrying capacity, recombination, and high cost. Although gene delivery efficiency of nonviral systems is less than that by viral systems, DNA delivery via the nonviral route has become a popular and powerful tool for elucidating gene structure, regulation, and function ***(4,5)***. Nonviral DNA delivery systems consist of cationic polymers and lipids that can interact electrostatically with negatively charged plasmid DNA and form nano-sized complexes and that facilitate cellular uptake ***(2,6–9)***.

We have shown that chitosan is a good candidate for a gene delivery system because of its positive charge upon dissolution, nontoxicity, and biodegradability ***(10–13)***. This chapter provides methods of gene delivery using plasmid DNA nanocomplexes with chitosan, and addresses the straightforward assays for transgene expression using flow cytometry and immunoblotting techniques with respect to the molecular weight of chitosan and surface charge of nanocomplexes.

2. Materials

2.1. Preparation of Plasmid DNA Nanocomplexes with Chitosan

1. Chitosan of 33 kDa (Taehoon Bio, Korea) and 10 kDa (Marine Bioengineering Co. Ltd., China) molecular weight.
2. Polyethylimine (Sigma, St. Louis, MO, USA).
3. A plasmid (pEGFP-N2, 4.7 kbp, Clontech, Mountain View, CA) containing the human cytomegalovirus (CMV) promoter and enhanced green fluorescent protein (GFP) gene was amplified in *Escherichia coli* and purified using GenElute HP Plasmid Maxprep Kits (Sigma).
4. Particle size analyzer with dynamic and electrophoretic light scattering (Nicomp 380/ZLS, Particle Sizing Systems Inc., CA).
5. Transmission electron microscope (Jeol, Japan).
6. Carbon-coated 400 mesh copper grids (Canemco Inc. & Marivac Inc., Canada).
7. Phosphotungstic acid: prepare a 2.0% solution in distilled water.

2.2. Cell Culture

1. Human embryonic kidney (HEK 293) cells (ATCC) are grown in an incubator at 37 °C with 5% CO_2.
2. Dulbecco's Modified Eagle's Medium (DMEM) (Mediatech Inc., Herndon, VA) supplemented with 10% fetal bovine serum (HyClone, Logan, UT) and 1 % penicillin streptomycin (Mediatech Inc.).

2.3. Micrographs of Transfected Cells

1. Microscope (ECLIPSE TE300 Inverted Microscope, Nikon, Japan).
2. WICF Image J program (NIH, USA).

2.4. Flow Cytometry

1. Trypsin/EDTA (Mediatech Inc.).
2. Propidum iodide (PI) (BD Biosciences, San Jose, CA, USA).
3. Phosphate-buffered saline solution (Mediatech Inc.).
4. 5-ml round-bottom tube, 12×75 mm style (BD Falcon, San Jose, MA, USA).

2.5. Immunoblotting

2.5.1. SDS-Polyacrylamide Gel Electrophoresis (SDS-PAGE)

1. Separating gel monomer solution (12%): 3.35 ml of deionized water, 2.5 ml of 1.5 M Tris–HCl (pH 8.8), 100 μl of 10% (w/v) SDS stock, 4.0 ml of acrylamide/Bis (30% stock). Degas the mixed solution under vacuum for 15 min.
2. Stacking gel monomer solution (4%): 3.05 ml of deionized water, 1.25 ml of 0.5 M Tris–HCl, pH 6.8, 50 μl of 10% (w/v) SDS, 650 μl of acrylamide/bis (30% stock). Degas the mixed solution under vacuum for 15 min.
3. Thirty percent acrylamide/bis solution, 29:1 (3.3% C) (Bio-Rad Laboratories, Hercules, CA, USA).
4. Ammonium persulfate: prepare 10% solution in water.
5. *N,N,N,N'*-Tetramethyl-ethylenediamine (TEMED, Bio-Rad Laboratories).
6. Running buffer (10× Tris/Glycine/SDS Buffer, Bio-Rad Laboratories): Dilute 100 ml with 900 ml of water.

2.5.2. Western Blotting

1. Lysis buffer: prepare a stock buffer containing 50 mM Hepes pH 7.5, 150 mM NaCl, 1 mM EDTA, 10% glycerol, 0.5% Nonidet P40. Before use, complete lysis buffer is made by adding proteinase inhibitor cocktail (1×), phosphate inhibitor cocktail (1×), and 1 mM dithiothreitol (DTT).
2. Transfer buffer: 25 mM Tris base, 0.2 M glycine, 0.05% SDS, 20% methanol (pH 8.5).
3. Blocking buffer: 1× TBS, 0.1 % Tween 20 with 5% w/v nonfat dry milk.
4. Tris-buffered saline with Tween (TBS-T): 1× TBS with 0.1% Tween 20.
5. Primary antibody: Rabbit anti- GFP polyclonal antibody (Chemicon, CA).
6. Secondary antibody: Anti-rabbit immunoglobulin G (IgG) conjugated to horseradish peroxide (HRP) (Cell Signaling, Danvers, MA).
7. Biotinylated Protein Ladder (Cell Signaling).
8. Anti-Biotin, HRP-conjugated (Cell Signaling).
9. Actin (C-2) mouse monoclonal IgG (Santa Cruz Biotechnology, Santa Cruz, CA, USA).
10. Goat anti-mouse IgG-HRP conjugated (Santa Cruz Biotechnology).
11. Enhanced chemiluminescent substrate (Pierce, Rockford, IL, USA).
12. Immuno-Blot PVDF membrane (Bio-Rad Laboratories).
13. Kapak SealPack pouches (Kapak, Minneapolis, MN, USA).
14. Western blotting stripping buffer (Pierce).

3. Methods

3.1. Preparation and Characterization of Nanocomplexes

1. Prepare a stock solution of chitosan (2mg/ml) by dissolving chitosan in 0.05% acetic acid and filtering the solution using a 0.45-μm filter (*see* **Note 1**).
2. Dilute 200 μl of chitosan stock solution in 1 ml of acetate buffer (pH 6.5) (*see* **Note 2**).
3. Prepare chitosan/DNA nanocomplexes by adding various amount of pDNA solution to the diluted chitosan solution to give weight ratios of chitosan to DNA (1:1 to 5:1). Incubate the nanocomplexes for 10 min and vortex for 30 s.
4. Measure the particle size and zeta potential of chitosan/pDNA nanocomplexes using Nicomp380/ZLS (Particle Sizing Systems Inc.), which combines dynamic light scattering and electrophoretic light scattering. An example of the results is shown in **Fig. 1** (*see* **Note 3**).
5. For examination by transmission electron microscopy, one drop of nanocomplex suspension is deposited on carbon coated 400 mesh copper grids. The grids are stained with phosphotungstic acid solution (2%) for 2 min, and excess phosphotungstic acid solution is removed with filter paper. The grids are allowed to dry for a further 20 min and examined with a transmission electron microscope. An example of the results is shown in **Fig. 2**.

3.2. Transfection

1. In a 6-well tissue culture plate (Corning Incorporated, NY), seed 0.7×10^6 HEK293 cells in 2 ml of growth medium (DMEM) supplemented with 10% of serum.
2. Incubate the cells at 37 °C in a CO_2 incubator until the cells are ~60% confluent (*see* **Note 4**).
3. For each transfection, dilute 2.0 μg of plasmid DNA into 100 μl of serum-free medium (DMEM) (*see* **Note 5**).
4. Place 2 to 10 μg of chitosan in 100 μl of serum-free medium (DMEM).
5. Combine the two solutions, vortex gently, and incubate at room temperature for 2 min.
5. Add 800 μl of serum-free medium to the tube containing DNA nanocomplexes.
6. Remove the cell growth medium and wash the cells with 1 ml of serum-free medium.
7. Gently overlay 1 ml of medium containing DNA nanocomplexes into each well and incubate at 37 °C in a CO_2 incubator for 6 h.
8. Replace the DNA containing medium with 2 ml of DMEM containing 10% FBS and incubate cells at 37 °C in a CO_2 incubator for a total of 24–48 h.

3.3. Flow Cytometry

1. Wash cells with PBS and aspirate PBS.
2. Collect cells using trypsin/EDTA and low speed centrifugation at 290 × g for 5 min. Wash cell pellets with cold PBS and resuspend cell pellet in PBS at a concentration of 1×10^6 cells/ml.

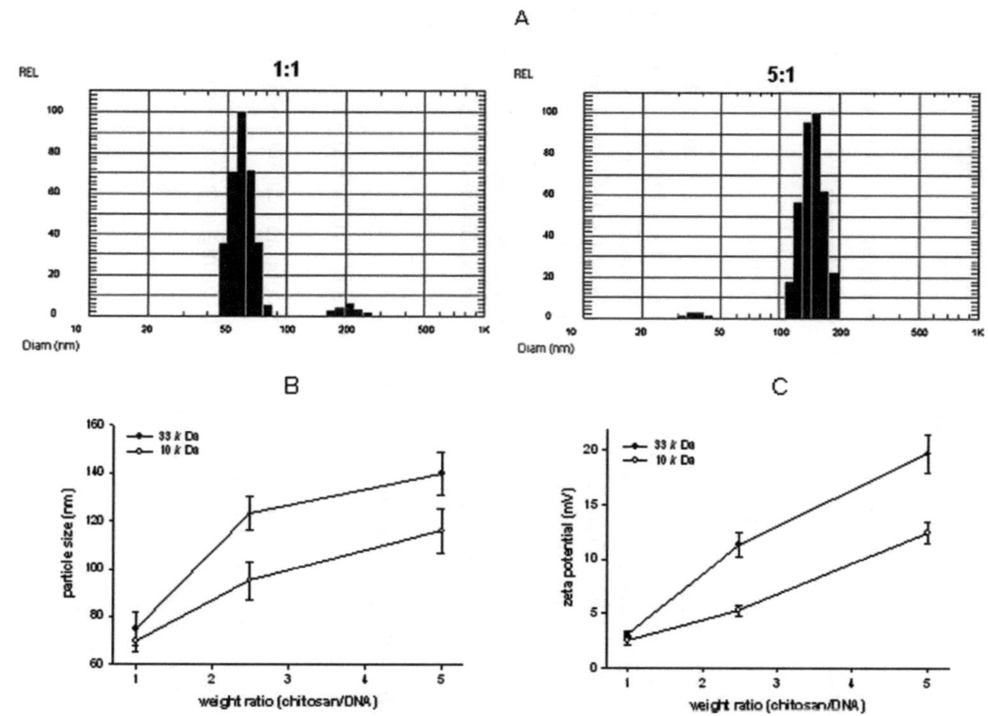

Fig. 1. Characterization of chitosan/DNA nanocomplexes. (**A**) Particle size of chitosan (33 kDa)/pDNA nanocomplexes at a weight ratio of 1:1 and 5:1; (**B**) The effect of weight ratio on the particle size of the chitosan/DNA nanocomplexes; (**C**) The effect of weight ratio on the zeta potential of chitosan/DNA nanocomplexes. The nanocomplexes were prepared at pH 6.5. Both the particle size and zeta potential of chitosan/DNA nanocomplexes increased with increasing weight ratio of chitosan to DNA. (Zeta potential refers to the electrostatic potential generated by the accumulation of ions at the surface of the colloidal particle that is organized into an electrical double layer.)

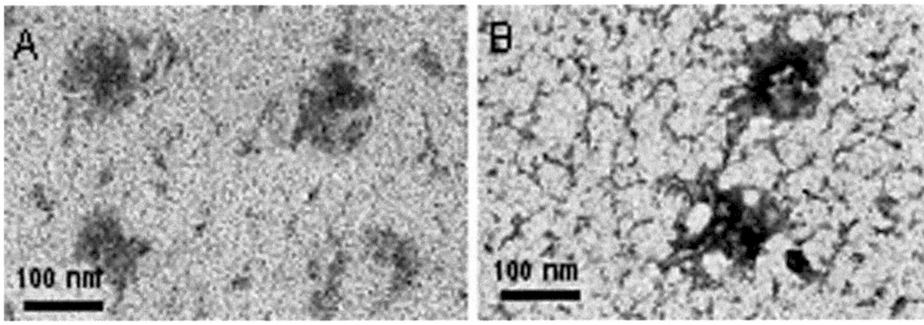

Fig. 2. TEM micrographs of pDNA nanocomplexes with chitosan, 10 kDa (**A**) and 33 kDa (**B**). Nanocomplexes were prepared at a weight ratio of 5:1 (chitosan/pDNA).

3. Add 1 μl of PI (2 mg/ml) to a tube containing approximately 1×10^6 cells and incubate on ice for 5 min in the dark before analysis.
4. For cell viability (**Fig. 3**), analyze by using flow cytometry with excitation at 488 nm and collect the fluorescence emission at 610 nm. An example is shown in **Fig. 3**. For the transfection efficiency, collect the fluorescence emission with a 510–530-nm long pass filter. Transfected cells are scored for GFP-positive cells with appropriate gating and controls using the green channel FL-1H. A total of 1.5×10^4 events were counted for each sample. The percentage of positive events is calculated as the events within the gate divided by total number of events then subtracting percentage of control samples. An example is shown in **Fig. 4**.

3.4. Fluorescent Micrographs for Transfected Cells

1. Cell culture plates are viewed under a microscope and brightfield images are captured.
2. Examine cell culture plates under a fluorescent microscope and capture brightfeild images. An example is shown in **Fig. 6**.
3. Capture fluorescent images of the same field of cells using a green interference filter (*see* **Note 6**). An example is shown in **Fig. 5**.

3.5. Immunoblotting Analysis

3.5.1. Cell Lysis

1. Harvest cells: Wash the cultured cell with cold 1× PBS twice. Lyse cells by adding 200 μl of complete lysis buffer on ice for 30 min.
2. Scrape the cell lysate using a cell scraper and transfer to 1.5-ml tube. Centrifuge at 12,000×*g* for 10 min at 4 °C.
3. Collect supernatant and measure the protein concentration using BCA Protein Assay Kit (Pierce). Store proteins at –80 °C until use. An example is shown in **Fig. 6**.

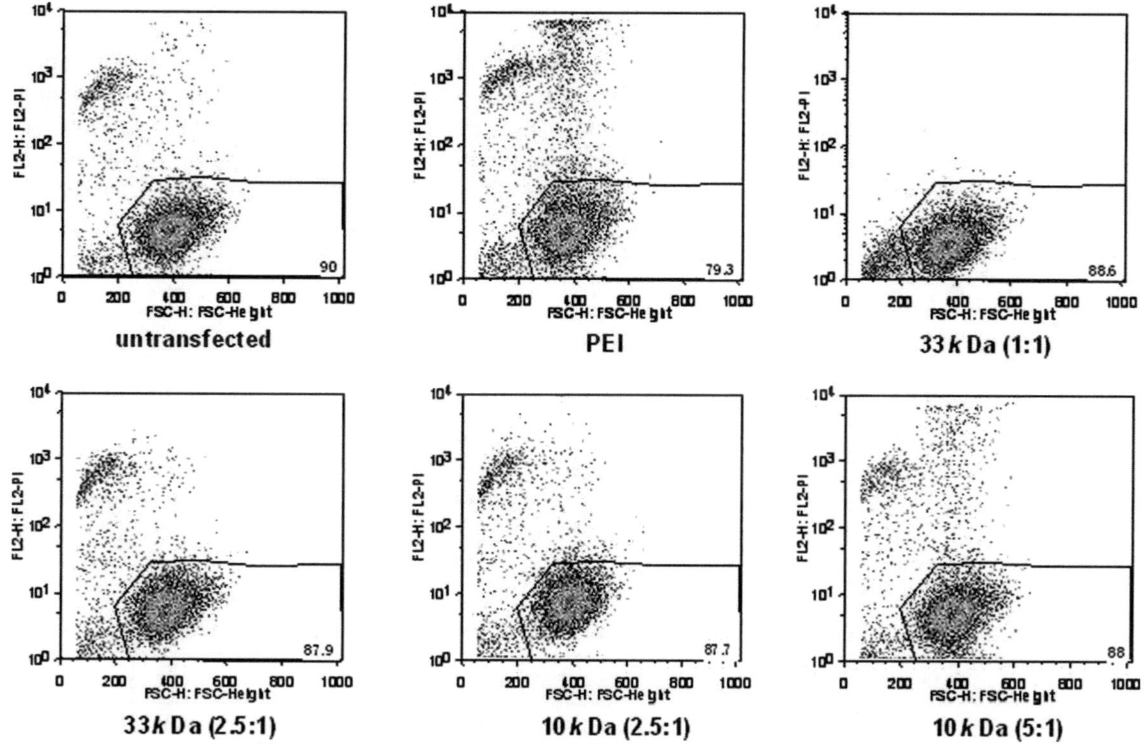

Fig. 3. Flow cytometry for cytotoxicity of chitosan/DNA nanocomplexes in HEK 293 cells. Flow cytometry (FACScan, BD Bioscience, San Jose, CA, USA) was carried out using PI as a viability marker at 48 h post-transfection. Untransfected cells are used as a negative control. A small fraction of cells (~10%) was damaged during the cell harvesting. A little cytotoxicity is observed with the cells transfected by a conventional polymer PEI. Chitosan/pDNA nanocomplexes did not exhibit cytotoxicity at all weight ratios (1:1~5:1).

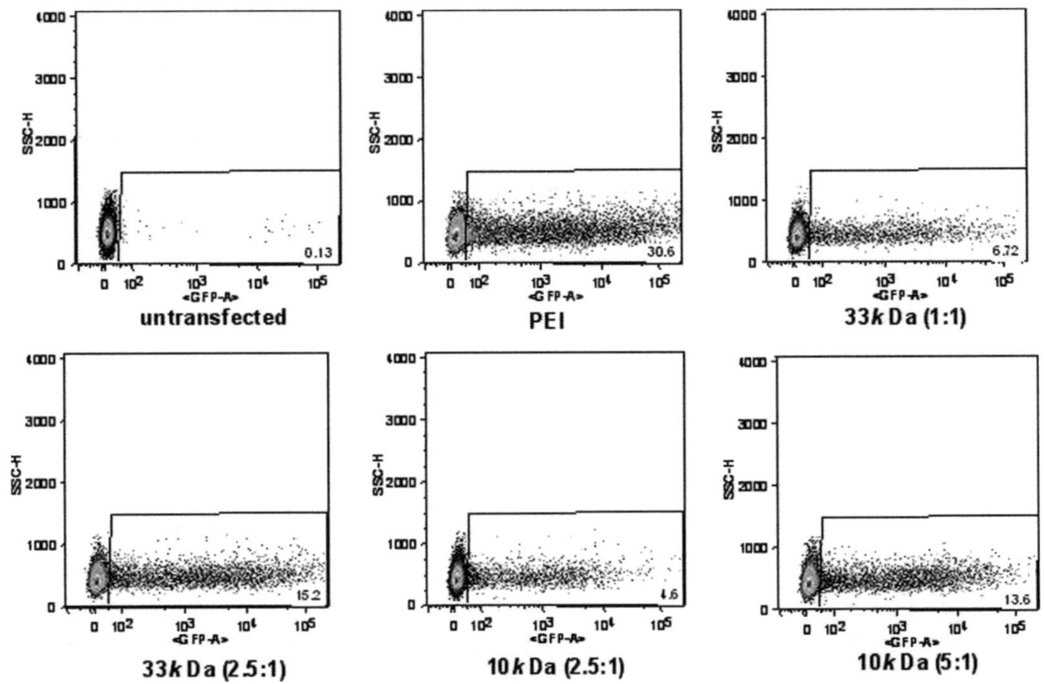

Fig. 4. Flow cytometry for transfection efficiency of chitosan/DNA nanocomplexes in HEK 293 cells. Flow cytometry (FACScan, BD Bioscience, San Jose, CA, USA) was carried out at 48 h post-transfection. Untransfected cells are used as a negative control. Conventional polymer PEI, a positive control, exhibited ~30% transfection efficiency that is higher than that of chitosan. Chitosan of 33k Da mediated the higher transgene expression than that of 10 kDa chitosan. Transfection efficiency of chitosan/pDNA nanocomplexes depends on the weight ratio. The results in **Figs. 3** and **4** indicate that chitosan is a non-toxic and useful non-viral vector for gene delivery. pH of transfection medium and presence of serum may influence the transfection efficiency of chitosan.

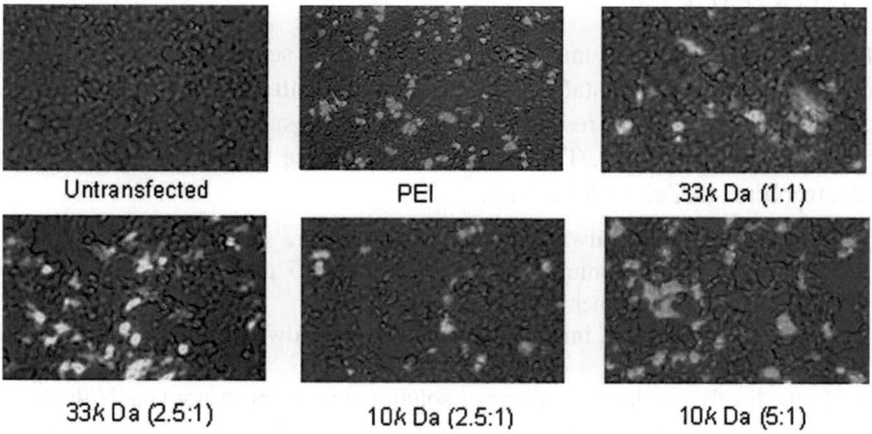

Fig. 5. Fluorescent micrographs of HEK293 cells transfected with chitosan/plasmid DNA nanocomplexes at different weight ratios. Micrographs (200×) were obtained on an ECLIPSE TE300 Inverted Microscope (Nikon, Japan) at 40 h post-transfection. Fluorescent images of the same field of cells at the same magnification were obtained with the identical exposure time. Brightfield images were merged with fluorescent images using a WICF Image J program (NIH, USA). The most GFP–expressing cells are observed when cells are transfected with PEI, a positive control. The level of transgene expression of chitosan depends on the molecular weight and weight ratio. The results support the transfection efficiency as shown in **Fig. 4**.

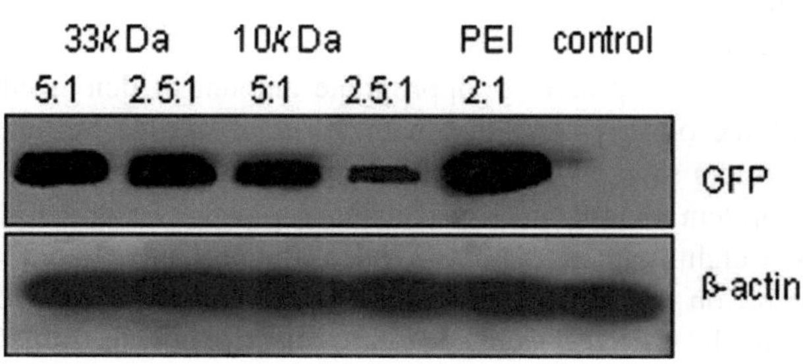

Fig. 6. Immunoblotting analysis of green fluorescent protein expressed in HEK293 cells transfected with pDNA expressing GFP. Proteins were extracted from the cells 50 h after transfection and were processed for assay of GFP by Western blotting. The values indicate the weight ratio of chitosan to plasmid DNA. No GFP expression is observed with the untransfected cells. PEI, a positive control, shows the highest level of GFP expression. In the comparison of GFP expression mediated by chitosan, higher gene expression was observed with cells transfected with higher molecular weight chitosan. Chitosan of molecular weight 10 kDa showed a higher transfection efficiency at a weight ratio 5:1. The results are in good agreement with the result of flow cytometry.

3.5.2. SDS–PAGE

Discontinuous polyacrylamide gels consist of a separating (lower) gel and stacking (upper) gel. The stacking gel acts to concentrate large sample volumes and results in better band resolution than using the same volume of separating gel without a stacking gel. These instruction assume the use of Mini-Protean® II Electrophoresis Cell (BIO-RAD).

1. Assemble the gel sandwich with a 1.5-mm spacer.
2. Add 50 μl of ammonium sulfate solution and 5 μl of TEMED to degassed separating gel monomer solution.
3. Pour the gel solution into the assembled gel sandwich using a plastic transfer pipette.
4. Immediately overlay the monomer solution with isopropanol (*see* **Note 7**).
5. Allow the gel to polymerize for 45 min to 1 h. Pour off the isopropanol and rinse the top of the gel with distilled water. Dry the area above the separating gel with a filter paper.
6. Place a comb in the gel sandwich and tilt it at ~10 ° angle (*see* **Note 8**).
7. Add 25 μl of APS and 5 μl of TEMED to the degassed stacking gel monomer solution and pour down the spacer until all comb teeth are covered by solution. Allow the gel to polymerize for 30 min.
8. Remove the comb by pulling it straight up gently and rinse the wells completely with distilled water.
9. Assemble the gel unit completely.
10. Prepare the running buffer by diluting 100 ml of 10× tris/glycine/SDS buffer with 900 ml of water and add the running buffer to the lower buffer chambers of the gel unit.
11. Dilute the 15 μg of protein in appropriate amount of deionized water to give a final volume of 20 μl and mix with 20 μl of sample buffer (2×). Heat the prepared protein samples at 90 °C for 5 min.
12. Load the protein sample in a well using a pipette. Load 10 μl of prestained molecular weight marker and Biotinylated protein ladder into a well.
13. Place the lid on the top of the lower buffer chamber. Apply power (200 V) to the Mini-PROTEIN II cell and allow electrophoresis for 45–60 min (*see* **Note 9**).

3.5.3. Western Blotting

1. Cut PVDF membrane and filter paper a little larger than the gel and soak them in the transfer buffer (*see* **Note 10**).
2. Prepare the gel sandwich for transfer of protein to membrane: One pre-wetted fiber pad is laid on the gray side of the gel holder cassette. A sheet of filter paper is laid on the fiber pad. Place the gel on filter paper. The pre-wetted PVDF membrane is laid on the gel. Complete the sandwich by placing a sheet of filter paper and fiber pad. Remove any air bubbles by using a glass tube to roll air bubbles out (*see* **Note 11**).

3. Close the cassette firmly and lock it with the latch.
4. Place the cassette in the electrode module.
5. Place the electrode module and a frozen Bio-Ice cooling unit in the buffer tank and fill the tank with transfer buffer completely.
6. Put on the lid, plug the cables into the power supply and run the blot. Transfer can be accomplished at 100 V for 1 h at 4 °C.
7. Once the transfer is complete, disassemble the blotting sandwich and remove the membrane carefully.
8. Incubate the blotting membrane in 25 ml of blocking buffer on a rocking platform at room temperature for 1 h.
9. Prepare 10 ml of primary antibody (rabbit anti-GFP polyclonal antibody) at a 1:800 dilution. Put the primary antibody solution and the blotting membrane in a Kapak SealPack pouch and seal it using an Impulse Sealer. Incubate the membrane overnight at 4 °C with gentle agitation (*see* **Note 12**).
10. Wash the blotting membrane with 25 ml of TBS-T for 5 min three times.
11. Prepare fresh secondary antibody (anti-rabbit IgG conjugated to HRP) at a 1:300 dilution and anti-biotin antibody solution at a 1:1000 dilution in 10 ml of blocking buffer. Incubate the membrane with the secondary antibody at room temperature for 1 h on a rocking platform.
12. Wash the blotting membrane with 25 ml of TBS-T for 5 min three times.
13. Prepare the chemiluminescent substrate solution. Mix the same volume of Luminol/Enhancer Solution and Stable Peroxide Solution in the dark.
14. Incubate the blotting membrane in a working substrate solution for 5 min in the dark.
15. Remove the blotting membrane from substrate solution and place it in plastic wrap. Remove the excess liquid and air bubbles between the membrane and surface of the plastic wrap. Keep light away from the membrane.
16. Place the blotting membrane in an X-ray film cassette with the protein side facing up. Place the film on the top of the blotting membrane. Exposure time may be varied to achieve the optimal results.
17. Develop film using appropriate developing solution and fixer (**Fig. 6**).

3.5.4. Stripping and Reprobing Blots for β-Actin

Once satisfactory results are obtained from the protein blot, the blotting membrane is stripped and reprobed for β-actin. This provides a loading control that validates equivalent loading of protein in the procedure.

1. Strip the blotting membrane using 20 ml of Western Blot Stripping Buffer at 37 °C for 30 min. Wash the membrane with TBS-T for 5 min three times.
2. The blotting membrane can be reprobed with anti-actin (C-2) mouse monoclonal IgG (1:300 in blocking buffer) and secondary antibody, goat anti-mouse IgG-HRP as above.

4. Notes

1. Chitosan is soluble at pH less than 6.5 because the pKa the chitosan is around 6.5. The high concentration of chitosan/DNA nanocomplexes in a buffer causes aggregation. For the particle size and zeta potential measurement, the pH should be identical because they are highly related to the pH of the solution.
2. The weight ratio of chitosan to DNA for best results varies by molecular weight of chitosan. In general, the lower molecular weight chitosan requires a higher weight ratio to form stable nanocomplexes with DNA and give high transfection efficiency. However, a very high weight ratio may cause cytotoxicity and diminish the transfection efficiency because of slow degradation and slow DNA release.
3. The particle size analyzer uses dynamic light scattering to obtain the particle size as well as size distribution. Zeta potential refers to the electrostatic potential generated by the accumulation of ions at the surface of the colloidal particles that are organized into an electrical double-layer. The main reason to measure zeta Potential is to predict the colloidal stability, which is related to the interaction between particles.
4. Cell culture confluency of 60–70% is optimal for transfection of HEK293 cells. It usually takes 20–24 h, but time will vary by cell types. As the transfection efficiency may be dependent on the culture confluency, it is critical to maintain a standard seeding protocol.
5. Opti-MEM Reduced Serum Medium, serum-free media, or tissue culture grade deionized water can also be used. The Volume of medium can be changed proportionally when different amounts of chitosan are employed.
6. Air bubbles are undesirable in the SealPak during the incubation with the primary antibody.
7. The advantage of using isopropanol is that the overlay solution can be applied rapidly with a pipette because very little mixing will occur. Do not allow isopropanol to remain on the gel more than 1 h, or the top of the gel will dehydrate. When water is used to overlay, it should be applied more slowly to prevent mixing with separating gel solution.
8. Tilting the teeth at 10 ° angle will prevent air from being trapped under the comb teeth while the stacking gel solution is poured.
9. The recommended conditions for optimal resolution is 200 V with constant voltage setting. During the run, the initial current ~60 mA slowly drops to 30 mA. The usual run time is 45 min.
10. PVDF membrane is an ideal support for immunoassays of blotted proteins. PVDF membrane exhibits better binding efficiency of blotted materials in the presence of SDS in the transfer buffer. Before use, PVDF must first be wetted in the 100% methanol and then soaked in the transfer buffer.
11. Membrane, filter paper and fiber pads should be soaked in the transfer buffer for 15 min. Air bubbles formed in the gel sandwich should be removed by using a glass tube to gently roll air bubbles out for good protein transfer.

12. Gene expression in HEK 293 cells can be observed ~10 h after transfection. Exposure time should be fixed to keep the intensity of green fluorescence in the images. The rate of gene expression varies by the transfection reagent. The faster onset of gene expression is observed with the lower molecular weight chitosan than the high molecular weight chitosan because of easier dissociation of nanocomplexes and faster degradation of chitosan.

Acknowledgments

The authors acknowledge the financial support by and VA Merit Review grant awarded to SSM, and support of the Joy McCann Culverhouse and Mabel and Ellsworth Simmons endowments and thank Stephanie Medley for assistance in preparation of the manuscript.

References

1. Mansouri, S., Lavigne, P., Corsi, K., Benderdour, M., Beaumont, E., and Fernandes, J. C. (2004) Chitosan-DNA nanoparticles as non-viral vectors in gene therapy: strategies to improve transfection efficacy. *European Journal of Pharmaceutics and Biopharmaceutics* **57**, 1–8.
2. Luo, D., and Saltzman, W. M. (2000) Synthetic DNA delivery systems. *Nature Biotechnology* **18**, 33–37.
3. Lee, M., Nah, J. W., Kwon, Y., Koh, J. J., Ko, K. S., and Kim, S. W. (2001) Water-soluble and low molecular weight chitosan-based plasmid DNA delivery. *Pharmaceutical Research* **18**, 427–31.
4. Ii'ina, A. V., and Varlamov, V. P. (2005) Chitosan-based polyelectrolyte complexes: A review. *Applied Biochemistry and Microbiology* **41**, 5–11.
5. Koping-Hoggard, M., Varum, K. M., Issa, M., Danielsen, S., Christensen, B. E., Stokke, B. T., and Artursson, P. (2004) Improved chitosan-mediated gene delivery based on easily dissociated chitosan polyplexes of highly defined chitosan oligomers. *Gene Therapy* **11**, 1441–52.
6. De Laporte, L., Rea, J. C., and Shea, L. D. (2006) Design of modular non-viral gene therapy vectors. *Biomaterials* **27**, 947–54.
7. Ishii, T., Okahata, Y., and Sato, T. (2001) Mechanism of cell transfection with plasmid/chitosan complexes. *Biochimica Et Biophysica Acta-Biomembranes* **1514**, 51–64.
8. Patil, S. D., Rhodes, D. G., and Burgess, D. J. (2005) DNA-based therapeutics and DNA delivery systems: A comprehensive review. *Aaps Journal* **7**, E61–E77.
9. Koping-Hoggard, M., Tubulekas, I., Guan, H., Edwards, K., Nilsson, M., Varum, K. M., and Artursson, P. (2001) Chitosan as a nonviral gene delivery system. Structure-property relationships and characteristics compared with polyethylenimine in vitro and after lung administration in vivo. *Gene Therapy* **8**, 1108–21.

10. Zhang, W. D., Yang, H., Kong, X. Y., Mohapatra, S., San Juan-Vergara, H., Hellermann, G., Behera, S., Singam, R., Lockey, R. F., and Mohapatra, S. S. (2005) Inhibition of respiratory syncytial virus infection with intranasal siRNA nanoparticles targeting the viral NS1 gene. *Nature Medicine* **11**, 56–62.
11. Kumar, M., Kong, X., Behera, A. K., Hellermann, G. R., Lockey, R. F., and Mohapatra, S. S. (2003) IFN-g-pDNA Nanoparticle (CIN) Therapy for Allergic Asthma. *Genetic Vaccines Therapy* **1**, 3.
12. Kumar, M., Behera, A. K., Lockey, R. F., Zhang, J., Bhullar, G., De La Cruz, C. P., Chen, L. C., Leong, K. W., Huang, S. K., and Mohapatra, S. S. (2002) Intranasal gene transfer by chitosan-DNA nanospheres protects BALB/c mice against acute respiratory syncytial virus infection. *Human Gene Therapy* **13**, 1415–25.
13. Mohapatra, S. S. (2003) Mucosal gene expression vaccine: a novel vaccine strategy for respiratory syncytial virus. *Pediatric Infectious Disease Journal* **22**, S100–S104.

9

PEG–PEI Copolymers for Oligonucleotide Delivery to Cells and Tissues

Gordon J. Lutz, Shashank R. Sirsi, and Jason H. Williams

Summary

Inefficient delivery of antisense oligonucleotides (AO) to target cell nuclei remains as the foremost limitation to their usefulness. Copolymers of cationic poly(ethylene imine) (PEI) and polyethylene glycol (PEG) are extremely well-studied compounds that markedly improve the in vitro and in vivo delivery of AOs to cells and tissues. By varying the Mw of PEI, as well as the nature of PEG shielding, PEG-PEI-AO nanoparticulates can be prepared with a dynamic range of size, surface charge, and stability. Each of these properties in-turn influences the transfection capacity of the PEG-PEI-AO polyplexes. In addition, PEG-PEI copolymers are readily functionalized for enhanced efficacy and specificity of cellular and tissue targeting. The synthesis and functionalization of PEG-PEI copolymers is remarkably simple and requires very little specialized equipment. Thus, PEG-PEI copolymers represent a tractable and adaptable oligonucleotide delivery system that can be customized and optimized to the investigators' specific application. This chapter describes the step-by-step synthesis of several PEG-PEI copolymers that are specifically formulated to provide effective delivery of AOs using both in vitro and in vivo applications. We describe the preparation of the PEG-PEI-AO polyplexes and provide examples showing transfection of cultured cells in vitro, as well as skeletal muscles in vivo using both local and systemic delivery.

Key Words: Nanopolymers; oligonucleotides; poly(ethylene imine); transfection; drug delivery.

From: *Methods in Molecular Biology, vol. 433: Volume 1: Production and In Vivo Applications of Gene Transfer Vectors*
Edited by: J. M. Le Doux © Humana Press, Totowa, NJ

1. Introduction

The use of antisense oligonucleotides (AOs) has increased dramatically over the past decade, in part due to significant advances in oligonucleotide chemistry that have greatly increased the stability, specificity, and functionality of AOs. For example, AOs comprised of phosphorothioate bases containing 2′O-methyl substitutions are extremely stable compounds that function within the cell nucleus to modulate pre-mRNA splicing. Various 2′O-methyl AOs have been shown to modulate alternative splicing, correct aberrant splicing, and promote both exon skipping and exon inclusion. Thus, 2′O-methyl AOs represent a powerful class of compounds with broad utility in both basic and translational research, with outstanding potential for use as therapeutic agents for various diseases. However, despite the major advances in AO chemistry and functionality, the delivery of AOs to target cells nuclei, especially in vivo, remains as the main limitation to their usefulness.

As illustrated in **Fig. 1**, the amine-rich cationic polymer poly(ethylene imine) (PEI) is an efficient nucleotide carrier that binds to the negatively charged

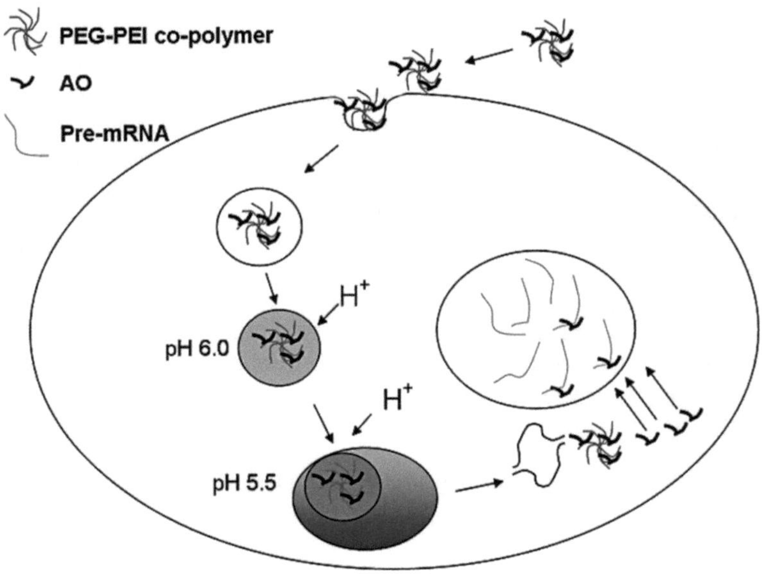

Fig. 1. Illustration of the presumed mechanisms of PEG-PEI-AO polyplex transfection. The positive surface charge of the polyplex stimulates nonspecific receptor-mediated endocytosis. Endosomal escape is facilitated by the "proton sponge" mechanism, involving protonation of PEI amine groups and osmotic swelling. The polyplex dissociates into "free AO" and copolymer before translocation of the AO into the cell nucleus, although the precise nature of the dissociation process remains unresolved.

phosphate backbone of nucleotides and negatively charged elements of cell membranes, facilitating endocytotic uptake of PEI–nucleotide complexes into cells ***(1–8)***. Protonation of the amine groups on PEI within endosomal compartments (the "so called" proton sponge effect) is thought to cause osmotic lysis, and release of the endosomal contents ***(6,7,9–11)***. Once released, AO may enter the nucleus by a process that likely requires dissociation from the polymer carrier. The functionality of PEI as a nucleotide carrier is significantly improved by incorporating a nonionic linear polymer polyethylene glycol (PEG) into PEG–PEI copolymers. The main purpose of PEGylation is to lower the cytotoxicity of PEI by shielding its surface charge ***(3,5,12)***. Overall, by varying the MW and the arrangement of PEG and PEI, PEG–PEI copolymers represent a flexible nucleotide delivery system with controllable size and adjustable unpackaging properties. Although PEG–PEI copolymers have been primarily used for plasmid delivery, other studies have provided strong evidence that PEGylated PEI can be adapted to function as an effective carrier for cellular delivery of small oligonucleotides ***(13–22)***.

In this chapter, we describe the step-by-step synthesis of three unique versions of PEG–PEI copolymers that are specifically formulated for transfection of AOs into cultured cells or for local and systemic delivery of AOs in vivo. Specific examples of the use of PEG-PEI-AO polyplexes for in vitro and in vivo delivery of AOs to skeletal muscle fibers are provided.

2. Materials

2.1. Synthesis of PEG–PEI Copolymers

2.1.1. Chemicals, Reagents, and Supplies

1. Methoxypoly(ethylene glycol)-5000 (mPEG5000) and mPEG550 (Sigma, St. Louis, MO, USA).
2. PEI25000 and PEI2000 (Sigma).
3. Hexane-1,6-diisocyanate (HMDI) (Sigma).
4. Et2O.
5. Hexanes (Sigma).
6. $CHCl_3$ (Sigma).
7. NHS-rhodamine (Pierce Biotechnology, Rockford, IL, USA).
8. Dialysis tubing (MWs 10000 and 1000, Spectrum Laboratories, Rancho Dominguez, CA, USA).
9. Nitrogen gas.

2.1.2. Instruments and Equipment

1. Reflux condenser.
2. Buchner funnel.
3. Filter paper (Fine grade Q2, Fisher Scientific, Rochester, NY, USA).

4. Rotary Evaporator (e.g., RV 06-ML, IKA Works, Wilmington, NC, USA).
5. Vacuum oven (e.g., Isotemp 281A, Fisher Scientific).
6. Lyophilizer (e.g., VirTis, Model 10-100V, Gardiner, NY, USA).

2.2. Preparation of PEG-PEI-AO Polyplexes

1. In our examples, we describe the use a 20 mer 2′O-methyl AO that is 5′ end-labeled with 6-FAM moiety. This AO has been used specifically to induce exon skipping of a mutated exon in dystrophin pre-mRNA, resulting in the expression of dystrophin. However, any 2′O-methyl AO (typically 18-25-mer) can be used, with the sequence chosen to modulate splicing of the specific gene of interest.
2. Bath sonicator (1510, Branson, Danbury, CT, USA).

2.3. Stability of PEG-PEI-AO Polyplexes–Polyanion Competition Assay

2.3.1. Chemicals, Reagents, and Supplies

1. Heparin (American Pharmaceutical Partners, Inc., Schaumburg, IL, USA).
2. Agarose (Sigma).
3. TAE: 40 mM Tris–acetate, pH approx. 8.3, containing 1 mM EDTA.
4. Loading buffer (Sigma).

2.3.2. Instruments and Equipment

1. Mini gel electrophoresis unit (B1A, Owl Separation Systems, Portsmouth, NH, USA).
2. Power supply.
3. Gel scanner with appropriate software for densitometry measurements.

2.4. Transfection of Cultured Cells with PEG-PEI-AO Polyplex

1. Growth medium: DMEM with 1% gentamycin (with and without 10% FBS).
2. Differentiation medium: DMEM with 2% horse serum or OptiMem.
3. Sterile 1× PBS.
4. Cell culture grade H_2O.
5. PEG-PEI-AO polyplex (prepared as in **Subheading 3.2.**) in sterile PBS.

2.5. IM and IV Delivery of PEG-PEI-AO Polyplexes

2.5.1. Chemicals, Reagents, and Supplies

1. Sterile insulin syringes: 27–28 G/0.5" and 31 G/0.5"; both with permanently attached needles.

2. 0.9% sterile saline.
3. 70% ethanol.
4. Opthalmic lubrication (Henry Schein, Melville, NY, USA).
5. Sterile swabs.
6. Isopropyl alcohol swabs.

2.5.2. Instruments and Equipment

1. Narrow blade hair trimmers.
2. Heated surface/heating pad.
3. Heat lamp.
4. Tail vein restraint (Plas-Labs, Inc, Lansing, MI, USA) or fabricated from a 50-ml conical tube.

3. Methods

3.1. Synthesis of PEG–PEI Copolymers

The nomenclature used for copolymers indicates the Mw of PEI and PEG, along with the number of PEG chains grafted per PEI (shown as subscript). For example, the copolymer PEI25K$(PEG5K)_{10}$ has 10 PEG chains of 5000 Da grafted to PEI of 25000 Da.

First we provide the protocol for the synthesis of the high Mw PEI25K$(PEG5K)_{10}$ copolymer, which is most useful for transfection of AOs into cultured cells in vitro. In **Subheadings 3.1.2** and **3.1.3**, we extend this to low Mw PEI2K-based copolymers that are more suitable for in vivo delivery of AOs.

3.1.1. High M_w PEI25K$(PEG5K)_{10}$

PEG–PEI copolymers are prepared using a simple two-step procedure as previously described ***(3,23)***. This involves complexation of activated mPEG and PEI. The synthesis is described to prepare about 3 g of final copolymer, although this is easily scaled and the exact yield will vary depending on the number of washes and desired purity of the sample.

1. Activate mPEG5000 with hexane-1,6-diisocyanate (HMDI), based on a previously reported procedure ***(3)***. Dissolve 10 g of mPEG5000 in 20 ml of $CHCl_3$ and add to 40 ml of HMDI (~200× molar excess HMDI) in a 100 ml round-bottom flask, and reflux under nitrogen for 16 h.
2. Recover resultant amine-reactive mPEG-HMDI by precipitation in 500 ml of hexanes (*see* **Note 1**). Isolate precipitated polymer by vacuum filtration, using a Buchner funnel and fine filter paper, and then wash multiple times with hexanes to remove excess HMDI.

3. Re-dissolve residue (mPEG-HMDI) in a minimal amount of $CHCl_3$ (10–15 ml) and re-precipitate into hexanes (as in **step 2**) three times.
4. Dry the mPEG-HMDI under vacuum for 48–72 h and store at –20°C in sealed vials under N_2.
5. Dissolve 1 g of PEI25000 and 2.06 g of activated mPEG5000 each in 50 ml of $CHCl_3$, combine in a 250 ml round-bottom flask, and reflux for 16 h under N_2. Choose the gram quantities of reactants to give PEG : PEI molar ratio of 10 (*see* **Table 1**).
6. Concentrate the copolymers using a rotary evaporator, precipitate into Et2O, dry under vacuum, and store at –20°C in sealed vials under N_2.
7. Prepare polymer stock solutions (typically 10 μg/μl) fresh each month; dissolve the copolymer in RNase-free diH20, vortex briefly, sonicate for 15 min, and then store at 4°C.

3.1.2. Low M_w PEI2K(PEG550)$_{10}$

The synthesis is exactly as described for the high M_w copolymer in **Subheading 3.1.1**, with the following differences (*see* also **Note 2**).

[**Step 1**] Dissolve 12 g of mPEG550 in 30 ml of $CHCl_3$ and add to 60 ml of HMDI (~17× molar excess) in a 250-ml round-bottom flask.

[**Steps 2–4**] Precipitate activated mPEG-HMDI in cold hexanes (–30°C). Perform re-precipitation from $CHCl_3$ four times.

[**Step 5**] React 0.84 g of water-free PEI2K with 3 g of activated mPEG550-HMDI (*see* **Note 3**).

[**Step 6**] Obtain copolymers by evaporation of the solvent and dry the products under high vacuum until weight constancy.

Table 1
Composition and Mw of PEG-PEI Copolymers

Copolymer	PEI M_w[a]	PEG M_w	PEG : PEI	Copolymer M_w
PEI25K(PEG5K)$_{10}$	25000	5000	10	75,000
PEI2K(PEG5K)$_{10}$	2000	5000	10	52,000
PEI2K(PEG550)$_{10}$	2000	550	10	7500

[a] The M_w values for PEI refer to the values provided by the supplier, and the corresponding M_ws listed for copolymers are based on these values. Recently, the M_w of commercially available branched PEI25K has been measured using SEC-MALLS and found to be substantially higher than indicated by the suppliers (*3,29*). Therefore, the PEG : PEI ratio listed for the high M_w PEI25K-based copolymer may underestimate the actual ratio.

3.1.3. Low M_w PEI2K(PEG5K)$_{10}$

The synthesis is exactly as for high M_w copolymer in **Subheading 3.1.1**, with the following differences.

[**Step 5**]. Dissolve 0.1 g of PEI2K and 2.6 g of activated mPEG5000 each in 50 ml of $CHCl_3$ and reflux for 16 h under N_2.

3.1.4. Rhodamine-Labeling of PEG–PEI Copolymers

Rhodamine labeling of PEG–PEI copolymers is useful for fluorescence tracking studies and is a relatively simple procedure involving covalent attachment of *N*-hydroxysuccinimide (NHS)-rhodamine to amine groups on PEI *(8)*. An example of rhodamine labeling of high M_w PEI25K(PEG5K)$_{10}$ at a ratio of 14 rhodamine's per mole PEG-PEI is as follows:

1. Add 10 mg of PEI25K(PEG5K)$_{10}$ (from stock solution) to 1× PBS (pH 8.0) to final volume of 1 ml.
2. Add 1 mg of NHS-rhodamine to 100 µl of DMSO.
3. Mix together the NHS-rhodamine and copolymer solutions and incubate on ice for 2 h.
4. Remove un-reacted NHS-rhodamine by dialyzing for 48 h against large volumes of di-H_2O (1–4 l; changed after 24 h).
5. Lyophilize the dialyzed solution, place in storage vials, flush the vials with N_2 gas (10 s) and then cap and seal to prevent oxidation during storage. Store at –20°C.

3.2. Preparation of PEG-PEI-AO Polyplexes

Figures 2 and 3 show the macromolecular properties of three unique PEG-PEI-AO polyplexes used as examples in this chapter (*see* **Note 4**). All polyplexes were prepared at a N : P ratio of 5, where N represents moles of amine on PEI, and P represents moles of phosphate on AO *(7)*. Preparation of PEG-PEI-AO polyplexes at a desired N : P ratio requires the following information:

1. M_w of the AO.
2. Number of phosphate groups on the AO (equal to 1 less than the number of base pairs).
3. M_w of the copolymer.
4. Number of amine groups per mole of copolymer.

Also *see* **Note 5** for further information.

Table 2 shows the amount of copolymer needed to prepare PEG-PEI-AO polyplexes (N : P = 5) containing 100 µg of AO. The AO used in these examples is a 6-FAM-AO (20 mer) with a MW of 7.4 kDa.

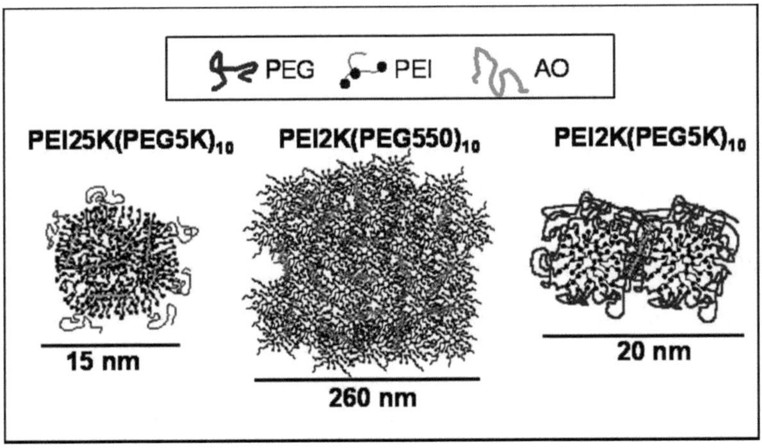

Fig. 2. Schematic illustrations of the three PEG-PEI-AO polyplexes described in this chapter. Grafting of high M_w PEG5K to either high M_w PEI25K (left) or low M_w PEI2K (right) provides enough steric repulsion to prevent aggregation, resulting in tiny nanoparticulates when complexed with AO. In contrast, when low M_w PEG550 is grafted to PEI2K (middle), significantly larger aggregate particles are formed. Reproduced from Glodde et al. *(23)* with permission.

Example of the preparation of 100 μl of PEI25K(PEG5K)$_{10}$-AO polyplex solution at a AO concentration of 1 μg/μl.

1. Pipette 10 μl of AO stock solution (10 μg/μl) and 72.6 μl of sterile PBS into an epi tube.
2. Add 17.4 μl of PEI25K(PEG5K)$_{10}$ copolymer stock solution (10 μg/μl) to the AO solution while mixing.

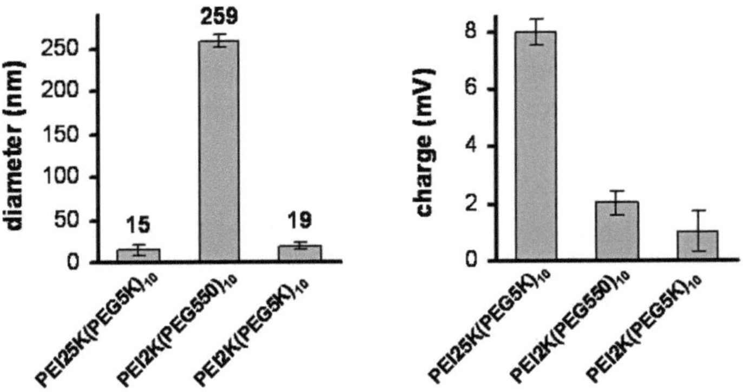

Fig. 3. Hydrodynamic diameter (left) and surface charge (right) of the three PEG-PEI-AO polyplexes shown in **Fig. 2**, as measured by dynamic light scattering and zeta potential, respectively. On the basis of data from Glodde et al. *(23)* with permission.

Table 2
Example Calculation of Amount of Copolymer Used to Form PEG-PEI-AO Polyplexes Containing 100 μg of the AO

Copolymer	M_W (Da)	~Amine groups/mol	Copolymer (μg)
$PEI25K(PEG5K)_{10}$	75,000	580	174
$PEI2K(PEG5K)_{10}$	52,000	46	1510
$PEI2K(PEG550)_{10}$	7,500	46	220

Example of the preparation of 100 μl of $PEI25K(PEG5K)_{10}$-AO polyplex solution at a AO concentration of 1 μg/μl.

3. Sonicate using a standard bath sonicator for 30 min at room temperature.
4. Incubate on ice for 30 min, and use within ~1 h (*see* **Note 6**).

3.3. Stability of PEG-PEI-AO Polyplexes-Polyanion Competition Assay

The association–dissociation properties (i.e., stability) of PEG-PEI-AO polyplexes is a critical determinant of their transfection capacity. In an optimal polyplex delivery system, the electrostatic charge association between the AO and copolymer must be strong enough to keep the polyplex together during transit to, and uptake into, the target cells. However, the stability must not be too strong as to prohibit release of "free AO," so that AO can be translocated to the nucleus. We recently showed that a simple polyanion competition assay is a useful way to assess the relative stability of the various polyplexes. In this assay, polyplexes were incubated with varying amounts of a competing polyanion heparin, electrophoresed on agarose gels, and the intensity of the "free AO" band (i.e., released AO) was quantified *(23)*. We found that the three polyplex formulations described in this chapter provided extremely stable particles (*see* **Fig. 4**). We reasoned that the stability was one of the salient features explaining their excellent transfection capacity. In contrast, copolymers that contained higher numbers of PEG chains were extremely unstable (*see* **Fig. 4**), and correspondingly their transfection efficiency was poor *(22)*. Thus, we caution that when experimenting with novel polyplex formulations, investigators should examine the stability of their polyplexes using this assay (*see* **Note 7**).

The protocol for the polyanion competition assay is as follows:

1. Prepare agarose gels (1.5% in TAE buffer).
2. Prepare polyplex solutions (using a FAM-labeled AO) as described in **Subheading 3.2** except 1–2 μg of AO should be used and final volumes should be 30 μl.
3. Add heparin to the polyplex solution to a final concentration of 0–0.2 μg/μl and incubate for 10 min.

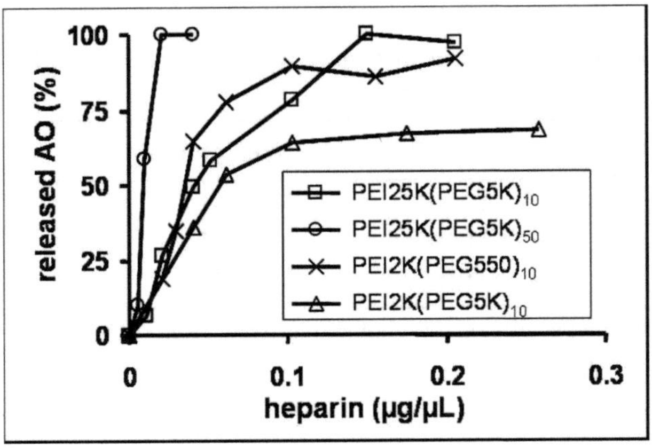

Fig. 4. Stability of PEG-PEI-AO polyplexes is measured using a heparin-based polyanion competition assay. Each of the three copolymers described in this chapter form very stable complexes with AO, requiring significantly more heparin to displace the AO than the unstable PEI25K(PEG5K)50-AO polyplex. Modified from Glodde et al. *(23)* with permission.

4. Add 2 µl of loading buffer and briefly vortex the samples.
5. Load samples onto agarose gels and electrophorese at 100 V for 25 min.
6. Image the gel using a gel scanner and quantify the intensity of the "free AO" band. The simplest quantification method is to place the gel on a UV light box and obtain a polaroid that is subsequently scanned on a conventional flatbed scanner. Scanned images (Tiff files) are then analyzed for band density using free-ware such as NIH ImageJ. Alternatively, UV scanners with bundled software packages are suitable for obtaining the optical density (densitometry) of the gels.
7. At a given heparin concentration, the percentage of AO dissociated from the polyplex is obtained by normalizing against the AO band measured from a sample containing AO only on the same gel.

3.4. Transfection of Cultured Cells with PEI25K(PEG5K)$_{10}$-AO Polyplexes

The macromolecular properties of polyplexes formulated with high M_w PEI25K(PEG5K)$_{10}$ copolymers are most favorable for transfection of cultured cells. We found that PEI25K(PEG5K)$_{10}$-AO polyplexes are composed of extremely stable and small (~15 nm) particles and exhibit a relatively high positive surface charge (*see* **Figs. 2–4**). Due mainly to their substantial surface charge, high M_w PEG-PEI-AO polyplexes are inherently more potent than their low M_w counterparts *(24–27)*. Recently, we showed using a dual fluorescence tracking approach with confocal microscopy (*see* **Fig. 5**) that

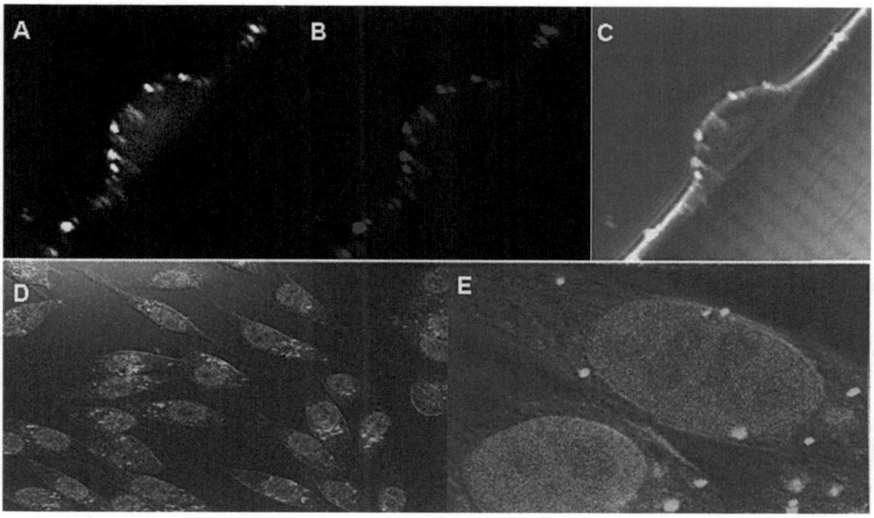

Fig. 5. Fluorescence tracking experiments document nuclear delivery of AO after transfection of cultured cells with high M_w PEI25K(PEG5K)$_{10}$-AO polyplex. (**A–C**) Isolated myofiber 24 h after transfection with a polyplex containing rhodamine-tagged PEI25K(PEG5K)$_{10}$ complexed with 6-FAM-tagged AO. The images show (**A**) rhodamine-tagged PEI25K(PEG5K)$_{10}$, (**B**), 6-FAM-AO, and (**C**) overlay with phase. The rho-PEI25K(PEG5K)$_{10}$ copolymer is localized in the perinuclear space, while 6-FAM-AO is clearly translocated without copolymer into the nucleus (see *(21)* for details). Modified from Sirsi et al. *(21)* with permission. (**D–E**) PC3 cells 24 h after transfection with PEI25K(PEG5K)$_{10}$ complexed with 6-FAM-AO. Nearly 100% of cells are transfected, with AO accumulating in the cell nuclei.

PEI25K(PEG5K)$_{10}$ copolymers provided effective AO delivery to myonuclei of mature skeletal muscle fibers *(21)*. Our tracking experiments further indicated that the copolymer-AO polyplex dissociated, presumably in the perinuclear space, allowing "free AO" to be translocated to the myonuclei, while the copolymer was mainly excluded from entering the nucleus (*see* **Fig. 5**). Using a similar fluorescence tracking approach, we also showed PEI25K(PEG5K)$_{10}$-AO polyplexes exhibited nearly 100% transfection efficiency and nuclear localization of the AO in PC3 cells (unpublished data, *see* **Fig. 5**). As expected, the in vitro transfection capacity of the high M_w PEI25K(PEG5K)$_{10}$-AO polyplexes was substantially greater than for low M_w polyplexes (unpublished data). Importantly, the high M_w polyplexes did not exhibit any overt signs of toxicity *(21)*.

Protocol for transfection of mouse primary myofibers with PEI25K (PEG5K)$_{10}$-AO polyplex

1. Pin the flexor digitorum brevis (FDB) muscle by tendons in a sylgard-lined 35 mm culture dish and incubate for 3 h at 37°C (5% CO2) in culture media containing DMEM, 10% FBS, 50 μg/ml gentamycin sulfate, and 0.5 mg/ml type I collagenase (*see* **Note 8**).
2. Following the collagenase digestion, gently triturate muscles with fire polished 9" glass pipettes *(28)*. Pipette dissociated fibers onto 15-mm glass cover slips coated with Matrigel (1:6 dilution) in 12-well plates containing 500 μl of DMEM (serum-free), and incubate overnight at 37°C.
3. Replace culture media in each well with 500 μl of DMEM with 10% FBS and 50 μg/ml gentamycin.
4. Prepare the PEI25K(PEG5K)$_{10}$-AO polyplex as described in **Subheading 3.2** except the final polyplex concentration will be 0.1 μg/μl of AO. For a 12-well plate prepare about 1.2 ml of polyplex containing 12 μg of AO.
5. Remove 100 μl of media from each well and add it to 100 μl of polyplex solution. Mix the solution thoroughly and replace the total volume of 200 μl back into the well.
6. Place 12-well plate into CO2 incubator. Replace each well with fresh media after 24 h.
7. For fluorescence tracking, fix cells with 2% paraformaldehyde for 10 min and subsequently 4 % paraformaldehyde for another 20 min. Counterstain with 1.2 μg/ml Hoechst dye (Molecular Probes) and mount with Vectashield (H-1000).

3.5. Utilization of Low M_w PEG–PEI Copolymers for In Vivo Delivery of AOs

Recent studies from our laboratory showed that low M_w PEI2K-PEG-AO polyplexes have macromolecular properties that make them most well-suited for in vivo applications *(22,23)*. Physiochemical analysis showed that both PEI2K(PEG550)$_{10}$-AO and PEI2K(PEG5K)$_{10}$-AO polyplexes are composed of extremely stable nanoparticles with only a minimal positive surface charge (*see* **Figs. 2–4**). An important difference between these polyplexes was that when PEG5K was grafted to PEI2K, polyplex diameter was extremely small (~10 nm) with no apparent aggregation (*see* **Figs. 2 and 3**). Apparently, the long 5K PEG chains impart steric repulsive forces that deter aggregation *(23)*. In contrast, when PEG550 was grafted to PEI2K, polyplexes appeared as much larger aggregate particles (~250 nm).

The minimal positive surface charge of low M_w PEI2K-based polyplexes is important for in vivo applications, as it minimizes protein adsorption and stimulation of macrophages and monocytes, thereby increasing circulation time in the bloodstream. The low surface charge also minimizes interactions between the polyplex and the negatively charged components in the tissue interstitium. Thus, these relatively inert polyplexes have a much more favorable biodistribution profile than their high M_w counterparts (*see* **Note 9**).

We recently evaluated AO-mediated dystrophin expression in skeletal muscles of dystrophin-null (*mdx*) mice after IM injection of AO complexed with low M_w PEI2K(PEG550)$_{10}$ copolymers *(22)*. Single IM injections of this aggregated polyplex produced dystrophin-positive fibers broadly distributed throughout the muscle cross-section (*see* **Fig. 6**) and showed nearly eightfold improvement in transduction efficiency relative to AO alone. Thus, we recommend the low M_w PEI2K(PEG550)$_{10}$-AO polyplex formulation for IM delivery or localized delivery to other tissues.

In pilot experiments, low M_w PEI2K(PEG5K)$_{10}$-AO polyplex was injected systemically in the tail vein of 8-week-old *mdx* mouse, using a repeat delivery regimen consisting six consecutive injections (1 mg AO each) at 1-week intervals. Inspection of limb muscles at 2 weeks after the final injection showed significant induction of dystrophin-positive fibers (unpublished, data not shown). Thus, for IV delivery, we recommend using the low M_w PEI2K(PEG5K)$_{10}$-AO polyplexes, which are composed of tiny (~15 nM), stable, and almost neutral nanoparticles (*see* **Note 10**).

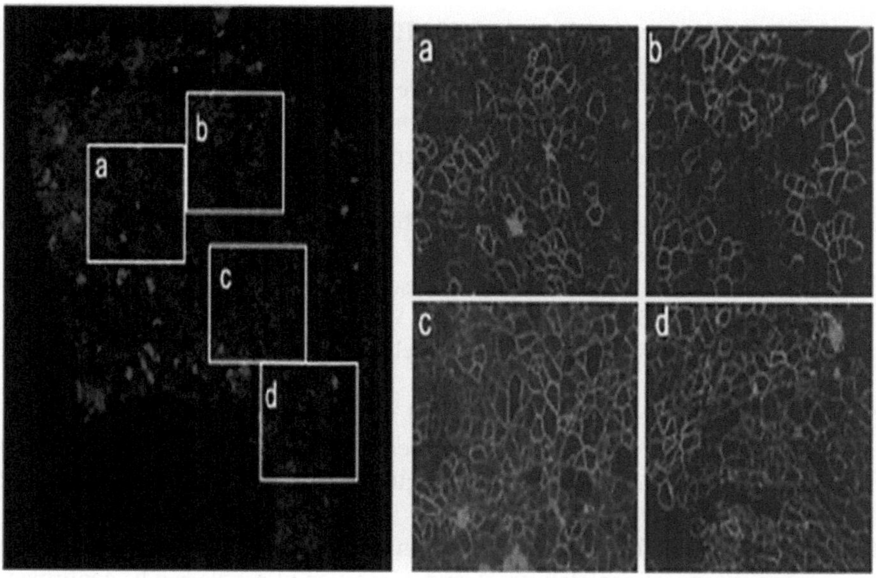

Fig. 6. Dystrophin immunolabeling of an entire transverse section of TA muscle from dystrophin-null *mdx* mouse at 3 weeks after intramuscular injection of 20 μg AO complexed with PEI2K(PEG550)$_{10}$. High magnification images of four different regions of the transverse section are shown (labeled a–d). Dystrophin-positive fibers are found to be distributed throughout the muscle. Reproduced from Williams et al. *(22)* with permission.

Protocol for intramuscular injection of PEI2K(PEG550)$_{10}$-AO in adult mice

1. Prepare the PEI2K(PEG550)$_{10}$-AO polyplex as described in **Subheading 3.2**, using sterile saline as the diluent to bring to the appropriate volume.
2. Load 15–30 μl of polyplex into a 31 G 5/16" length insulin syringe with permanently attached needle and allow it to come to room temperature. Generally, 5–10 μg of AO is sufficient for a single injection.
3. Anesthetize mice by intraperitoneal injection of a ketamine-xylazine mixture at an induction dose of 20–100 mg/kg.
4. It is necessary to then place the anesthetized mouse on a heated surface (a simple slide warmer; Fisher Scientific) to maintain body temperature. Apply ophthalmic lubrication to both eyes with a sterile swab to ensure moisture during anesthesia.
5. Shave the hair surrounding the injection site for visual confirmation of the appropriate muscle(s) and wipe the skin surface with isopropyl alcohol swabs to disinfect the area.
6. Insert the needle through the skin along the mid-belly portion of the muscle and slowly expel the volume of polyplex, leaving the needle in place for 1 min to minimize leakage out the needle track.
7. Keep mice on the heated surface until awakening from anesthesia.

Protocol for intravenous injection of PEI2K(PEG5K)$_{10}$-AO in adult mice

1. Prepare the PEI2K(PEG5K)$_{10}$-AO polyplex as described in **Subheading 3.2**, using sterile saline as the diluent to bring to the appropriate volume. Use 200–300 μl volumes for conventional tail vein delivery or 1.6–3 ml (10% of bodyweight) for hydrodynamic based delivery.
2. Warm mouse in cage using heat lamp taking precaution not to overheat. This will promote blood vessel dilation.
3. Place mouse in a commercial tail-vein restrainer (Plas-labs) or homemade version fabricated from a 50-ml conical tube.
4. Wipe tail with 70% ethanol to further help dilate blood vessels and disinfect the area.
5. Holding the tail between the thumb and index finger, insert needle (27 or 28 G insulin syringe, 0.5" length with permanently attached needle) bevel down at the distal end of the tail. Insert the full length of the needle visualizing it under the skin while keeping it parallel to tail.
6. Expel the volume in one constant motion in 4–7 s for hydrodynamic delivery or slower for conventional delivery. Correct insertion of needle can be confirmed by visualizing blanching of the vein during the injection. If the needle is not in the vein, there will be significant pressure resisting the injection. The needle should be removed, wiped with 70% ethanol, and reinserted at a location more proximal along the tail.
7. Blot the injection site with a sterile swab to stop any bleeding and return mouse to cage.

4. Notes

1. Precipitation of activated mPEG can be difficult, and if done incorrectly may result in excess un-reacted HMDI in the final product, decreasing polymer solubility in water. During precipitation of activated mPEGs, it is important to add the concentrated mPEG dropwise into the hexanes while mixing. If a titration setup is used, the mPEG should be added to the hexanes before the chloroform evaporates off. The activated mPEG product should be a consistent powder. If large clumps appear, dissolve the polymer in small amounts of chloroform and re-precipitate into hexanes. Low M_w mPEGs are more difficult to precipitate and should be done at low temperatures to sustain higher yields, using either a liquid nitrogen bath or adding dry ice directly to the hexanes.
2. Copolymers made using PEG5K consist of white fluffy powders, whereas low M_w PEI2K(PEG550)$_{10}$ copolymers appear as thick viscous oils, which are more difficult to precipitate. For this reason, it is recommended that the copolymer be recovered by evaporation of the solvent, which may take several days depending on the amount of the copolymer and efficiency of the vacuum pump.
3. As a starting material, PEI2K should not contain any water, which may interfere with the reaction between activated mPEG and amine groups on PEI. If stock PEI2K is shipped in 50% (w/v) water, it should be lyophilized before use.
4. Functionalization of PEG–PEI copolymers for enhanced AO delivery and cell/tissue-specific targeting is an active area of research ***(13–20)***. In this chapter, we presented relatively simple formulations of the PEG–PEI copolymers. Investigators are encouraged to experiment with attaching various cell targeting ligands to the copolymers, which can be done using relatively simple conjugations such as EDC coupling.
5. We routinely use N : P ratio of 5 for all polyplex formulations. For high M_w PEI25K(PEG5K)$_{10}$–AO polyplexes we found that lower N : Ps resulted in the formation of aggregated and unstable particles. Higher N : Ps do not appear to increase the transfection efficiency of these polyplexes in culture, although this has not been systematically investigated. Likewise, we have not systematically investigated the effect of N : P on transfection capacity of low M_w PEI2K-based polyplexes, and it is quite possible that either higher or lower N : Ps may improve transfection depending on the context in which they are used.
6. It is important to use polyplex solutions within 1–2 h after preparation. Prolonged incubation times can lead to aggregation of particles in solution and subsequently affect the transfection properties of the polyplex.
7. The polyanion competition assay should include five to seven different heparin concentrations (ranging from 0–0.2 μg/μl) to produce a full curve. The assay is moderately dependent on incubation time, which should be kept relatively constant to insure more consistent results. For AOs that do not contain a fluorophore, ethidum bromide staining (0.5 μg/ml) of the gel can be used to visualize the AO bands.
8. Cell culture polyplex transfection conditions must be optimized for each cells type. Conditions that may influence polyplex transfection capacity include media

type, serum versus no serum, and transfection duration. Investigators may choose to begin using conditions that were previously established using commercially available transfection reagents.

9. The M_w of PEG was found to be the main determinant of polyplex size, through its influence on particle aggregation. Generally, greater steric repulsion provided by long 5 K PEG chains minimizes particle aggregation, whereas short PEG550 chains may result in aggregated nanoparticulates. In the context of low M_w PEI2K, it is not yet clear whether aggregated or non-aggregated polyplexes provide better results, although most of our work has been done with the aggregated PEI2K$(PEG550)_{10}$-AO polyplexes. It is possible that these aggregate particles yield better results because more AO may be delivered per endocytotic event. The aggregate structures may also provide better protection of AOs from nuclease/RNase digestion.
10. The copolymers described herein may be used for enhanced delivery of other small nucleic acid compounds including siRNAs, LNAs, mixmers, and gapmers.

References

1. Bieber, T., Meissner, W., Kostin, S., Niemann, A., and Elsasser, H. P. (2002) Intracellular route and transcriptional competence of polyethylenimine-DNA complexes. *J.Control Release.* **82**(2–3), 441–454.
2. Suh, J., Wirtz, D., and Hanes, J. (2003) Efficient active transport of gene nanocarriers to the cell nucleus. *Proc. Natl. Acad. Sci. U. S. A.* **100**(7), 3878–3882.
3. Petersen, H., Fechner, P. M., Fischer, D., and Kissel, T. (2002) Synthesis, characterization, and biocompatibility of polyethylenimine-*graft*-poly(ethylene glycol) block copolymers. *Macromolecules.* **35**, 6867–6874.
4. Petersen, H., Fechner, P. M., Martin, A. L., Kunath, K., Stolnik, S., Roberts, C. J., Fischer, D., Davies, M. C., and Kissel, T. (2002) Polyethylenimine-graft-poly(ethylene glycol) copolymers: influence of copolymer block structure on DNA complexation and biological activities as gene delivery system. *Bioconjug. Chem.* **13**(4), 845–854.
5. Kichler, A. (2004) Gene transfer with modified polyethylenimines. *J. Gene Med.* **6**(Suppl 1), S3–10.
6. Thomas, M. and Klibanov, A. M. (2003) Non-viral gene therapy: polycation-mediated DNA delivery. *Appl. Microbiol. Biotechnol.* **62**(1), 27–34.
7. Boussif, O., Lezoualc'h, F., Zanta, M. A., Mergny, M. D., Scherman, D., Demeneix, B., and Behr, J. P. (1995) A versatile vector for gene and oligonucleotide transfer into cells in culture and in vivo: polyethylenimine. *Proc. Natl. Acad. Sci. U. S. A.* **92**(16), 7297–7301.
8. Godbey, W. T., Wu, K. K., and Mikos, A. G. (1999) Tracking the intracellular path of poly(ethylenimine)/DNA complexes for gene delivery. *Proc. Natl. Acad. Sci. U. S. A.* **96**(9), 5177–5181.
9. Hara-Chikuma, M., Yang, B., Sonawane, N. D., Sasaki, S., Uchida, S., and Verkman, A. S. (2005) ClC-3 chloride channels facilitate endosomal acidification and chloride accumulation. *J. Biol. Chem.* **280**(2), 1241–1247.

10. Sonawane, N. D., Szoka, F. C., Jr., and Verkman, A. S. (2003) Chloride accumulation and swelling in endosomes enhances DNA transfer by polyamine-DNA polyplexes. *J. Biol. Chem.* **278**(45), 44826–44831.
11. Akinc, A., Thomas, M., Klibanov, A. M., and Langer, R. (2004) Exploring polyethylenimine-mediated DNA transfection and the proton sponge hypothesis. *J. Gene Med.* **7**(5), 657–663.
12. Sung, S. J., Min, S. H., Cho, K. Y., Lee, S., Min, Y. J., Yeom, Y. I., and Park, J. K. (2003) Effect of polyethylene glycol on gene delivery of polyethylenimine. *Biol. Pharm. Bull.* **26**(4), 492–500.
13. Kunath, K., von, Harpe A., Petersen, H., Fischer, D., Voigt, K., Kissel, T., and Bickel, U. (2002) The structure of PEG-modified PEI influences biodistribution and pharmacokinetics of their complexes with NF-kappaB decoy in mice. *Pharm. Res.* **19**(6), 810–817.
14. Brus, C., Petersen, H., Aigner, A., Czubayko, F., and Kissel, T. (2004) Physicochemical and biological characterization of polyethylenimine-graft-poly(ethylene glycol) block copolymers as a delivery system for oligonucleotides and ribozymes. *Bioconjug. Chem.* **15**(4), 677–684.
15. Schiffelers, R. M., Ansari, A., Xu, J., Zhou, Q., Tang, Q., Storm, G., Molema, G., Lu, P. Y., Scaria, P. V., and Woodle, M. C. (2004) Cancer siRNA therapy by tumor selective delivery with ligand-targeted sterically stabilized nanoparticle. *Nucleic Acids Res.* **32**(19), e149.
16. Fischer, D., Osburg, B., Petersen, H., Kissel, T., and Bickel, U. (2004) Effect of poly(ethylene imine) molecular weight and pegylation on organ distribution and pharmacokinetics of polyplexes with oligodeoxynucleotides in mice. *Drug Metab. Dispos.* **32**(9), 983–992.
17. Vinogradov, S. V., Batrakova, E. V., and Kabanov, A. V. (2004) Nanogels for oligonucleotide delivery to the brain. *Bioconjug. Chem.* **15**(1), 50–60.
18. Jeong, J. H., Kim, S. W., and Park, T. G. (2003) A new antisense oligonucleotide delivery system based on self-assembled ODN-PEG hybrid conjugate micelles. *J.Control Release.* **93**(2), 183–191.
19. Vinogradov, S. V., Bronich, T. K., and Kabanov, A. V. (1998) Self-assembly of polyamine-poly(ethylene glycol) copolymers with phosphorothioate oligonucleotides. *Bioconjug. Chem.* **9**(6), 805–812.
20. Vinogradov, S., Batrakova, E., Li, S., and Kabanov, A. (1999) Polyion complex micelles with protein-modified corona for receptor-mediated delivery of oligonucleotides into cells. *Bioconjug. Chem.* **10**(5), 851–860.
21. Sirsi, S. R., Williams, J, and Lutz, G. J. (2005) Poly(ethylene imine)-polyethylene glycol copolymers facilitate efficient delivery of antisense oligonucleotides to nuclei of mature muscle cells of *mdx* mice. *Hum. Gene Ther.* **16**(11), 1307–1317.
22. Williams, J. H., Sirsi, S. R., Latta, D., and Lutz, G. J. (2006) Induction of dystrophin expression by exon skipping in *mdx* mice following intramuscular injection of antisense oligonucleotides complexed with PEG-PEI copolymers. *Mol. Ther.* **14**(1), 88–96.

23. Glodde, M, Sirsi, S. R., and Lutz, G. J. (2006) Physiochemical properties of low and high molecular weight peg-grafted poly(ethylene imine) copolymers and their complexes with oligonucleotides. *Biomacromolecules.* **7**(1), 347–356.
24. Kursa, M., Walker, G. F., Roessler, V., Ogris, M., Roedl, W., Kircheis, R., and Wagner, E. (2003) Novel shielded transferrin-polyethylene glycol-polyethylenimine/DNA complexes for systemic tumor-targeted gene transfer. *Bioconjug. Chem.* **14**(1), 222–231.
25. Ogris, M., Steinlein, P., Carotta, S., Brunner, S., and Wagner, E. (2001) DNA/polyethylenimine transfection particles: influence of ligands, polymer size, and PEGylation on internalization and gene expression. *AAPS PharmSci.* **3**(3), E21.
26. Ogris, M., Walker, G., Blessing, T., Kircheis, R., Wolschek, M., and Wagner, E. (2003) Tumor-targeted gene therapy: strategies for the preparation of ligand-polyethylene glycol-polyethylenimine/DNA complexes. *J. Control Release.* **91** (1–2), 173–181.
27. Ogris, M., Brunner, S., Schuller, S., Kircheis, R., and Wagner, E. (1999) PEGylated DNA/transferrin-PEI complexes: reduced interaction with blood components, extended circulation in blood and potential for systemic gene delivery. *Gene Ther.* **6**(4), 595–605.
28. Rosenblatt, J. D., Lunt, A. I., Parry, D. J., and Partridge, T. A. (1995) Culturing satellite cells from living single muscle fiber explants. *In Vitro Cell Dev. Biol. Anim.* **31**(10), 773–779.
29. Kunath, K., von Harpe, A., Fischer, D., Petersen, H., Bickel, U., Voigt, K., and Kissel, T. (2003) Low-molecular-weight polyethylenimine as a non-viral vector for DNA delivery: comparison of physicochemical properties, transfection efficiency and in vivo distribution with high-molecular-weight polyethylenimine. *J. Control Release.* **89**(1), 113–125.

10

Non-Viral Gene Delivery with Cationic Liposome–DNA Complexes

Kai K. Ewert, Ayesha Ahmad, Nathan F. Bouxsein, Heather M. Evans, and Cyrus R. Safinya

Summary

A large amount of research activity worldwide is currently directed towards developing lipid- or polymer-based, non-viral gene vectors for therapeutic applications. This strong interest is motivated by their low toxicity, ease of production, ability to transfer large pieces of DNA into cells, and lack of immunogenic protein components. Cationic liposomes (CLs) are one of the most powerful non-viral vectors. In fact, CL-based vectors are among the prevalent synthetic carriers of nucleic acids currently used in human clinical gene therapy trials as well as in cell transfection applications for biological research. Our understanding of the mechanisms of action of CL–DNA complexes is still in its infancy. However, the relevance of a few crucial parameters, such as the lipid/DNA charge ratio (ρ_{chg}) and the membrane charge density of lamellar complexes (σ_M), is well established. To arrive at true comparisons of lipid performance, one must optimize both these parameters using a reproducible, reliable transfection assay. In this chapter, we aim to provide the reader with detailed procedures for liposome formation and transfection. It is our hope that the use of such optimized protocols will improve the comparability of transfection data obtained with novel lipids.

Key Words: Cationic liposome; cationic lipid; transfection; lipofection; non-viral; gene therapy; gene delivery.

1. Introduction

There currently is a large amount of interest in the development and improvement of new methods to deliver genes, i.e., DNA, to cells, motivated

From: *Methods in Molecular Biology, vol. 433: Volume 1: Production and In Vivo Applications of Gene Transfer Vectors*
Edited by: J. M. Le Doux © Humana Press, Totowa, NJ

mainly by the promises of gene therapy. Most current methods of gene therapy start with a vector, containing the therapeutic gene, which has the ability to transfer the gene into cells. This vector is then applied in either ex vivo or in vivo transfer methods ***(1)***. In the ex vivo method, cells derived from patient tissue are transfected in vitro, possibly selected for successful gene transfer, and then returned to the patient. The in vivo method may involve either systemic delivery, e.g., by i.v. injection, or local application of the vector, e.g., by injection into tumors ***(2,3)***.

Gene delivery vectors may be divided into two major classes. Viral vectors are based on replication-deficient engineered viruses, which include retroviruses and adenoviruses among several others ***(4)***. Their main advantage is high gene transfer efficiency, both in vitro and in vivo. There are, however, safety concerns associated with the use of viral vectors, which have been highlighted by a few recent setbacks ***(5,6)***. In a prominent example, 3 of 11 patients developed a leukemia-like disease caused by insertional mutagenesis. This marred the first clinical success of gene therapy, which had managed to correct X-linked severe combined immunodeficiency (SCID-X1) with an engineered retrovirus vector ***(7–9)***.

Synthetic vectors, which are based on lipids, polymers, peptides, or combinations of these ***(1,10–14)***, are gaining importance as a safe alternative to viral vectors ***(15)***. They have other advantages as well: their preparation is facile and straightforward and several parameters can be tuned to optimize transfection results; they need not contain immuno-stimulating peptide or protein components; last but not least, they do not impose a size limit on their genetic cargo. In fact, while viral vectors have a maximum carrying capacity of about 40,000 base pairs, cationic liposome (CL) vectors have successfully been employed to deliver human artificial chromosomes, with a size between 6 and 10 million base pairs, to mammalian cells ***(16)***. Currently, more than 20% of open clinical trials of gene therapy worldwide use synthetic vectors, with CL carriers employed in about one-third of those trials ***(17)***.

1.1. Liposomes

Liposomes or vesicles are closed shells of lipid bilayers, which may form spontaneously when a lipid film is exposed to water ***(18,19)***. The liposomes used for gene delivery typically contain at least two types of lipid, one cationic and one neutral. Small liposomes of a fairly uniform size distribution are the preferred starting material for the preparation of CL–DNA complexes. These small liposomes are typically prepared by hydrating a lipid film and treating the resulting aqueous solution (which will contain a variety of structures, including large and small liposomes as well as multilamellar vesicles) with ultrasound or extruding it several times through filters with a small (200 nm) pore size. For effective liposome formation, the temperature at which hydration of the lipid film is performed needs to exceed the lipid's gel-to-liquid-crystal transition temperature, T_C. Above its T_C, the hydrophobic chains of a lipid are in a liquid rather than in an ordered state, resulting in fluid, flexible membranes.

1.2. Structures of CL–DNA Complexes

The first application of CLs as gene vectors resulted from a landmark study by Felgner and collaborators *(20)*. It was originally thought that DNA would simply wrap around CLs to form the complexes. In reality, the equilibrium structures of CL–DNA complexes result after major rearrangements, as determined by synchrotron X-ray diffraction. To date, three equilibrium structures of CL–DNA complexes have been discovered. These are shown in **Fig. 1**.

The lamellar (L_α^C) phase of CL–DNA complexes is the most widely found. It consists of stacked lipid bilayers with DNA intercalated in between *(21–24)*. As an example, lipid mixtures of DOTAP (2,3-dioleoyl-1-trimethylammonium propane, a monovalent cationic lipid) and neutral DOPC (1,2-dioleoyl-*sn*-glycero-phosphatidylcholine) form lamellar DNA complexes throughout the entire composition range (DOTAP mole fraction).

The commonly used neutral co-lipid DOPE (1,2-dioleoyl-*sn*-glycero-phosphatidylethanolamine) can lead to the formation of a different structure, which is shown in **Fig. 1** (middle) *(25)*. In this inverted hexagonal (H_{II}^C) phase, the DNA molecules are enveloped in inverse lipid micelles and assembled on a

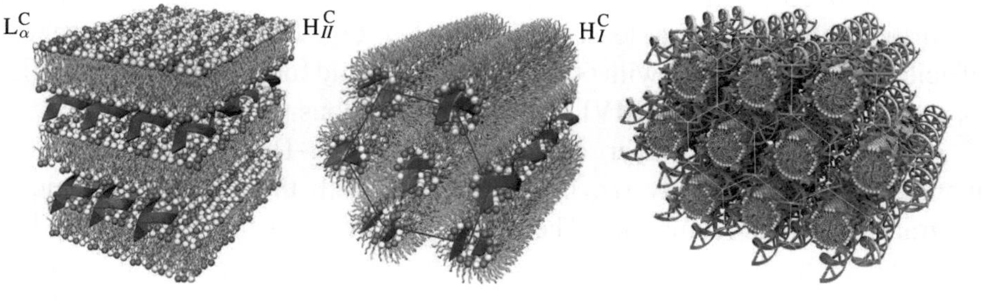

Fig. 1. The three known equilibrium structures of cationic liposome (CL)–DNA complexes. (Left) The lamellar, L_α^C, phase of CL–DNA complexes with alternating lipid bilayers and DNA monolayers. (Middle) The inverted hexagonal, H_{II}^C, phase of CL–DNA complexes with DNA chains coated by inverse micelles and arranged on a hexagonal lattice. Adapted from **ref.** *(32)*. (Right) In the hexagonal, H_I^C, phase of CL–DNA complexes, cylindrical lipid micelles are assembled into a hexagonal lattice, with DNA arranged in the interstices and forming a honeycomb lattice. Schematics of the L_α^C and H_{II}^C phases reprinted with permission from **ref.** *(25)*. Schematic of the H_I^C phase reprinted with permission from **ref.** *(26)*. Copyright 2006 American Chemical Society.

hexagonal lattice. The main reason for the formation of this different structure is that DOPE is a cone-shaped (rather than cylinder-shaped) lipid, which confers a negative spontaneous curvature to membranes.

Very recently, we have shown that a novel cationic lipid with a large, highly charged headgroup (16+), termed MVLBG2, gives rise to a third structure of CL–DNA complexes when combined with DOPC. In this structure, named H_I^C, cylindrical lipid micelles are assembled into a hexagonal lattice, with DNA arranged in the interstices to form a honeycomb lattice ***(26)***.

1.3. Important Parameters of CL–DNA Complexes

CL–DNA complexes spontaneously form when solutions of DNA and small CLs (of 50–100 nm diameter) are mixed. This self-assembly process is driven by entropy gained from the release of counterions, which were tightly bound to the highly charged DNA and liposomes ***(27,28)***, into solution, as the charges on DNA and liposomes compensate each other ***(29,30)***.

A few compositional parameters of CL–DNA complexes of established importance affect their transfection efficiency (TE; a measure of the vector's ability to deliver the gene into cells) strongly. One of these is the lipid/DNA charge ratio (ρ_{chg}) ***(31)***. For all lipids investigated in our laboratory to date, TE increases with ρ_{chg} up to a saturation value; this behavior is independent from the ratio of cationic to neutral lipid in the membrane. The investigated lipids cover headgroup charges from +1 to +16 and varied headgroup structures. The onset of the saturation depends on the cationic lipid. For example, as shown in **Fig. 2**, ρ_{chg} = 3 lies in the saturated regime for DOTAP, whereas a group of recently synthesized dendritic lipids required at least ρ_{chg} = 4.5 ***(32)***. **Fig. 2** displays the transfection efficiencies of complexes with 60 mol% cationic lipid for DOTAP (+1), MVLG2 (+4), MVLBG1 (+8), and MVLBG2 (+16) at various values of ρ_{chg}.

A second key parameter affecting TE of CL–DNA complexes is the membrane charge density (σ_M), which varies with the ratio of cationic and neutral lipid in the membrane. The membrane charge density provides a lipid-independent measure of how cationic a membrane is, because it is defined simply as the cationic charge per unit area. For example, two membranes, each containing the same molar fraction of a cationic lipid, may exhibit very different values of σ_M if the two cationic lipids carry a different charge (assuming their headgroup areas are the same). At the same time, σ_M of two membranes containing very different molar fractions of cationic lipid may be similar, if the lipids bear very different charges. To calculate the membrane charge density, one needs to know the effective charge of the cationic lipid in DNA complexation [which can be obtained using an ethidium bromide based assay ***(33)***] and the lipid's headgroup area, which we use as a fitting parameter.

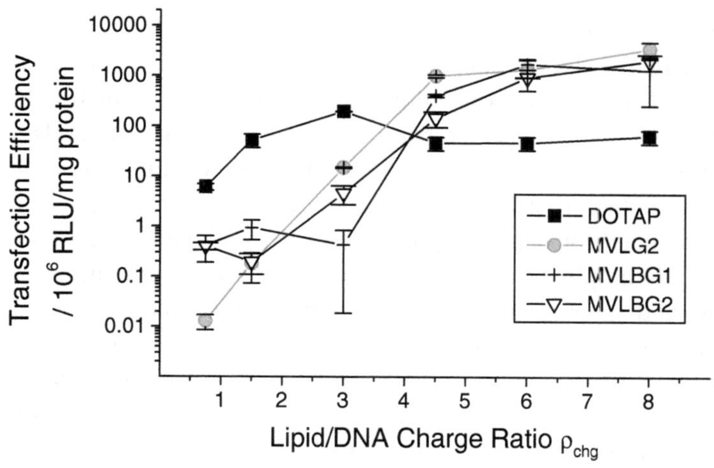

Fig. 2. Transfection efficiency (TE), determined according to the protocol described below, of DNA complexes containing 60 mol% cationic lipid at various lipid/DNA charge ratios for DOTAP (+1), MVLG2 (+4), MVLBG1 (+8), and MVLBG2 (+16). The amount of DNA is constant for all data points. Reprinted with permission from **ref.** *(32)*. Copyright 2006 American Chemical Society.

The remarkable usefulness of the parameter σ_M is illustrated in **Fig. 3**. The plot on the left shows TE for DNA complexes of several lipids, with headgroup charges ranging from +1 to +5, as a function of the cationic lipid/DOPC molar ratio *(33,34)*. The amount of DNA was kept constant for all data points. All cationic lipids exhibit a maximum in TE as a function of lipid composition: at 65 mol% for MVL2 (+2), 70 mol% for MVL3 (+3), 50 mol% for MVL5 (+5), 55 mol% for TMVL5 (+5), and 90 mol% for DOTAP (+1). This result is of note with a view to literature results which often only compare one or two ratios of cationic and neutral lipid: although the optimized TE is similar for all lipids, this TE appears at different molar ratios. Thus, testing only a few ratios is inadequate to fully assess the potential of a new lipid. The optimal molar ratios result in a TE up to three orders of magnitude larger than that of complexes that transfect poorly.

The plot on the right of **Fig. 3** shows the same TE data, now plotted versus σ_M. Remarkably, a notable simplification takes place and all the data points merge onto a single curve. This demonstrates that the membrane charge density is a universal parameter and a predictor of TE for lamellar (L_α^C) CL–DNA complexes. The resulting universal curve reveals an optimal charge density of $\sigma_M{}^* = 17.0 \pm 0.1 \times 10^{-3}$ (e/Å^2) *(33)*.

X-ray diffraction shows that DOTAP as well as the MVLs form lamellar (L_α^C) cationic lipid/DOPC–DNA complexes. Notably, the TE of DOTAP/DOPE-containing complexes, which exhibit the H_{II}^{C} phase in the low-σ_M region labeled Regime I in **Fig. 3**, is independent of σ_M. This deviation from the universal curve is indicative of a distinctly different transfection mechanism for the inverted hexagonal phase, which has also been confirmed by other methods *(35)*.

Considering the data for DOTAP/DOPE–DNA complexes, it may be tempting to conclude that DOPE is a generally better choice of co-lipid which eliminates the need for optimizing σ_M. However, independence of TE from σ_M is a property of complexes in the H_{II}^{C} phase, not of DOPE-containing complexes. New lipids, and in particular multivalent lipids with their larger headgroups (which can even be large enough to impose the H_{I}^{C} structure in mixtures with DOPC), may result in lamellar rather than inverted hexagonal DNA complexes when mixed with DOPE. Even with DOTAP, complexes containing larger fractions of cationic lipid exhibit the lamellar phase. Furthermore, DOPE has turned out to be unsuitable for in vivo applications where cholesterol, which promotes lamellar complexes, has gained importance. It is thus important to dispel the widespread belief that complexes containing DOPE rather than DOPC as the neutral lipid are always higher transfecting *(36–38)*. We have repeatedly shown that this assessment is not true if lipid composition is optimized, at which point DOPC-containing (lamellar, L_{α}^{C}) complexes transfect as well as the best DOPE-containing (H_{II}^{C}) complexes *(33–35)*.

The mechanistic implications of the data shown in **Fig. 3** and detailed mechanisms for transfection with both lamellar and inverted hexagonal CL–DNA

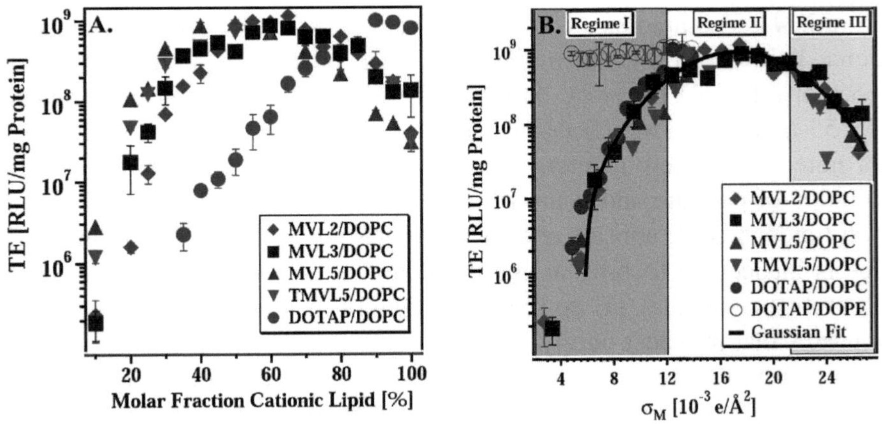

Fig. 3. **(A)** Transfection efficiency (TE) as a function of molar fraction of cationic lipid for cationic liposome (CL)/DOPC–DNA complexes prepared with MVL2 (+2, diamonds), MVL3 (+3, squares), MVL5 (+5, triangles), TMVL5 (+5, inverted triangles), and DOTAP (+1, circles). **(B)** The same TE data plotted against the membrane charge density, σ_M. The solid line is a Gaussian fit to the data. Data for DOTAP/DOPE complexes (open circles, H_{II}^{C} phase) are also shown. Reprinted with permission by John Wiley & Sons, Inc. from **ref.** *(33)*.

complexes have been discussed elsewhere ***(33,35,39–41)***. In practice, to arrive at true comparisons of lipid performance, one must optimize crucial parameters such as ρ_{chg} and σ_M for each lipid using a reproducible, reliable transfection assay. In the protocols below, we aim to provide the reader with detailed procedures for liposome formation and cell transfection. We expect from the reader a general knowledge and competence in the culture of adherent cell lines as described, e.g., in **ref.** *(42)*. We hope that these optimized protocols will help yield transfection data for novel lipids that allow for true comparisons of different lipids and formulations.

In the following, we detail the liposome preparation and cell transfection protocol that has evolved in our group. Lipids are mixed in chloroform solution for homogenous mixing, dried to prepare a lipid film, and subsequently hydrated and sonicated to form small liposomes. These are combined with plasmid DNA encoding luciferase and the resulting complexes transferred onto cells. After incubation, cells are harvested and the amount of expressed luciferase protein is measured. Luciferase, the protein generating the firefly's bioluminescence, allows measurement of the expressed protein levels by means of a light-emitting assay, providing for a very large dynamic range.

2. Materials

2.1. Liposome Preparation

1. Small glass vials (2 ml or 4 ml; Fisher Scientific, Pittsburgh, PA, USA) (*see* **Notes 1** and **2**).
2. Larger glass vials or measuring flasks for solvent stock solutions (Fisher) (*see* **Note 2**).
3. Plastic syringes and Teflon membrane filters (0.2 μm, Whatman, Florham Park, NJ, USA).
4. Low-conductivity water (18 MΩ/cm; from Millipore, Billerica, MA, USA).
5. Neutral (e.g., DOPC or DOPE) and cationic (e.g., DOTAP) lipid(s) (Avanti Polar Lipids, Alabaster, AL, USA). Store at –20 °C. (*see* **Note 3**).
6. Chloroform and methanol (ACS grade, from EM Science, Gibbstown, NJ, USA). Inhalation, ingestion and skin contact should be avoided. They should be handled in a fume hood.
7. Ultrasound generator (e.g., Vibracell from Sonics & Materials, Newtown, CT, USA).
8. Incubator (at 37 °C).

2.2. Cell Transfection

1. 150-mm Cell culture dish.
2. 1.5-ml Centrifuge tubes ("Eppendorf tubes").
3. 50-ml Centrifuge tubes.

4. Hemocytometer (Sigma, St. Louis, MO, USA).
5. 24-well culture plates (2 cm^2/well).
6. pGL3 Control Vector (Firefly Luciferase) Plasmid DNA (Promega, Madison, WI, USA) (*see* **Note 4**).
7. L-Cells (mouse fibroblast cells; from ATCC, Manassas, VA, USA; ATCC number: CCL-1) (*see* **Note 5**).
8. Serum-free medium: DMEM (Dulbecco's Modified Eagle's Medium; Invitrogen, Carlsbad, CA, USA).
9. Supplemented medium: 5% Fetal Bovine Serum (Invitrogen) in DMEM.
10. 1×Phosphate-buffered saline (PBS; Invitrogen).
11. Trypsin buffer (or TrypLE; both Invitrogen).
12. Liposome solutions (prepared as described in **Subheading 3.1.**) (*see* **Note 6**).
13. Passive Lysis Buffer (5×; Promega).
14. Luciferase Assay System (Promega).
15. Protein concentration assay reagent (BioRad, Hercules, CA, USA).

3. Methods

3.1. Liposome Preparation

1. For each neutral and cationic lipid, weigh out an appropriate amount of lipid in a larger glass vial (*see* **Note 7**).
2. Dissolve each lipid in chloroform at a concentration of 0.6 mM (*see* **Notes 8–11**).
3. In small glass vials, combine the appropriate volumes of organic solutions of neutral and cationic lipid(s) to make the desired lipid mixtures (*see* **Notes 12** and **13**).
4. Remove the solvent, using a stream of dry nitrogen gas in a fume hood (*see* **Note 14**).
5. Once this has yielded a film on the surface of the vial and no visible liquid remains, place the vial in a desiccator attached to an oil pump and incubate in a vacuum overnight (*see* **Note 15**).
6. To the dried lipid films, add the amounts of water required to prepare a solution of the desired concentration. The typical concentration of liposomes for transfection purposes is 0.6 mM.
7. Close the vials tightly and incubate them at 37 °C (or above T_C of the used lipid) for 4–8 h. Longer times (8 h) are usually advantageous for facile liposome formation, but ester bonds in the lipids may be hydrolyzed upon more prolonged incubation in water, especially at elevated temperature (*see* **Note 3**).
8. Place the vials in a water bath (*see* **Note 16**) and tip-sonicate the lipid solution for 10 min or until clear (*see* **Notes 17–20**).
9. Aspirate the solution with a syringe and transfer it to a new container (glass vial or 1.5 ml centrifuge tube) by passing the solution through a filter (0.2 μm pores) (*see* **Note 21**).
10. Store the resulting liposome solutions at 4 °C (*see* **Note 6**).

3.2. Cell Transfection

3.2.1. Day 1: Preparing Cells in 24-Well Plates

Starting with cells that are almost confluent on a 150-mm dish, release the adherent cells and reseed them in a 24-well plate:

1. Aspirate the medium from a 150-mm dish containing an almost confluent layer of cells (*see* **Note 22**).
2. Wash with 1× PBS and aspirate PBS (*see* **Note 23**).
3. Briefly coat the cells with 2 ml of trypsin buffer (*see* **Note 24**) and incubate the dish at 37 °C for no more than 5 min.
4. Transfer the cells to a 50-ml centrifuge tube using 15 ml of supplemented DMEM (*see* **Note 25**).
5. Determine the cell density using a hemocytometer (*see* **Note 26**).
6. Add 500 μl of supplemented DMEM to each well of a 24-well plate that is to be used.
7. Calculate the required volume of cell suspension (*see* **Note 27**) and seed the 24-well plate(s) at 85,000 cells per well, ensuring an even distribution of cells in the well (*see* **Note 28**).
8. Several plates may be seeded at a given time. The plates are then incubated under the conditions used for maintenance of the cells for about 20 h (*see* **Note 29**).

3.2.2. Day 2: Transfection

1. Check the cells in the 24-well plates (*see* **Note 30**).
2. Calculate the volumes of liposome stock solutions required for all values of ρ_{chg} and σ_M to be investigated, starting from a constant amount of 1 μg pGL3 DNA (*see* **Note 31**). Add the lipid solution(s) to 1.5-ml centrifuge tubes (*see* **Note 32**).
3. Prepare the required amount of a solution of pGL3 plasmid DNA at 1 μg/250 μl in serum-free DMEM (*see* **Note 33**).
4. Dilute the liposome solutions to 250 μl with serum-free DMEM. After the lipid samples have been incubated for 5 min (never longer than 30 min!) (*see* **Note 34**), add 250 μl of the prepared DNA solution. Invert the closed tube 10 times to mix and centrifuge at 3000 rpm (604 g) for 30 s to collect all the solution in the bottom of the tube.
5. Incubate the samples at room temperature for 30 min to allow the CL–DNA complexes to form.
6. Remove a 24-well plate from the incubator. Processing 12 wells at a time, aspirate the medium and wash the cells once with 1× PBS (*see* **Notes 22, 23,** and **35**).
7. After gently mixing the CL–DNA complex solution by repeatedly filling and emptying the pipette, add 200 μl to each well (*see* **Note 36**).
8. Repeat **steps 6** and **7** for more wells as desired.

9. Incubate the cells with the complex solution for 6 h under the conditions used for their maintenance (*see* **Note 37**). The plates should be left undisturbed during this time (*see* **Note 38**).
10. Processing 12 or 24 wells at a time, aspirate the complex solution, wash the cells once with 1× PBS, and add 500 μl of supplemented medium (*see* **Note 35**).
11. Incubate the cells for 20 h, i.e., the time required for one full cell cycle to occur, under the conditions used for their maintenance (*see* **Note 39**).

3.2.3. Day 3: Harvesting of Cells; Determining Luciferase and Total Protein Content

1. Freshly prepare a sufficient amount of 1× Passive lysis buffer (PLB): 150 μl of PLB for each well (*see* **Note 33**).
2. After aspirating the medium, wash the cells once with 1× PBS. Then add 150 μl of PLB to each well (*see* **Note 40**). After processing all wells, place the plates in a freezer overnight.
3. Thaw the plates for 30–45 min (*see* **Note 41**).
4. Completely transfer the content of each well into a 1.5-ml tube and centrifuge for 5 min at 5000 rpm (1677 g) (*see* **Note 42**).
5. Add 20 μl of the supernatant to a 12×75 mm disposable glass culture tube (*see* **Note 43**). Place the tubes in the luminometer and collect the light output readings using the luciferase assay.
6. Using a ultraviolet visible (UV/VIS) spectrophotometer, determine the protein concentration of the supernatant with the protein assay reagent as described by the reagent manufacturer (*see* **Note 44**). The light output readings from the luciferase assay are normalized with this number. The total cellular protein content is also a crude indicator of cell viability.

4. Notes

1. The desired final volume of liposome solution determines the size of the vial. In order for sonication of the liposome solution to be possible, a 4-ml vial needs to contain at least 200 μl of solution, whereas only about 50 μl are required in a 2-ml vial. These numbers should be confirmed by testing with plain water. Large volumes of liposome solutions can be prepared using round-bottom flasks and a rotary evaporator.
2. The vials should have Teflon-lined screw cap closures to minimize solvent evaporation.
3. Common lipids such as DOPC, DOPE, or DOTAP are available as powders or convenient chloroform solutions. They need to be warmed to room temperature before weighing; otherwise, water will condense on the lipid and in the container.
4. We propagate this plasmid in *Escherichia coli* as prescribed by Promega and purify it using a Qiagen (Valencia, CA, USA) Giga kit (following the directions for a "high yield" plasmid). The resulting aqueous solution may be stored for

months in a refrigerator and for years at –20 °C. The DNA concentration is measured spectrophotometrically at 260 nm (an absorbance of 1 at a path length of 1 cm corresponds to 50 μg/ml DNA).

5. For reliable assessment of the usefulness of a (new) lipid, it is important to test it in several different cell lines, e.g., originating from different species. Examples of commonly used cell lines are COS-1, HeLa, NIH-3T3.
6. Lipids with ester bonds are subject to hydrolysis upon prolonged storage in aqueous suspension. Therefore, liposome solutions should be used within 1–2 months. They should be re-sonicated before use if stored for over a week or if opaque.
7. Lipid powders can be hygroscopic and/or glue-like and thus tricky to weigh out. A method that has worked well in our laboratory is to use two spatulas: the first (with a wide blade) to remove some lipid from its container, and the second (thin blade, narrow tip) to transfer the lipid from the first spatula into the vial. Dissolving the lipid is facilitated by lightly depositing the lipid rather than smearing it onto the wall of the vial. Hygroscopic lipids need to be handled speedily or under a nitrogen atmosphere. In this case, it is imperative to adjust the concentration by the amount of solvent that is added rather than by weighing out a specific amount of lipid.
8. The exact concentration of lipid is not crucial at this step, but should be at or above that of the desired concentration of the liposome solution (0.6 mM), so that the entire lipid film (see **steps 4** and **5**) will be wetted. Having identical molar concentrations for all lipids greatly facilitates the calculations for making lipid mixtures.
9. Chloroform/methanol mixtures may be required to dissolve lipids with higher headgroup charge or Poly(ethylene glycol) (PEG) chains (2:1, 4:1, 9:1 are typical ratios). In this case, it is best to stepwise add the solvents, starting with part of the calculated amount of methanol, then adding part of the calculated amount of chloroform, etc. This allows an empirical determination of the optimal solvent composition while the solution is being prepared. Obviously, it is crucial to keep track of the total amount of solvent added.
10. If feasible, organic solvents (especially chloroform) should be dispensed only using glass pipettes. If pipettes with plastic tips (only polypropylene should be considered) are used, the time that the solvent is in contact with the plastic must be minimized. In addition, it is crucial to fill and empty the tip once or twice before aspirating the amount to be measured. Otherwise, the vapor pressure of the solvent pushes part of it out of the tip.
11. The chloroform or chloroform/methanol solutions may be stored at –20 °C under argon in glass (never plastic) containers with Teflon-lined closure. Marking the level of liquid on the container is advisable, because solvent evaporation is the main reason for the limited shelf life (1–2 months) of the solutions.
12. It may not be necessary to prepare liposome solutions for all desired ratios of neutral/cationic lipid. Mixing of two liposome solutions to achieve an intermediate concentration just prior to addition to DNA may be successful. However,

this has first to be tested (i.e., must give same results, in structural characterization as well as in transfection) for a given lipid combination by preparing the desired concentrations both ways. For example, we have successfully prepared samples in the DOTAP/DOPC system covering the whole composition range in 5% or 10% steps using only liposome stock solutions of 100, 70, 40, and 10 or 100, 80, 60, 40, and 10 mol% DOPC.

13. These calculations are most conveniently performed and recorded using a spreadsheet program.
14. A stream of dry nitrogen is directed into the vial(s) via Pasteur pipettes or plastic pipette tips of appropriate size. If preparing a series of lipid compositions, constructing a setup with multiple outlets is helpful. Care must be taken to adjust the speed of the nitrogen stream so it does not spill the solution. It also must not be too gentle a stream, which would lead to a thick film with little surface area at the bottom of the vial. To achieve the goal of forming a thin film, one may have to adjust the position of the vial with respect to the pipette over time.
15. The vials may remain open at this stage or the caps placed very loosely on the vials.
16. The water bath prevents heat buildup in the vial. The vial has to be secured very tightly to prevent it from falling into the water bath because of vibrations generated by the sonication.
17. The sonicator tip has to be submerged in the lipid suspension. It should be as close to the meniscus of the suspension as possible for optimum performance but must not touch the wall of the glass vial. The amplitude of the ultrasound should be increased slowly and maintained at the highest level possible without causing the suspension to splash and spill. The sonicator tip should be cleaned with ethanol before processing the next sample.
18. If the sonicator tip touches the wall of the vial, the glass may break. Because movement may occur during sonication, it is best to monitor the process. The tip touching the glass wall is usually apparent from a change in pitch of the sound generated by sonication.
19. The noise pollution due to sonication can be minimized by encasing the whole setup in a cupboard or box.
20. Alternate methods for the preparation of small liposomes are using a high-power ultrasound bath and extrusion (which is easily scaled up). For more information, see, for example, the websites of Avanti Polar Lipids (http://www.avantilipids.com) and Northern Lipids Inc. (http://www.northernlipids.com).
21. This step sterilizes the solution and removes debris stemming from the sonicator tip. Very small amounts of lipid solution may be transferred to 1.5-ml tubes and briefly centrifuged to collect the debris at the bottom of the tube.
22. We use a setup with a large washing flask connected to a water aspirator. The waste needs to be autoclaved before it can be disposed of. (b) Cells should never be allowed to reach confluency for the sake of reproducibility of the data.

23. All media and solutions that are applied to the cells need to first be warmed to room temperature or 37 °C in a water bath.
24. After adding the trypsin buffer to the cells and rocking the dish back and forth to cover the cells, immediately aspirate the trypsin buffer to minimize the amount that will remain on the cells.
25. Tilting the cell culture dish after removing it from the incubator will indicate whether the cells have successfully been dislodged from the surface (a slightly opaque film should slowly slide to the bottom of the flask). Using a pipette, the added DMEM should be aspirated and re-added to the flask a few times to completely remove the cells from the surface and ensure a homogeneous mixture.
26. Directions for using the hemocytometer: Pipet 20 μl from a 50-μl aliquot of the cell suspension onto the hemocytometer (between coverslip and grid). Count the number of cells in each of the nine large squares of the cytometer and take the average of that number. The result, multiplied by 10,000, gives the number of cells per ml of the suspension.
27. A typical numerical example: to yield 85,000 cells (n) per well, 56.7 μl (V) of a cell suspension with 1.5×10^6 cells/ml (c) needs to be added to each well: $V = n/c$.
28. It is good practice to gently agitate the cell stock about every 12 wells to prevent the cells from settling in the centrifuge tube. To achieve even distribution of cells, one should circle the pipette over the well while the cell suspension is added. More importantly, once cells have been added to half or all of the plate, the plate is rocked first back and forth and then side to side. This movement should be quick and vigorous, but not so abrupt as to spill the content of the wells. It is vital to *never* swirl the plate, even avoiding rotational movement as the plates are transferred back to the incubator.
29. Again, the cells should not be allowed to reach full confluency. For reproducibility, it is important to perform experiments at similar levels of confluency, because this parameter affects the measured TE.
30. Cells should be homogeneously distributed in each well with around 80–90% confluency. If there appears to be any problem or irregularity with the cells, it is best to discard the plates and restart the experiment by seeding new 24-well plates.
31. This description assumes that every experimental data point will be performed in duplicate. To perform triplicates, scale each lipid sample up to a final volume of 700 μl, calculating the amount of lipid required for 1.4 μg DNA. For each well, 200 μl of the lipid solution will be used eventually, but preparing a slight excess is necessary to account for pipetting inaccuracies and losses. All these calculations and pipetting schemes are most conveniently performed and recorded using a spreadsheet program.
32. It is prudent practice to agitate the liposome stock solution just prior to using it to ensure a homogenous solution.

33. To account for inaccuracies and losses in pipetting, it is sensible to prepare 20% more of this solution than the amount calculated from the number of samples.
34. Because of the salt content of DMEM, aggregation/fusion of liposomes due to decreased electrostatic repulsion will set in. The exact timescale will depend on the lipid mixture, which determines the zeta potential of the liposomes. In our experience, incubating for 5 min or less has avoided aggregation-related effects (a certain incubation time is unavoidable when processing a larger number of wells). Aggregation/fusion of small liposomes in the DOTAP/DOPC system starts to be significant at times of 30 min or more.
35. The reason for only processing 12 wells at a time is to prevent the cells from drying out. An alternative method (likely more suitable when getting familiar with the procedure) is to process two (or three if performing experiments in triplicate) wells at a time, aspirating the medium, washing with PBS, aspirating the PBS, and then adding the 200 μl of complex solution.
36. Accomplish this step as quickly and as gently as possible. When pipetting the complex solution into the well, the tip of the pipette should be in contact with the side of the well rather than the bottom of the well (the cells) to prevent dislodging cells when adding the complexes.
37. This time (the time required for an optimum amount of complexes to be taken up by the cells) was determined by a series of experiments measuring TE as a function of incubation time, where TE first increases and then saturates or decreases (due to serum starvation of the cells). A time of 6 h universally works well for common cell lines as investigated in our and other laboratories.
38. Even very small hydrodynamic forces, such as those generated by taking out a plate, looking at it, and rocking it back and forth a bit, can have a notable (favorable but irreproducible) effect on complex uptake and, thus, TE.
39. Incubating for a full cell cycle allows for the machinery of the cell to transport the transferred DNA into the nucleus. Waiting for longer than one cell cycle is not advised because it may lead to cell crowding or daughter cells that do not retain the transferred DNA.
40. It is now no longer necessary to prevent the cells from drying out by processing only a part of the wells at a time.
41. Also thaw the luciferase assay kit at this time.
42. Complete removal of the contents of the well is ensured by repeatedly (about 10 times) aspirating and releasing the content with a pipette (set to about 200 μl volume). Centrifugation pellets insoluble cell components.
43. This applies if using a Berthold Autolumat Plus luminometer. If using a different luminometer, follow the directions supplied with the instrument and the luciferase assay.
44. This can be done, for example, by mixing 20 μl of supernatant and 800 μl of protein assay reagent in 1-ml disposable cuvettes. These readings can also be done more efficiently in 96-well plates on a UV/VIS plate reader using 2 μl of the supernatant with 198 μl of reagent.

Acknowledgment

Our work is supported by the National Institutes of Health und grant number GM-59288. Our laboratory is indebted to Prof. C. E. Samuel and Dr. C. X. George for the introduction to and continued support regarding cell culture and biological protocols. We also thank past and present graduate students who have helped to develop the presented protocols: Dr. Nelle L. Slack, Dr. Alison J. Lin, and Alexandra Zidovska.

References

1. Friedmann, T. (1997) Overcoming the obstacles to gene therapy. *Sci. Am.* **276**, 96–101.
2. Nabel, G., Nabel, E., Yang, Z., Fox, B., Plautz, G., Gao, X., Huang, L., Shu, S., Gordon, D. and Chang, A. (1993) Direct gene transfer with DNA–liposome complexes in melanoma: expression, biologic activity, and lack of toxicity in humans. *Proc. Natl. Acad. Sci. U.S.A.* **90**, 11307–11311.
3. Rinehart, J., Hersh, E., Issell, B., Triozzi, P., Buhles, W. and Neidhart, J. (1997) Phase 1 trial of recombinant human interleukin-1-beta (rhIL-1-beta), carboplatin, and etoposide in patients with solid cancers: Southwest Oncology Group Study 8940. *Cancer Invest.* **15**, 403–410.
4. Kay, M. A., Glorioso, J. C. and Naldini, L. (2001) Viral vectors for gene therapy: the art of turning infectious agents into vehicles of therapeutics. *Nature Med.* **7**, 33–40.
5. Marshall, E. (2002) Gene therapy on trial. *Science* **288**, 951–952.
6. Raper, S. E., Chirmule, N., Lee, F. S., Wivel, N. A., Bagg, A., Gao, G. P., Wilson, J. M. and Batshaw, M. L. (2003) Fatal systemic inflammatory response syndrome in a ornithine transcarbamylase deficient patient following adenoviral gene transfer. *Mol. Genet. Metab.* **80**, 148–158.
7. Cavazzana-Calvo, M., Hacein-Bey, S., de Saint Basile, G., Gross, F., Yvon, E., Nusbaum, P., Selz, F., Hue, C., Certain, S., Casanova, J. L., Bousso, P., Le Deist, F. and Fischer, A. (2000) Gene therapy of human severe combined immunodeficiency (SCID)-X1 disease. *Science* **288**, 669–672.
8. Hacein-Bey-Abina, S., von Kalle, C., Schmidt, M., Le Deist, F., Wulffraat, N., McIntyre, E., Radford, I., Villeval, J. L., Fraser, C. C., Cavazzana-Calvo, M. and Fischer, A. (2003) A serious adverse event after successful gene therapy for X-linked severe combined immunodeficiency. *N. Engl. J. Med.* **348**, 255–256.
9. Hacein-Bey-Abina, S., von Kalle, C., Schmidt, M., McCormack, M. P., Wulffraat, N., Leboulch, P. et al. (2003) LMO2-associated clonal T cell proliferation in two patients after gene therapy for SCID-X1. *Science* **302**, 415–419.
10. Huang, L., Hung, M.-C. and Wagner, E. (eds.) (1999) *Non-Viral Vectors for Gene Therapy*. Academic Press, San Diego, CA.

11. Huang, L., Hung, M.-C. and Wagner, E. (eds.) (2005) *Non-Viral Vectors for Gene Therapy, (Advances in Genetics Vol. 53)*. Elsevier, San Diego, CA.
12. Mahato, R. I. and Kim, S. W. (eds.) (2002) *Pharmaceutical Perspectives of Nucleic Acid-Based Therapeutics*. Taylor & Francis, London.
13. Miller, A. D. (1998) Cationic liposomes for gene therapy. *Angew. Chem. Int. Ed.* **37**, 1768–1785.
14. Niidome, T. and Huang, L. (2002) Gene therapy progress and prospects: nonviral vectors. *Gene Ther.* **9**, 1647–1652.
15. Ferber, D. (2001) Gene therapy: safer and virus-free? *Science* **294**, 1638–1642.
16. Harrington, J. J., van Bokkelen, G., Mays, R. W., Gustashaw, K. and Willard, H. F. (1997) Formation of de novo centromeres and construction of first-generation human artificial microchromosomes. *Nat. Genet.* **15**, 345–355.
17. Edelstein, M. L., Abedi, M. R., Wixon, J. and Edelstein, R. M. (2004) Gene therapy clinical trials worldwide 1989–2004 – an overview. *J. Gene Med.* **6**, 597–602.
18. Lasic, D. D. (1993) *Liposomes: From Physics to Applications*. Elsevier, San Diego, CA.
19. See also the website of Avanti Polar Lipids: http://www.avantilipids.com/
20. Felgner, P. L., Gadek, T. R., Holm, M., Roman, R., Chan, H. W., Wenz, M., Northrop, J. P., Ringold, G. M. and Danielsen, M. (1987) Lipofection: a highly efficient, lipid-mediated DNA-transfection procedure. *Proc. Natl. Acad. Sci. U.S.A.* **84**, 7413–7417.
21. Rädler, J. O., Koltover, I., Salditt, T. and Safinya, C. R. (1997) Structure of DNA–cationic liposome complexes: DNA intercalation in multilamellar membranes in distinct interhelical packing regimes. *Science* **275**, 810–814.
22. Salditt, T., Koltover, I., Rädler, J. O. and Safinya, C. R. (1998) Self-assembled DNA–cationic lipid complexes: two-dimensional smectic ordering, correlations, and interactions. *Phys. Rev. E* **58**, 889–904.
23. Lasic, D. D., Strey, H., Stuart, M. C. A., Podgornik, R. and Frederik, P. M. (1997) The structure of DNA–liposome complexes. *J. Am. Chem. Soc.* **119**, 832–833.
24. Koltover, I., Salditt, T. and Safinya, C. R. (1999) Phase diagram, stability and overcharging of lamellar cationic lipid–DNA self-assembled complexes. *Biophys. J.* **77**, 915–924.
25. Koltover, I., Salditt, T., Rädler, J. O. and Safinya, C. R. (1998) An inverted hexagonal phase of cationic liposome–DNA complexes related to DNA release and delivery. *Science* **281**, 78–81.
26. Ewert, K. K., Evans, H. M., Zidovska, A., Bouxsein, N. F., Ahmad, A. and Safinya, C. R. (2006) A columnar phase of dendritic lipid-based cationic liposome–DNA complexes for gene delivery: hexagonally ordered cylindrical micelles embedded in a DNA honeycomb lattice. *J. Am. Chem. Soc.* **128**, 3998–4006.
27. Manning, G. S. (1978) Limiting laws and counterion condensation in polyelectrolyte solutions. I. Colligative properties. *J. Chem. Phys.* **51**, 924–933.
28. Le Bret, M. and Zimm, B. H. (1984) Distribution of counterions around a cylindrical polyelectrolyte and Manning's condensation theory. *Biopolymers* **23**, 287–312.

29. Bruinsma, R. (1998) Electrostatics of DNA cationic lipid complexes: isoelectric instability. *Eur. Phys. J. B* **4**, 75–88.
30. Harries, D., May, S., Gelbart, W. M. and Ben-Shaul, A. (1998) Structure, stability, and thermodynamics of lamellar DNA–lipid complexes. *Biophys. J.* **75**, 159–173.
31. Lin, A. J., Slack, N. L., Ahmad, A., Koltover, I., George, C. X., Samuel, C. E. and Safinya, C. R. (2000) Structure–function studies of lipid–DNA nonviral gene delivery systems. *J. Drug Target.* **8**, 13–27.
32. Ewert, K. K., Evans, H. M., Bouxsein, N. F. and Safinya, C. R. (2006) Dendritic cationic lipids with highly charged headgroups for efficient gene delivery. *Bioconjug. Chem.* **17**, 877–888.
33. Ahmad, A., Evans, H., Ewert, K., George, C. X., Samuel, C. E. and Safinya, C. R. (2005) New multivalent lipids reveal bell-curve for transfection versus membrane charge density: lipid–DNA complexes for gene delivery. *J. Gene Med.* **7**, 739–748.
34. Ewert, K., Ahmad, A., Evans, H. M., Schmidt, H.-W. and Safinya, C. R. (2002) Efficient synthesis and cell-transfection properties of a new multivalent cationic lipid for nonviral gene delivery. *J. Med. Chem.* **45**, 5023–5029.
35. Lin, A. J., Slack, N. L., Ahmad, A., George, C. X., Samuel, C. E. and Safinya, C. R. (2003) Three-dimensional imaging of lipid gene carriers: membrane charge density controls universal transfection behavior in lamellar cationic liposome–DNA complexes. *Biophys. J.* **84**, 3307–3316.
36. Farhood, H., Serbina, N. and Huang, L. (1995) The role of dioleoyl phosphatidylethanolamine in cationic liposome mediated gene transfer. *Biochim. Biophys. Acta* **1235**, 289–295.
37. Hui, S., Langner, M., Zhao, Y., Ross, P., Hurley, E. and Chan, K. (1996) The role of helper lipids in cationic liposome-mediated gene transfer. *Biophys. J.* **71**, 590–599.
38. Rao, N. M. and Gopal, V. (2006) Cationic lipids for gene delivery *in vitro* and *in vivo*. *Expert Opin. Ther. Pat.* **16**, 825–844.
39. Ewert, K., Slack, N. L., Ahmad, A., Evans, H. M., Lin, A. J., Samuel, C. E. and Safinya, C. R. (2004) Cationic lipid–DNA complexes for gene therapy: understanding the relationship between complex structure and gene delivery pathways at the molecular level. *Curr. Med. Chem.* **11**, 133–149.
40. Ewert, K., Ahmad, A., Evans, H. M. and Safinya, C. R. (2005) Cationic lipid–DNA complexes for non-viral gene therapy: relating supramolecular structures to cellular pathways. *Expert Opin. Biol. Ther.* **5**, 33–53.
41. Ewert, K., Ahmad, A., Evans, H. M., Slack, N. L., Lin, A. J., Martin-Herranz, A. and Safinya, C. R. (2005) Lipoplex structures and their distinct cellular pathways, in *Advances in Genetics, Vol. 53: Non-Viral Vectors for Gene Therapy* (Huang, L., Hung, M.-C., Wagner, E., eds.). Elsevier, San Diego, CA, pp. 119–155.
42. Freshney, R. I. (2000) *Culture of Animal Cells*. Wiley-Liss, New York, NY.

11

Applications of Lentiviral Vectors in Noninvasive Molecular Imaging

Abhijit De, Shahriar Shah Yaghoubi, and Sanjiv Sam Gambhir

Summary

Noninvasive imaging of molecular-genetic and cellular processes is an effective way to determine the location(s), magnitude, and time variation of action of gene products used for many therapeutic strategies. Lentiviral vectors provide effective means for the delivery, integration, and expression of transgenes in cultured mammalian cells as well as in vivo. Therefore, the combination of lentiviral vector-mediated therapeutic and imaging-targeted reporter gene delivery to various target organs holds promise for the future treatment of diseases. In this chapter, we provide protocols for developing lentiviral vectors that can be utilized for noninvasive monitoring/imaging of reporter gene expression. We have described the procedures to perform cellular assays and animal imaging based on positron emission tomography (PET), optical bioluminescence, and fluorescence reporter genes. The protocols described here are standardized for mouse models, which can also be adapted for other small animal models (e.g., rats).

Key Words: Lentiviral vector; bioluminescence; fluorescence; micro positron emission tomography (microPET); molecular imaging.

1. Introduction

Molecular imaging is a rapidly emerging biomedical research field that provides a visual estimation in space and time of normal as well as abnormal cellular processes at a molecular or genetic level in intact living subjects. Recent developments in the biological sciences have enabled the scientific community with unprecedented opportunities in understanding the molecular

From: *Methods in Molecular Biology, vol. 433: Volume 1: Production and In Vivo Applications of Gene Transfer Vectors*
Edited by: J. M. Le Doux © Humana Press, Totowa, NJ

mechanisms of disease, allowing for development of innovative therapies at the genetic/molecular level. Today, in vivo molecular imaging makes it possible to understand the disease pathogenesis in real time and in a noninvasive manner from intact microenvironments within the body. Imaging assessment, in combination with innovative-targeted therapies, would allow assessment of therapeutic response in real time, long before any clinical symptoms occur. To this end, several strategies have emerged for achieving important goals in biomedical research: (1) to develop noninvasive in vivo imaging methods that reflect specific cellular and molecular processes, (2) to monitor multiple molecular events near-simultaneously, (3) to follow trafficking and targeting of cells, (4) to optimize drug and therapeutic transgene delivery, (5) to image drug efficacy at molecular and cellular level, and (6) to assess disease progression at a molecular pathological level ***(1)***.

There are two major imaging strategies that are being employed in biomedical research: direct and indirect imaging, of which the latter is predominant in studies with pre-clinical animal models. The direct imaging strategy is based on imaging the target directly, usually with a target-specific probe ***(2)***. Indirect molecular imaging may involve multiple components. One type of indirect imaging that is now being widely used is based on reporter gene imaging, which involves single or multiple reporter genes and their specific reporter probe partners ***(1,3)***. The advantage of reporter gene imaging is the ability to develop and validate indirect imaging strategies more rapidly and at considerably lower cost than direct imaging strategies. This is possible because a well-characterized reporter gene–reporter probe pair can be linked to either a gene-specific promoter or to a gene sequence as a bi-cistronic vector to image many different biological and molecular-genetic processes ***(4,5)***. The reporter gene can encode for a protein that can produce light signal in the presence of a substrate [e.g., firefly luciferase (*Fluc*) in the presence of D-luciferin], or can trap a radiolabeled substrate [e.g., 9-(4-[^{18}F]-fluoro-3-hydroxy-methylbutyl) guanine (FHBG)] for the HSV-1sr39TK reporter ***(6,7)***. Alternatively, the reporter gene product can be a receptor or transporter that irreversibly traps the probe in transduced cells during the period of image acquisition [e.g., 3-(2′-^{18}F-fluoroethyl)-spiperone (FESP) for the dopamine-2 receptor (D2R)] ***(8,9)***. Recently, PET reporter gene/probe (PRG/PRP) systems based on membrane transporters has also been described [e.g., [^{124}I]-iodide for sodium/iodide symporter (NIS) reporter gene] ***(10,11)***. On the other hand, fluorescent reporter genes do not need a probe, but need external light illumination to measure the gene expression.

For reporter gene imaging in living subjects, positron emission tomography (PET), single photon emission computed tomography (SPECT), magnetic resonance imaging (MRI), and optical imaging are well standardized and are being used extensively in small animal models ***(4,12)***. However, to overcome the shortcomings of each modality, one may find a multimodality approach very useful for detecting the expression of multiple reporter genes from single vector. Combining two or three different technologies (e.g., PET with optical), one cab obtain a unified vector that has the advantage of speed and ease of validating approaches in small animals that in turn can be translated to humans. With recent advances and innovations in vector design, lentiviral expression vectors are playing a very effective role in delivering and expressing combined reporter genes of interest to almost any mammalian cell including nondividing cells and target organs in model organisms ***(13–15)*** (**Fig. 1**). By packaging the lentiviral vector as viral particles, one can obtain highly efficient transduction of the expression constructs even with the most difficult-to-transfect cells, such as primary, stem, and differentiated cells ***(16–19)***. Most of the lentiviral vectors currently used for gene delivery and gene therapy approaches are HIV-derived vectors. The *cis*- and *trans*-acting factors of lentiviruses can be separated while preserving their functions, making them as effective and safer viral vector to use in biological research ***(20)***. When delivered as viral particle, the transgene is integrated into the host genome, providing stable, long-term expression of the target gene: therefore one can study viral distribution including the site of target gene delivery.

In this chapter, protocols for PET, optical bioluminescent, and fluorescent reporter-based gene-imaging techniques that help to monitor lentiviral gene delivery are provided so that reporter gene expression can be studied in vivo. For this purpose, we choose one reporter gene–reporter probe combination under each category, such as radionuclide-based microPET imaging or bioluminescence (BL)-based optical imaging, etc. We primarily discuss protocols standardized for mouse models, which can be adapted for other models. Animals larger than the rodents would be difficult to image using optical methods due to poor penetration of light, and therefore non-optical methods, such as PET or MRI, would be more appropriate for those situations.

2. Materials

2.1. Imaging Instrumentation

1. An inverted fluorescent microscope with an attached camera can be very useful when fluorescent reporter genes are introduced in the lentiviral cDNA, helping one to identify transduction efficiency and imaging cells after viral transduction.
2. Fluorescence-activated cell sorting (FACS): After lentiviral transduction, when a fluorescence reporter gene is used, positive expressers can be easily sorted out using FACS and the data can be analyzed for deciding transduction efficiency and/or titering the viral supernatant. For example, the BD FACSvantage SE system (Becton Dickinson, San Jose, CA) can be used for both multicolor FACS analysis and cell-sorting purposes.

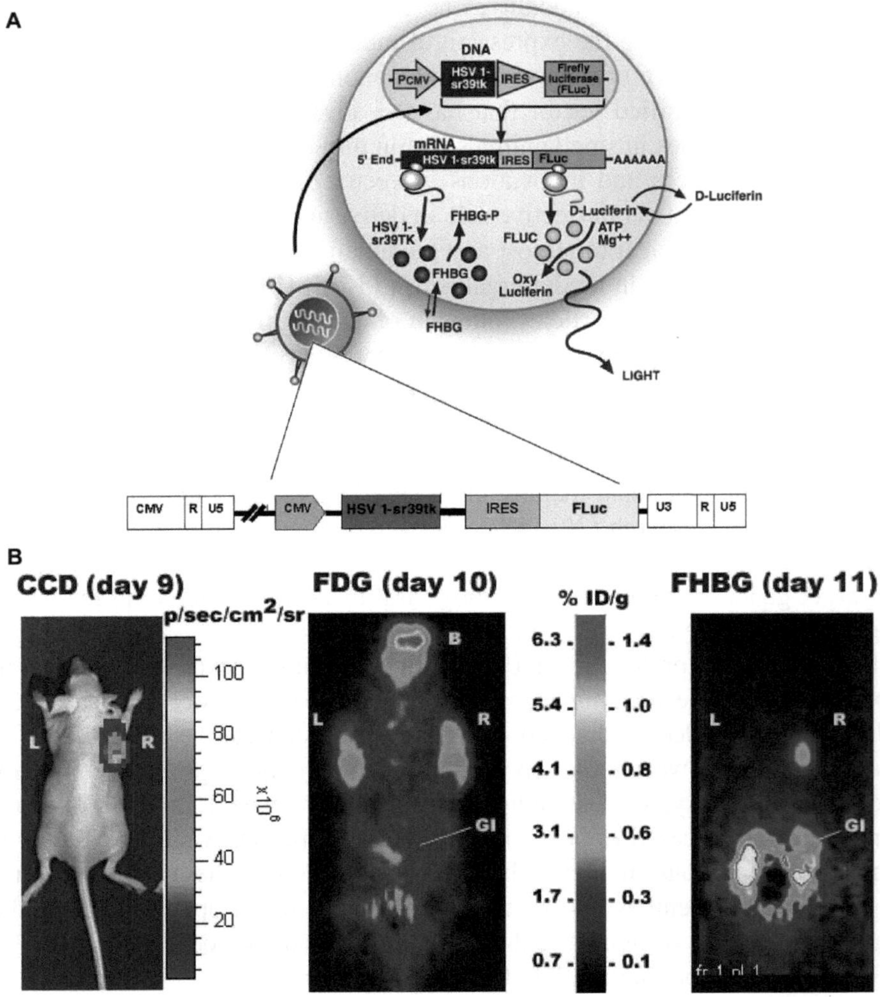

Fig. 1. (A) Diagrammatic representation showing an internal ribosomal entry site (IRES)-based bi-cistronic lentivector used to link the activity of two reporter genes. As shown in this example, a positron emission tomography (PET) reporter gene (HSV1-*sr39tk*) is linked to a bioluminescent reporter gene (*Fluc*) by an encephalomyocarditis virus (EMCV) IRES sequence. When these combined reporter genes are packaged within a lentivector for cell targeting, the virus delivers this DNA cassette to the host cell nucleus and the reporter genes get integrated within the host genome. As the host genome replicates, the messenger RNA (mRNA) for the reporter gene combination also replicates. Then the mRNA molecules are translated to proteins within the host cytoplasm. By using ^{18}F-fluoro-3-hydroxymethylbutylguanine (FHBG) as a PET reporter probe, one can quantitatively image the HSV1-1sr39TK activity within living subjects. The same is also true for the *Fluc* gene, whereby administration of D-luciferin substrate produces light signal that can be imaged using an optical charge-coupled

3. PET is a well-established imaging method used to map the distribution and concentration of injected radiolabeled tracers within a living organism. In PET imaging, reporter-gene-specific chemical molecules are labeled with radioisotopes that are positron emitters. This is in contrast to the SPECT imaging, where molecules are "tagged" with radioisotopes that directly emit gamma rays. The image acquired with a PET scanner after injecting the probe is a snapshot of the concentration of tracer throughout the imaging region. The advantage of PET over SPECT is the improved resolution due to the natural "collimation" provided by the physics. In PET, a "count" is recorded when *two* 511 KeV gamma rays emitted upon collision of a positron with an electron *simultaneously* trigger detectors on opposite sides of the sample. The line drawn through space between the two detectors gives the only possible locus of points from which the annihilation event could occur releasing the two gamma rays in opposite directions; thus, effective collimation is achieved. The creation of positron emitters that attach to biologically active molecules allows PET to probe the functional biochemistry of the living organism. Concord R4 microPET (Siemens, New York, NY) and Gamma Medica microSPECT-CT (Gamma Medica Inc., Northridge, CA) imaging systems can be used for tracking radionuclide-based reporter gene imaging.
4. Cooled charge-coupled device (CCD)-based Xenogen IVIS 200 biophotonic imager from Caliper (Alameda, CA) can be used for both BL and fluorescence imaging. The system excels at imaging light emitted from small animals such as mice or rats but can also be used to image light emitted directly from cells in 96-well plates. The system is capable of measuring the intensities of BL signals produced by luciferases and other bioluminescent reporters, as well as fluorescence signals produced by fluorescent proteins (GFP, RFP, etc.) and other fluorescent molecules. The output is generally a superimposed image of photographic and luminescence/fluorescence image captured sequentially.
5. CRI Maestro (Boston, MA) in vivo fluorescent imaging system can be used for multispectral acquisition and analysis, with multicolor flexibility and quantitative accuracy for measuring fluorescent probes with emission at visible light ranges as well as near-infrared ranges. This system employs unique technology to improve

Fig. 1. device (CCD) camera. **(B)** Optical CCD and microPET imaging of a mouse implanted with N2a cells stably expressing HSV1-*sr39tk* and *Fluc* reporter genes. Tumors were grown by injecting 5×10^6 N2a cells transduced in culture by lentivirus (CS-CMV*sr39tk*-I-*Fluc*) (R) and a control tumor with equal number of parental N2a cells (L). Optical CCD image for *Fluc* expression on day 9, microPET FDG scan to check tumor viability on day 10, and microPET FHBG scan on day 11 for HSV1-*sr39tk* gene expression. B, brain; GI, gastrointestinal tract; L, left tumor; R, right tumor. [Reprinted with modifications from **ref.** ***4*** with permission from Elsevier].

signal-to-noise by eliminating auto-fluorescence, allowing much smaller or fainter targets to be detected (*see* **Note 1**).

2.2. DNA Plasmid Vectors

1. HIV-1-based expression plasmid [e.g., pCS-CMV-GFP *(20)* or pHR-CMV-GFP*(21)* or pRRLsin18.cPPT.CMV.EGFP *(22)*] (*see* **Note 2**).
2. Packaging plasmid [e.g., pDelta R8.2 *(20)* or pMDLg/p RRE and pRSV-REV *(22)*].
3. Envelope plasmid [e.g., pMD2.G *(20,22)*].
4. Various commercially available reporter vectors such as pCMV-*Fluc* (Promega, Madison, WI) or pCMV-Rluc (Promega) or pCMV-GFP (Clonetech) or pCMV-DsRed2 (Clonetech, Palo Alto, CA) and their mutant variants can be used for amplifying reporter cDNA and cloning in the lentivectors. The CMV-HSV1sr39tk plasmid, used here as a PRG, was a gift from Dr. M. Black (Washington State University, WA).

2.3. Cell Lines

Human embryonic kidney fibroblast (293T, ATCC) or 293FT (Invitrogen, Grand Island, NY) cell lines are most commonly used for virus production. Minimum Essential Medium (Invitrogen) completed with 10% fetal bovine serum (FBS) and 1% penicillin/streptomycin (P/S) is used for culturing 293 cell lines. Dulbecco's Modified Eagle's Medium (DMEM) completed with 10% FBS, 1% P/S/glutamine (P/S/G), and 1% nonessential amino acids is used for 293FT cells.

2.4. Chemicals

1. ^{18}F-fluoro-2-deoxyglucose (FDG) *(23)*.
2. ^{18}F-fluoro-3-hydroxymethylbutylguanine (FHBG) *(6,7)*.
3. D-Luciferin potassium salt (Caliper, Alameda, CA). Dissolve measured amount of D-luciferin powder in Dulbecco's phosphate-buffered saline (DPBS) to prepare 15 mg/ml stock.
4. Coelenterazine (Prolume Ltd., Pinetop, AZ). Dissolve measured amount of coelenterazine powder in methanol to prepare 1 mg/ml stock (*see* **Notes 3–4**).
5. DPBS, without Mg^{++} and Ca^{++}.
6. Sodium phosphate buffer (pH 7.4).

3. Methods

Noninvasive and repetitive imaging of cellular events directly from small animal models has become an important tool to provide data about biochemical, genetic, or pharmacological processes in vivo. The techniques described below focus on the preparative phase and imaging methods of lentiviral expression from small animal imaging.

3.1. Preparation of Expression Lentiviral Vector

1. To prepare a lentiviral vector for mammalian expression, either sub-clone full-length cDNA fragments from another plasmid vector or amplify the cDNA by polymerase chain reaction (PCR) and then clone it. Because of limited restriction enzyme sites in lentiviral vectors, PCR-based cloning is the most convenient way for full-length cDNA cloning in lentivectors (*see* **Notes 5–10**).
2. Amplify the target cDNA fragments by designing a 5′-primer (containing a Kozak sequence and ATG codon) and a 3′-primer with unique restriction sites present in the multiple cloning site of the lentivector, but not present in the cDNA sequence. Amplify the cDNA fragment using high-fidelity long-distance PCR kit following manufacturer's recommended procedure, purify, and digest the amplified product with end-specific restriction enzyme(s) and purify the digested PCR product by agarose gel electrophoresis.
3. Digest the lentivector with the corresponding restriction enzymes used for preparation of cDNA fragments, and verify complete digestion of the vector by agarose gel electrophoresis. If a single restriction enzyme is used, dephosphorylate the digested DNA by re-digesting with calf intestinal phosphatase (CIP) for 30 min and elute the vector from agarose gel.
4. Proceed to ligation and screening of the ligated product after bacterial transformation using standard cloning procedures ***(24)***.

3.2. Packaging of the Lentiviral Expression Constructs into Pseudoviral Particles

For developing a viral particle carrying the genes of interest, it is required to package the lentiviral expression construct into pseudoviral particles. For this purpose, co-transfect the expression plasmid, packaging plasmid, and envelope plasmid in 293T/293FT cells as per recommendations.

1. Transfect 293T/293FT cells either by using classical calcium chloride transfection methods ***(20)*** or by using Lipofectamine™ 2000 complexes (Invitrogen) as per manufacturer's recommendations. For details of viral production, follow other methods described in this book (*see* **Note 11**).
2. Harvest virus-containing supernatants 72 h post transfection by removing medium to a 15-ml sterile, capped conical tube.
3. Centrifuge at 200 × *g* or 15 min at +4 °C to pellet cell debris.
4. Filter the viral supernatant through a sterile, 0.45-μm low-protein-binding filter after the low speed centrifugation to remove the cellular debris, using Millex-HV 0.45-μm PVDF filters from Millipore, Billerica, MA.
5. Pipet viral supernatants into cryovials in 1-ml aliquots. Store viral stocks in aliquots at –80 °C. Proceed to titering (for details, see other chapters in this book).

6. Alternatively, one can concentrate the virus by ultra-centrifugation using a swinging bucket rotor at 50,000 × *g* for 2 h and then re-suspend the pellet in media of choice.

3.3. Ex Vivo Cell Labeling Using Lentiviral Particles

For titering and verification of transgene expression as in vitro assays, target cells can be transduced with the lentiviral particles in culture. To perform this, do the following:

1. Split the target cell type a day before with 60–70% confluency and grow overnight.
2. Remove the media before adding virus.
3. After collecting the virus, transfer 3–5 ml of the media containing viral particles into the target cell plate. Add 8 μg/ml Polybrene (hexadimethrine bromide, Sigma, St. Louis, MO) into the virus-containing media before adding into the cells. Polybrene is added in medium during transduction experiments in order to facilitate vector penetration into cells. Depending upon the envelope pseudotyping or the nature of target cells, such components may not have any effect in increasing the transduction efficiency.
4. Incubate overnight at 37 °C incubator with 5% CO_2 to transduce cells.
5. After infection, dispose the virus containing media in 10% bleach solution for complete de-contamination. Wash the plate twice with PBS and then add normal growth media.
6. Assays can be performed after growing them for another 24 h to allow enough time for maturity of the reporter proteins. This is especially important where fluorescent proteins are used that require longer maturity times.
7. Repeat with the same assays after the cells past culture for few more passages to ensure stable gene expression in vitro (*see* **Note 12**).

3.4. In Vitro Reporter Assays

It is important to evaluate reporter gene expression in cultured cells prior to their use for imaging from animal models.

3.4.1. HSV1-TK/HSV1-sr39TK Enzyme Activity Assay

1. Prepare 20 ml of cell or tissue lysis buffer by mixing the following: 2 ml of 100 mM Tris–HCl (pH 7), 4 μl of 2-mercaptoethanol, 100 μl of Igepal CA-630, 2.5 μl 200 mM NaF, 15.4 ml of deionized water.
2. Harvest the cells by re-suspending them in 1 ml of 1× PBS and transfer to a 1.5-ml micro-centrifuge tube. Obtain cell pellets by centrifuging at a rate that will not harm them.

3. Add lysis buffer to the pellet at a volume approximately equal to 3× the volume of the cell pellet (*see* **Note 13**).
4. If frozen, quickly thaw the pellet in lysis buffer, vortex to detach the pellet, and keep on ice for 30 min. Otherwise, vortex to detach the pellet after the addition of the lysis buffer and then keep on ice for 30 min.
5. After lysis, spin the tubes for 20 min in a micro-centrifuge at 20,000 × *g* at 4 °C.
6. Determine the protein concentration in the supernatant. To do this, mix 5 μl of each sample with 1 ml of 1× Bradford reagent. Mix well and estimate the optical density (OD) at A_{595}, and calculate the protein concentration using standard curve determined from known protein standards.
7. Dilute sample in lysis buffer to obtain a concentration of 0.2 μg/μl and then transfer 5 μl (1 μg) to each of triplicate tubes.
8. Prepare 85 μl of HSV1-TK reaction buffer by mixing the following: 20 μl of 1 μCi/μl [^{3}H]-Penciclovir (^{3}H-PCV) (Moravek Biochemicals, Brea, CA), 40 μl of 0.5 M NaH_2PO_4 (pH 6.0), 20 μl of 0.1 M ATP (pH 7.0), 2 μl of 1 M magnesium acetate, 3 μl of deionized water.
   Depending upon sample numbers, scale up or down the total volume maintaining the proportion.
9. Add 3 μl of TK reaction buffer to the 5 μl of protein lysate in each tube and place the mixture in a 37 °C water bath for exactly 20 min (*see* **Note 14**).
10. After 20 min of incubation, place the tubes on ice and add 40 μl ice cold water to the reaction mix. Then place the tubes in 95 °C Heat Block for 2 min. The reaction will be stopped and the enzyme will be inactivated.
11. Do a quick 10 s max speed centrifugation to spin down the vapors on the walls of the tubes (*see* **Note 15**).
12. Blot filters (Whatman, DE81, 2.3 cm diameter) with 40 μl of reaction mixture. Phosphorylated Penciclovir is firmly attached to DE81 filters, due to its charge. Un-phosphorylated Penciclovir will be washed away from the filter in the next steps.
13. Let it dry for at least 1 min.
14. Count the total amount of TK Mix added to each lysate (3 μl) sample on DE81 filter (do not wash, but go through the process after washing) to figure out total count. Also, add the total amount of TK Mix added to each lysate sample (3 μl) to a DE81 filter and wash it. This will provide the background counts.
15. Add 4 ml of 1 M ammonium formate (final = 4 mM) and 1 ml of 10 mM thymidine (final = 10 μM) to 1 L of water.
16. Wash membranes three times in approximately 150 ml (per 20 filters) ammonium formate/thymidine for 5 min each.
17. Wash once for 5 min in 150 ml water.
18. Wash twice for 5 min each in 100–150 ml 95% ethanol.
19. Dry the filters on paper towel (~5 min).

20. Put filters for 1 h in 1 ml acid (0.1 M HCl/0.2 M KCl) in scintillation vials.
21. Add 10 ml scintillation fluid (Cytoscint ES), and let it equilibrate for at least an hour before counting, then count for tritium in a scintillation counter.
22. Data analysis: Figure out the efficiency factor of your scintillation counter. The Beckman Coulter model LS6500 in our laboratory has an efficiency factor of 59.2% for tritium and 78.7% for carbon-14. Divide the counts per minute (CPM) by the efficiency factor of the scintillation counter to get disintegrations per minute (DPM) and then divide that by 40/48 (as 40 μl of the 48 μl reaction mixture was blotted on DE81 filters) to obtain the total DPM from each reaction. Determine the average background of tritium on the filters and subtract that from the total DPM of each reaction (step A). Calculate the average total DPM in 3 μl of HSV1-tk reaction buffer from actual measurements (step B). Divide the value from step A by the μg of protein, reaction time, and the value from step B, then multiply by 100. The final value is the percentage of [^{3}H]-Penciclovir phosphorylated divided by the amount of protein in the reaction mixture, per minute of reaction time (*see* **Note 16**).

3.4.2. Luciferase Enzyme Reporter Assay

1. Harvest the cells after viral transduction by re-suspending them in 1 ml of 1× PBS and transfer to a 1.5-ml micro-centrifuge tube.
2. Obtain cell pellets by centrifuging at a rate that will not harm them.
3. Add 1× passive lysis buffer (Promega) to the pellet at a volume approximately equal to 3× the volume of the cell pellet (*see* **Note 17**).
4. If frozen, quickly thaw the pellet in lysis buffer, vortex to detach the pellet, and keep on ice for 20 min. Otherwise, vortex to detach the pellet after the addition of the lysis buffer and then keep on ice for 20 min.
5. After lysis, spin the tubes for 5 min in a micro-centrifuge at 20,000 × *g* at 4 °C.
6. Determine the protein concentration in the supernatant. To do this, mix 5 μl of each sample with 1 ml of 1× Bradford reagent. Mix well and estimate the OD at A_{595}, and calculate the protein concentration using standard curve determined from known protein standards.
7. Dilute sample in 1× passive lysis buffer and aliquot 20 μl of each sample in triplicate tubes.
8. For measuring firefly luciferase activity, take the tube containing 20-μl aliquots of each sample in micro-centrifuge tubes. Now, one at a time, add 100 μl of luciferase assay reagent II (LAR II) to each sample, mix by vortex, and count the photon output immediately for 10 s using a luminometer (e.g., Turner T20/20). For measuring *Renilla* luciferase activity, add 100 μl of 0.05 M sodium phosphate buffer (pH 7.0) containing 1 μg of coelenterazine to the 20 μl sample, mix by vortex, and measure using the same instrument and protocol.
9. Final result outputs are generally represented as relative light unit (RLU)/μg of protein for each sample.

3.5. Imaging Methods

3.5.1. MicroPET Imaging of HSV1-tk/HSV1-sr39tk Reporter Gene Activity in Mice with [^18F]FHBG

Investigators can choose radioactive molecular probes that can detect HSV1-TK or HSV1-sr39TK that are most appropriate for their study, or are available at their institution. Comparative studies have already shown that the detection sensitivities and specificities of different PRPs varies with wild-type (HSV1-TK) or mutant (HSV1-sr39TK) PRG in cell culture and in mice ***(6,25,26)***. We have chosen [^{18}F]FHBG, because with the HSV1-*sr39tk* reporter gene, it is the most sensitive combination ***(25)*** that has thus far been reported for imaging in animals. Additionally, [^{18}F]FHBG is a relatively poor substrate for mammalian thymidine kinases, resulting in a relatively high specificity and has been extensively tested for safety studies in humans, allowing rapid translation from pre-clinical imaging studies to clinical imaging studies ***(27)***. The procedures below describe use of a R4 microPET imaging system, because it is the one available at our institution and used for our imaging studies ***(28)***. Other small animal PET imaging instruments are available with varying sensitivities and spatial resolutions ***(29–33)*** (*see* **Note 18**).

3.5.1.1. Preparation and Injection of [^{18}F]FHBG

1. Synthesis of [^{18}F]FHBG has been described in detail in several research articles ***(6,7)***, and will not be described in this protocol. For administration into mice, the activity concentration should be greater than 1 μCi/μl and the percentage of ethanol in the final solution should be less than 10%.
2. Fifteen minutes prior to scanning, start anesthetizing the mouse using the method described in your approved study protocol. In our laboratory, we have used either 2% Isoflurane as gas anesthesia or 4:1 Ketamine (100 mg/ml)/Xylazine (100 mg/ml) injected intraperitoneally.
3. Draw [^{18}F]FHBG into an insulin syringe (1/2 cc). Measure approximately 200 μCi of [^{18}F]FHBG with a dose calibrator (should be routinely tested for accuracy) in a volume less than 200 μl and greater than 50 μl. Record the exact measurement and the measurement time.
4. Inject the entire amount of [^{18}F]FHBG into the tail vein (intravenous or i.v.) of the mouse. This procedure may require a lot of practice to avoid injection of [^{18}F]FHBG outside of the vein. To dilate the vein, use heat packs or lamps. Record the exact injection time.

3.5.1.2. Preparation for [^{18}F]FHBG microPET Scanning

1. For a static [^{18}F]FHBG scan, one should usually wait 60 min after the injection time to scan the mouse. This time is required for the clearance of the tracer from sites not expressing the PRG and for sufficient uptake and entrapment into the cells from the tissues expressing the PRG. For example, liver and kidneys have

high initial uptake, because they are involved in tracer clearance. One may allow more time for clearance, but it is important to keep the time between injection and scan constant for all scans in a given research study. During injection and waiting period, the mouse need not be anesthetized.
2. Fifteen minutes prior to scanning, start anesthetizing the mouse using the method described in your approved study protocol.
3. Ten minutes prior to scan time, start emptying the bladder by applying pressure on the bladder and soaking it to absorbent cotton without contaminating the area.
4. Position the unconscious mouse in the center of microPET's field of view (FOV) and then move the bed into the camera. The microPET R4 can usually scan an entire mouse at a single-bed position, because it has an axial FOV of 7.8 cm.

3.5.1.3. Image Acquisition, Reconstruction, and Analysis

1. Refer to your scanners' manual for instructions on image acquisition.
2. For a whole-body static scan, we usually use a 10-min acquisition time. The longer the acquisition time, the better will be the quality of acquired images.
3. Record the scan start time. Images are decay corrected to the scan start time and this time will be necessary for calculation of the injected dose at scan start time
4. Extra precautions to be taken so that the animal does not wake up or move during the scan.
5. The manufacturer of R4 provides software for the reconstruction of the images with the Flitered Back Projection (FBP) and ordered subsets expectation maximization (OSEM) algorithms *(34)*. The OSEM reconstruction algorithm usually yields images of better quality.
6. The manufacturer should have provided an image analysis software. In our laboratory, we have used the AMIDE software developed by us, available freely for download online *(35)*. This software allows visualization and quantitative analysis of images acquired by a Concorde Microsystems microPET as well as most other scanners.
7. Coronal, sagital, or transaxial images can be displayed. By providing the injected dose and the cylinder calibration factor, the AMIDE program can calculate the percentage of ID/g or standard uptake values (SUVs) in regions of interests (ROIs) (**Fig. 2A**).

3.5.2. Monitoring Therapeutic Gene Expression Using Optical Bioluminescence Imaging

In comparison with the radionuclide-based systems, advantage of optical BL reporter-based system is that they are at least an order of magnitude more sensitive than the radionuclide-based techniques at limited depths *(36)*. Furthermore, the direct and indirect costs are generally less than radionuclide-based techniques. However, there is significantly less spatial information obtained from BL imaging, and the source tissue depth could limit the signal

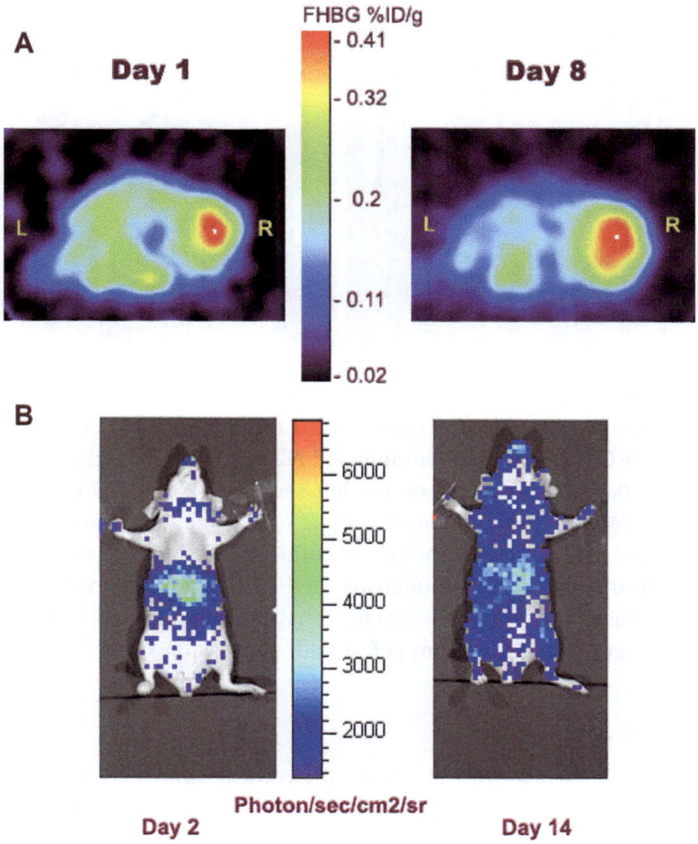

Fig. 2. (**A**) ^{18}F-fluoro-3-hydroxymethylbutylguanine (FHBG) microPET images of transverse sections of a mouse on days 1 and 8 with HSV1-*sr39tk* gene expression in a tumor xenograft. N2a parental cells were implanted subcutaneously on both shoulders and allowed to grow to a palpable size of ~5 mm diameter before lentivirus (CSCMV*sr39tk*-I-*Fluc*) was injected directly into the right tumor (R) only once. L represents the left tumor without any virus injection. (**B**) Optical charge-coupled device (CCD) image of a nude mouse injected intravenously (via tail-vein) with 50 μl of a concentrated stock of the CSCMV*sr39tk*-I-*Fluc* lentiviral vector (multiplicity of infection 5×10^6). After virus injection, bioluminescence imaging was performed on days 2 and 14 showing high expression of *Fluc* gene in liver and spleen, indicating the primary organ targets of VSVG-pseudotyped lentivirus. [Reprinted with modifications from **ref.** ***4*** with permission from Elsevier].

obtained from luminescence reporter system (**Fig. 2B**). Newer BL tomography systems are under development, but most current systems are based on a projection image with no depth of signal information (**Fig. 3**).

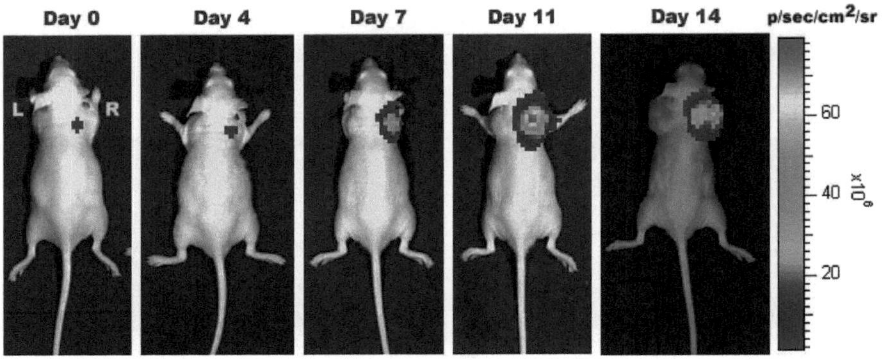

Fig. 3. Optical charge-coupled device (CCD) images of tumor xenografts showing *Fluc* gene expression over time. Tumors were implanted subcutaneously with 5×10^5 lentiviral (CS-CMV*sr39tk*-I-*Fluc*) infected N2a cells on the right shoulder (R) and same number of parental N2a cells on the left shoulder (L) on day 0. The mouse was repeatedly scanned with the CCD on days 0, 4, 7, 11, and 14 as tumors grew. Highest FLUC expression was observed on day 11, after which internal necrosis within the tumors likely resulted in lowered intensity of FLUC signal on day 14. All images shown are the optical CCD bioluminescence images superimposed on the visible light photographic image. [Reprinted from **ref. *4*** with permission from Elsevier].

3.5.2.1. Animal Anesthesia

The mice are placed into a clear plexiglass anesthesia box (2.5–3.5% isoflurane) that allows unimpeded visual monitoring of the animals, e.g., one can easily determine if the animals are breathing. The tube that supplies the anesthesia to the box is split so that the same concentration of anesthesia is pumped to the anesthesia manifold located inside the imaging chamber. After the mice are fully anesthetized, they can be transferred to the imaging chamber with their snout placed to the nose cones attached inside and close the door (*see* **Note 19**).

3.5.2.2. Preparation and Injection of D-Luciferin/Coelenterazine for In Vivo Bioluminescence Imaging

1. Prepare a fresh stock solution of D-luciferin at 15 mg/ml or 30 mg/ml in DPBS. Filter sterilize by passing through a 0.2-μm filter. Coelenterazine is dissolved in methanol and a stock solution of 1 mg/ml can be stored at –80 °C in working aliquots. For applications where higher coelenterazine concentrations (80–100 μg) are required for animal injections, dissolve the powder in 50% ethanol: 50% PEG mix. Just before experiments, take out a vial and prepare a working solution by adding 10 μg of the coelenterazine stock per 100 μl of DPBS.
2. Inject each mouse with 150 mg D-luciferin/kg body weight (e.g., for a 20 g mouse, inject 200 μl of 15 mg/ml to deliver 3 mg of D-luciferin), whereas for

coelenterazine, a dose of 1 mg/kg body weight is recommended (e.g., for a 20 g mouse, inject 200 μl of 1 μg/10 μl to deliver 20 μg of coelenterazine). However, for certain applications up to five times higher coelenterazine substrate can be used.

3. Inject D-luciferin solution via an intraperitoneal (i.p.) route and allow the substrate to distribute in awake animals for about 5–15 min. For coelenterazine, inject the substrate via the intravenous (i.v.) (tail-vein) route and image immediately (*see* **Notes 20–21**).

3.5.2.3. Intraperitoneal Injection of D-Luciferin

Preferred site: Lower left abdominal quadrant of the animal subject.
Needle size: 25 gauge, usually used with 1 cc syringe
Injection method: Manually restrained, abdomen side up, with cranial (head) end of animal pointed down. Needle should be bevel-side up and slightly angled when entering the abdominal cavity. Penetrate just through abdominal wall (about 4–5 mm). The tip of the needle should just penetrate the abdominal wall of the animal's left lower abdominal quadrant. Push the syringe content steadily over few seconds time.

3.5.2.4. Intravenous Injection of Coelenterazine

Preferred site: Lateral veins on animal tail.
Needle size: 30 gauge, usually used with 1 cc syringe.
Injection method: Manually restrained, under deep anesthesia place the mice on side. Injection: Tail-veins are superficially located on each side of the animal tail and are covered by tough skin. Needle should be bevel-side up and slightly angled when entering the lateral veins of tail. The tip of the needle should just penetrate the thick skin on the sides of the animal tail and then protrude the needle in parallel for few more millimeters within the vein. Push the substrate steadily over few seconds time (*see* **Note 22**).

3.5.2.5. Image Acquisition and Data Analysis Using Cooled CCD Camera

1. As per manufacturer's recommendation, initialize the IVIS system before starting image acquisition.
2. Place the animals inside the black box imaging chamber and close the door. Remember to posture the mice with the imaging surface facing the camera, e.g., if the implanted cells are on the back, place the animal exposing the dorsal side or if you are trying to image cells from lungs, place the animal on its back exposing the chest towards the camera. This is important to minimize the path length of the luminescent light through different tissues/organs.
3. Now, on the IVIS software, set different parameters, such as integration time, bin, FOV, and emission filter. Integration time can range from 10 s to 5 min

depending upon the cellular luciferase expression. Bin is generally set medium, which balances between sensitivity and resolution of the CCD camera. Depending upon number of mice scanned at a time, the FOV can be set at higher or lower range. For example to accommodate about five mice, set FOV at 25 cm^2, or for single mice, it can be set at 10 cm^2.

4. Acquire images by clicking the "acquire image" button on the displayed screen. At first, the camera acquires a photographic view of the subject followed by luminescence acquisition from the subject for the set period of time. After the acquisition is over, a superimposed image of the photographic view and the pseudocolor luminescence image will appear on the screen.
5. Save the image data in preferred locations. For data analysis, draw ROI by clicking on the "create ROI" button and then dragging and dropping the ROI on the target site of the displayed image. Now, click on the measure button and the measured photon values will be displayed in a new window. You can export this data to the Microsoft Excel program for further use and statistical analysis.

3.5.3. Indirect Monitoring of Therapeutic Gene Expression with Fluorescence Imaging

Despite the overabundance of visualization techniques, fluorescence imaging is emerging as an important alternative because of its operational simplicity, safety, and cost-effectiveness. Fluorescence imaging has recently become particularly interesting because of the advancements in fluorescent probe technology, including targeted fluorochromes as well as fluorescent "switches" sensitive to specific biochemical events. While past biological investigations using fluorescence have focused on microscopic examination of ex vivo, in vitro, or intravital specimens, techniques for macroscopic fluorescence imaging are now emerging for in vivo molecular imaging applications. Because photon penetration into living tissue is highly dependent on the absorption and scattering properties of tissue components, red and far-red shifted fluorescent reporters are preferred for animal imaging purpose ***(37)***. For lentiviral applications harboring fluorescent reporter genes, a fluorescent imaging system can be utilized to monitor reporter gene expression noninvasively from intact animals. In our facility, we use either IVIS200 or CRI Maestro system. The following protocol is for the Maestro imaging system.

3.5.3.1. Animal Anesthesia

Same as **Subheading 3.5.2.1**.

3.5.3.2. Image Acquisition

1. Load the Maestro system with the fluorophore-specific excitation and emission filter sets before starting image acquisition (refer to the Filter Selection

Guide in the *Maestro User's Manual*). Adjust the focal length manually if necessary.

2. Place the animals inside the imaging chamber and close the door. Remember to posture the mice with the imaging surface facing the camera, e.g., if the implanted cells are on the back, place the animal exposing the dorsal side or if you are trying to image cells from lungs, place the animal on its back exposing the chest towards the camera. This is important to minimize the path length of the excitation and emission light through different tissues/organs.
3. On the Maestro image acquisition software, select the "Acquire" tab. Click the "Live" button to see live streaming video of the image.
4. Set different parameters, such as integration time, excitation, and emission range. Select "Binning" and "Region-of-Interest (ROI)" options according to the desired image quality. The default is full-field and 2 × 2 binning. Image files will be smaller with smaller ROI and increased binning. Binning increases sensitivity to light but reduces image resolution.
5. In the "Wavelength and Exposure" box, enter the wavelength at which you expect to see an image of your specimen. This should correspond to the installed emission filter. Following acquisition, determine background signal by drawing ROI.
6. Open the door and focus the lens to get a sharp image of your specimen (*see* **Note 23**).
7. Close the imaging module door. "Autoexpose" the image again if you had to change the "Exposure" setting while focusing.
8. In the "Cube Wavelength Selection" box, select the "Preset Filter Setting" that corresponds to the installed filters. If you do not find a preset that works for your specimen, you can edit the Start, Step, and End settings manually. Refer to the *Maestro User's Manual* for instructions.
9. Switch off all interior lights and open the excitation light shutter to illuminate the specimen with the excitation light source.
10. Click the "Autoexpose Cube" button to automatically calculate the exposure settings for the image cube. Adjust the exposure time manually; if the image is too dark, increase the exposure time or if the image is too bright or saturated (indicated by solid red pixels), reduce the exposure time in the "Time (milliseconds)" box.
11. Click the "Acquire Cube" button to generate an image cube.
12. After Maestro has finished generating the image cube, a color representation of the cube displays in the image viewing area.
13. Click the "Save Cube" button on the toolbar to save the cube. Select a location and enter a file name (file name format suggestion: *project_sample_operator_date_time*).
14. Select a cube type option: *(1)* "Image Cubes" saves the cube in CRi format, which includes hardware and display settings for the cube. *(2)* "TIFF Cubes" saves the cube as a series of TIFF images with the assigned file name plus an appended number indicating the wavelength for each image in the cube.

3.5.3.3. Data Processing

1. In the Maestro application, select the "Spectral Processing" tab.
2. Open or acquire an image cube. To do this, click the "Load Cube" button on the toolbar. Cube format types include CRi format (*.im3) cubes and TIFF (*.tif) cubes. An RGB color representation of the cube will display in the Maestro image viewing area.
3. If desired, select a spectral display option from the Scaling combo box.
4. Click the "Real Component Analysis (RCA)" button near the bottom of the "Spectral Processing" panel. The RCA function automatically detects and separates each of the component signals in the cube (*see* **Note 24**).
5. In the "RCA" dialog box, you have the option of manually sampling the autofluorescence. Click the "Sample Spectrum" button and use the mouse pointer to draw a sampling line on an area of the specimen that contains only autofluorescence.
6. Click the "Find Component Images" button. Maestro identifies and displays the autofluorescence and fluorescence signals.
7. Select the signals to unmix. The autofluorescence is displayed in the upper left image. Click two times anywhere in the image to select it as the autofluorescence. Set its pseudocolor to white. Single-click on the other component image(s) to unmix, and select a pseudocolor for each.
8. Extract the pure fluorescence signal(s) from the autofluorescence by clicking the "Find spectra" button. The component signals display in the spectral graph. Occasionally, for weak signals, there may be too much baseline offset in the data. If you did not get the desired unmixing results, by selecting the "Fit Offset" option you can try to improve it.
9. Use the controls in the "Computed Spectra" box to specify each spectrum's name, pseudocolor, and location within the spectral library.
10. When finished, click the "Transfer to Library, Unmix and Close" button. This adds the spectra to the spectral library and unmixes the spectra in the cube.
11. New images will display in the image viewing area. There is a small component image for each selected spectrum, each with a colored border that corresponds to the pseudocolor of the spectrum. There is also a larger pseudocolored composite image displayed beside the original RGB image of the cube. The colors used to create this image are library pseudocolors. The result should be a clear display of the specific fluorescence signal, with obvious differentiation from autofluorescence (**Fig. 4**).
12. The resulting images can be saved. Also, save the protocol and/or spectral library for use throughout your experiment using either "Save Protocol" or "Save Spectral Library" from the File menu of the software.

3.6. Discussion

The field of noninvasive molecular imaging has been identified as a new discipline only within the past decade although it has its roots in Nuclear

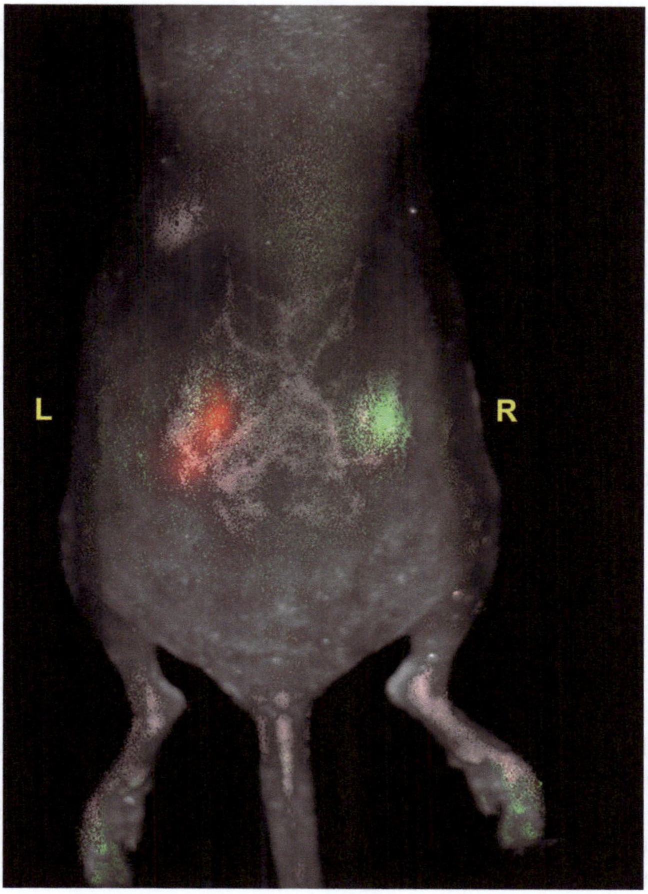

Fig. 4. Imaging fluorescent reporter gene expression from nude mice implanted with A375M human melanoma cells expressing RFP and GFP fluorescent reporters. Cells were transduced either by using lentivector carrying tri-fusion reporter (CS-CMV-*sr39tk-RFP-Rluc*) on L site or with a bi-fusion reporter (CS-CMV-*Fluc-GFP*) on R site. The image acquisition was performed using CRI Maestro fluorescent spectral imager allowing RFP and GFP fluorescent signal resolved spectrally from the same mouse.

Medicine, which is a much older field. The opportunities for molecular imaging research with animal models where animals bearing transduced xenografts and orthotopic and metastatic tumors, and animals transduced with viral vectors bearing specific reporter constructs currently provide the main resource for in vivo molecular imaging research. Among the various viral vectors in use, lentiviral-based gene transfer provides effective means for the delivery, integration, and long-term expression of transgenes in mammalian cells in culture as well as in vivo *(38)*. Studies have already shown that lentiviral vectors can be altered for delivery of multiple reporter genes *(39)*, linking a reporter gene with a therapy targeted gene *(4)*, directing transgene expression in specific tissue types *(40,41)* or even knocking down specific genes by RNA interference (RNAi) method *(42,43)*. With all these different modifications possible, lentiviruses are likely to play a greater role in gene therapy applications in future. In the context of reporter gene imaging, the approach is initially limited in patients due to the necessity of transducing target tissue with specific reporter constructs. However, the approval process for reporter systems may ease somewhat once a number of reporter gene–reporter probe combinations become established for clinical use. As safe and high-titer lentiviral vectors continue to develop, efficient delivery of transgenes into target cells, irrespective of their dividing state, will likely become more frequent, and the application of these vectors in different gene therapy trials including cancer trials should grow *(44–46)*.

4. Notes

1. There are various instruments available from different manufacturer's in each category and one can use any as per availability. However, the sensitivity and resolution of each instrument can vary as per manufacturer specifications. The examples stated here are entirely based on the instruments available and used in our small animal imaging facility.
2. Other Feline Immuno-deficiency Virus (FIV)- or Equine Infectious Anemia Virus (EIAV)-based expression vectors also work well. On the basis of the application, this is the vector to be modified and clone the therapeutic gene of interest and/or the reporter gene suitable for the right imaging instrument in use. The transgene vector construct contains several optimized genetic elements in order to fulfill biosafety criteria and to increase its transduction efficiency, such as a central polypurine tract (cPPT) to permit increased nuclear translocation efficiency of the viral pre-integration complexes (PICs) and a woodchuck hepatitis B post-transcriptional responsive element (WPRE) to permit increased messenger RNA stability, enhancing transgene expression levels.
3. PET radiotracers are high-energy positron emitters and therefore proper radioactive shielding and precautionary measures must be used to avoid unnecessary radioactive exposure while working with these materials. ^{18}F-labeled probes have short isotope half-life (^{18}F half-life is 110 min) and therefore require a cyclotron facility.
4. Both D-luciferin and coelenterazine are light sensitive and therefore the working solution should be prepared in amber tubes to protect them from light. These substrates are also susceptible to degradation and are oxidized at room temperature or higher and therefore should be stored at 4 °C for the entire duration of the work.

5. The lentivector does not contain an ATG initiation codon. Therefore, a translation initiation sequence must be incorporated in the insert cDNA if the cDNA fragment to be cloned does not already have an ATG codon. It is also recommended that a Kozak sequence (i.e., GCCACC) be included before the ATG for optimal translation.
6. For building lentivectors with tissue-specific expression, replace the internal CMV promoter with the tissue-specific promoter of choice (e.g., prostate-specific PSA) ***(40)*** or a U6/H1 promoter for RNAi applications ***(47,48)***.
7. For building bi-cistronic lentivectors, use of an encephalomyocarditis virus internal ribosomal entry site (EMCV-IRES) (Clonetech) sequence between the first and second cistron works well. The IRES element enables translation initiation within the bi-cistronic mRNA, permitting cap-dependent translation of the first cistron and cap-independent, IRES-mediated translation of the second cistron ***(49)***. The cap-dependent translation of the first cistron has been shown to be several-fold higher than cap-independent translation of the second cistron. Successful imaging studies in animals using IRES constructs have been performed and indicate that proportional co-expression can be achieved ***(4,50,51)***.
8. A fusion gene approach can also be attempted to fuse a reporter gene with the therapeutic target gene. The fusion protein is a hybrid of the two individual proteins that comes as a translational product of a single mRNA. The functional (therapeutic and reporter) characteristics of both components are retained in many cases but may require adjustments in the linker length ***(52,53)***.
9. Because of limited restriction enzyme sites in the majority of the lentivectors, these constructs can be made using any common mammalian expression plasmid (such as pcDNA3.1 (Invitrogen), validate the desired functions of the two genes and then sub-clone the cDNA fusion cassette downstream of the promoter of choice within the lentiviral vector ***(39)***.
10. Exclude polyA signal associated with the transgene while cloning in the lentivector expression vector.
11. Expression of the VSVG glycoprotein causes 293 cells to fuse, resulting in the appearance of multi-nucleated syncitia. This morphological change is normal and does not affect production of the lentivirus.
12. Because of pseudo-transduction of the reporter protein, it is recommended that the reporter assay be repeated after third passage of the transduced cells. Remember, proper care and precautionary measures should be taken while handling active virus particles. Remember that you are working with infectious virus at this stage. Follow the recommended guidelines for working with Biosafety Level-2 organisms. Working with active virus particles needs prior institutional approval under NIH and local institutional guidelines.
13. Cells can be stored frozen at this stage at –80 °C, immediately after addition of the lysis buffer.
14. It is important to keep all reaction mixture tubes on ice until the TK mix has been added to all tubes containing protein lysates, then place all the tubes in the

37 °C water bath at once. Also, take out all tubes at once and immediately put on ice. This is to ensure the time of reaction is constant for all samples.

15. Samples can be stored frozen at –20 °C overnight at this stage.
16. It is important to keep all reaction mixture tubes on ice until the TK mix has been added to all tubes containing protein lysates, then place all tubes in 37 °C water bath at once. Also, take out all tubes at once and immediately put on ice. This is to ensure time of reaction is constant for all samples ***(54)***.
17. Cells can be stored frozen at this stage at –80 °C, immediately after addition of the lysis buffer.
18. Training is required prior to working with radioisotopes. Lab coat and radiation badges and rings must be worn at all times when you are in the presence of [^{18}F]FHBG. To reduce exposure to ionizing radiation, increase your distance, as much as possible, from radiation sources and limit your exposure time to the duration necessary to complete the studies. Fluorine-18 is a positron emitter, with a half-life of 109.77 min. The collision of a positron with an electron results in the emission of two Gamma rays each with an energy of 511 KeV.
19. Deep anesthesia can be examined by tail pinch method. Make sure the stage temperature is set to 37 °C, so that the animal body temperature is maintained during the scan period.
20. A D-luciferin/coelenterazine kinetic study should be performed for each animal model to determine peak signal time after D-luciferin/coelenterazine administration. This is especially important for coelenterazine, because coelenterazine has flash kinetics and the signal peaks very quickly.
21. Up to 1 ml i.p. injection of a nonirritating solution is easily tolerated. Up to 200 μl i.v. injection of aqueous solution is easily tolerated.
22. Maintaining the animal on a 37 °C heating-pad helps i.v. injection. A successful tail-vein flushes out the red color of the vein while pushing the injection and a blood droplet oozes out after the needle is withdrawn.
23. If the live image becomes saturated with light when the door is open, it may be difficult to focus the image properly. Try auto-exposing the image again with the door open.
24. To unmix the cube by manually identifying the component signals, populate the color library with spectra and click the Manual Compute Spectra button. Refer to the *Maestro User's Manual* for instructions.

Acknowledgments

We are thankful to Manish Patel for helping us revising the protocols.

References

1. Massoud TF, Gambhir SS. (2003) Molecular imaging in living subjects: seeing fundamental biological processes in a new light. *Genes Dev* **17**, 545–580.

2. Miyagawa T, Oku T, Uehara H, et al. (1998) "Facilitated" amino acid transport is upregulated in brain tumors. *J Cereb Blood Flow Metab* **18**, 500–509.
3. Doubrovin M, Ponomarev V, Blasberg RG. (2004) PET-based reporter gene imaging. Assessment of endogenous molecular-genetic events. *IEEE Eng Med Biol Mag* **23**, 38–50.
4. De A, Lewis XZ, Gambhir SS. (2003) Noninvasive imaging of lentiviral-mediated reporter gene expression in living mice. *Mol Ther* **7**, 681–691.
5. Wang Y, Iyer M, Annala AJ, Chappell S, Mauro V, Gambhir SS. (2005) Noninvasive monitoring of target gene expression by imaging reporter gene expression in living animals using improved bicistronic vectors. *J Nucl Med* **46**, 667–674.
6. Alauddin MM, Conti PS. (1998) Synthesis and preliminary evaluation of 9-(4-[18F]-fluoro-3-hydroxymethylbutyl)guanine ([18F]FHBG): a new potential imaging agent for viral infection and gene therapy using PET. *Nucl Med Biol* **25**, 175–180.
7. Yaghoubi S, Barrio JR, Dahlbom M, et al. (2001) Human pharmacokinetic and dosimetry studies of [(18)F]FHBG: a reporter probe for imaging herpes simplex virus type-1 thymidine kinase reporter gene expression. *J Nucl Med* **42**, 1225–1234.
8. Gambhir SS, Barrio JR, Herschman HR, Phelps ME. (1999) Assays for noninvasive imaging of reporter gene expression. *Nucl Med Biol* **26**, 481–490.
9. Liang Q, Satyamurthy N, Barrio JR, et al. (2001) Noninvasive, quantitative imaging in living animals of a mutant dopamine D2 receptor reporter gene in which ligand binding is uncoupled from signal transduction. *Gene Ther* **8**, 1490–1498.
10. Boland A, Ricard M, Opolon P, et al. (2000) Adenovirus-mediated transfer of the thyroid sodium/iodide symporter gene into tumors for a targeted radiotherapy. *Cancer Res* **60**, 3484–3492.
11. Groot-Wassink T, Aboagye EO, Wang Y, Lemoine NR, Reader AJ, Vassaux G. (2004) Quantitative imaging of Na/I symporter transgene expression using positron emission tomography in the living animal. *Mol Ther* **9**, 436–442.
12. Ray P, Bauer E, Iyer M, et al. (2001) Monitoring gene therapy with reporter gene imaging. *Semin Nucl Med* **31**, 312–320.
13. Kafri T, van Praag H, Gage FH, Verma IM. (2000) Lentiviral vectors: regulated gene expression. *Mol Ther* **1**, 516–521.
14. Naldini L. (2006) Inserting optimism into gene therapy. *Nat Med* **12**, 386–388.
15. Naldini L, Blomer U, Gallay P, et al. (1996) In vivo gene delivery and stable transduction of nondividing cells by a lentiviral vector. *Science* **272**, 263–267.
16. Hamaguchi I, Woods NB, Panagopoulos I, et al. (2000) Lentivirus vector gene expression during ES cell-derived hematopoietic development in vitro. *J Virol* **74**, 10778–10784.
17. Kosaka Y, Kobayashi N, Fukazawa T, et al. (2004) Lentivirus-based gene delivery in mouse embryonic stem cells. *Artif Organs* **28**, 271–277.
18. Oka M, Chang LJ, Costantini F, Terada N. (2006) Lentiviral vector-mediated gene transfer in embryonic stem cells. *Methods Mol Biol* **329**, 273–281.

19. Woods NB, Mikkola H, Nilsson E, Olsson K, Trono D, Karlsson S. (2001) Lentiviral-mediated gene transfer into haematopoietic stem cells. *J Intern Med* **249**, 339–343.
20. Miyoshi H, Blomer U, Takahashi M, Gage FH, Verma IM. (1998) Development of a self-inactivating lentivirus vector. *J Virol* **72**, 8150–8157.
21. Zhang B, Xia HQ, Cleghorn G, Gobe G, West M, Wei MQ. (2001) A highly efficient and consistent method for harvesting large volumes of high-titre lentiviral vectors. *Gene Ther* **8**, 1745–1751.
22. Follenzi A, Sabatino G, Lombardo A, Boccaccio C, Naldini L. (2002) Efficient gene delivery and targeted expression to hepatocytes in vivo by improved lentiviral vectors. *Hum Gene Ther* **13**, 243–260.
23. Hamacher K, Coenen HH, Stocklin G. (1986) Efficient stereospecific synthesis of no-carrier-added 2-[18F]-fluoro-2-deoxy-D-glucose using aminopolyether supported nucleophilic substitution. *J Nucl Med* **27**, 235–238.
24. Sambrook J, Russell, D.W. (2001) *Molecular Cloning: A Laboratory Manual.* Cold Spring Harbor Laboratory Press, Cold Spring Harbor, New York.
25. Min JJ, Iyer M, Gambhir SS. (2003) Comparison of [18F]FHBG and [14C]FIAU for imaging of HSV1-tk reporter gene expression: adenoviral infection vs stable transfection. *Eur J Nucl Med Mol Imaging* **30**, 1547–1560.
26. Tjuvajev JG, Doubrovin M, Akhurst T, et al. (2002) Comparison of radiolabeled nucleoside probes (FIAU, FHBG, and FHPG) for PET imaging of HSV1-tk gene expression. *J Nucl Med* **43**, 1072–1083.
27. Yaghoubi SS, Couto MA, Chen CC, et al. (2006) Preclinical safety evaluation of 18F-FHBG: a PET reporter probe for imaging herpes simplex virus type 1 thymidine kinase (HSV1-tk) or mutant HSV1-sr39tk's expression. *J Nucl Med* **47**, 706–715.
28. Yaghoubi SS, Gambhir SS. (2006) PET imaging of herpes simplex Virus type 1 thymidine kinase (HSV1-tk) or mutant HSV1-sr39tk reporter gene expression in mice and humans using [18F]FHBG. *Nature Protocols* **1**, 3069–3075.
29. Chatziioannou AF. (2002) PET scanners dedicated to molecular imaging of small animal models. *Mol Imaging Biol* **4**, 47–63.
30. Chatziioannou AF. (2002) Molecular imaging of small animals with dedicated PET tomographs. *Eur J Nucl Med Mol Imaging* **29**, 98–114.
31. Phelps M. (2000) Inaugural article: positron emission tomography provides molecular imaging of biological processes. *Proc Natl Acad Sci USA* **97**, 9226–9233.
32. Phelps ME. (2000) PET: The merging of biology and imaging into molecular imaging. *J Nucl Med* **V41**, 661–681.
33. Townsend DW, Carney JP, Yap JT, Hall NC. (2004) PET/CT today and tomorrow. *J Nucl Med* **45 Suppl 1**, 4S–14S.
34. Hudson H, Larkin, RS. (1994) Accelerated image reconstruction using ordered subsets of projection data. *IEEE Trans Med Imaging* **13**, 601–609.
35. Loening AM, Gambhir SS. (2003) AMIDE: a free software tool for multimodality medical image analysis. *Mol Imaging* **2**, 131–137.

36. Contag CH, Bachmann MH. (2002) Advances in in vivo bioluminescence imaging of gene expression. *Annu Rev Biomed Eng* **4**, 235–260.
37. Frangioni JV. (2003) In vivo near-infrared fluorescence imaging. *Curr Opin Chem Biol* **7**, 626–634.
38. Verma IM, Somia N. (1997) Gene therapy—promises, problems and prospects. *Nature* **389**, 239–242.
39. Ray P, De A, Min JJ, Tsien RY, Gambhir SS. (2004) Imaging tri-fusion multimodality reporter gene expression in living subjects. *Cancer Res* **64**, 1323–1330.
40. Iyer M, Salazar FB, Lewis X, et al. (2004) Noninvasive imaging of enhanced prostate-specific gene expression using a two-step transcriptional amplification-based lentivirus vector. *Mol Ther* **10**, 545–552.
41. Iyer M, Salazar FB, Wu L, Carey M, Gambhir SS. (2006) Bioluminescence imaging of systemic tumor targeting using a prostate-specific lentiviral vector. *Hum Gene Ther* **17**, 125–132.
42. Sumimoto H, Miyagishi M, Miyoshi H, et al. (2004) Inhibition of growth and invasive ability of melanoma by inactivation of mutated BRAF with lentivirus-mediated RNA interference. *Oncogene* **23**, 6031–6039.
43. Tiscornia G, Singer O, Ikawa M, Verma IM. (2003) A general method for gene knockdown in mice by using lentiviral vectors expressing small interfering RNA. *Proc Natl Acad Sci USA* **100**, 1844–1848.
44. Firat H, Zennou V, Garcia-Pons F, et al. (2002) Use of a lentiviral flap vector for induction of CTL immunity against melanoma. Perspectives for immunotherapy. *J Gene Med* **4**, 38–45.
45. Metharom P, Ellem KAO, Schmidt C, Wei MQ. (2001) Lentiviral vector-mediated tyrosinase-related protein 2 gene transfer to dendritic cells for the therapy of melanoma. *Hum Gene Ther* **12**, 2203–2213.
46. Ohashi K, Park F, Kay MA. (2002) Role of hepatocyte direct hyperplasia in lentivirus-mediated liver transduction in vivo. *Hum Gene Ther* **13**, 653–663.
47. Ventura A, Meissner A, Dillon CP, et al. (2004) Cre-lox-regulated conditional RNA interference from transgenes. *Proc Natl Acad Sci USA* **101**, 10380–10385.
48. Xiong C, Tang DQ, Xie CQ, et al. (2005) Genetic engineering of human embryonic stem cells with lentiviral vectors. *Stem Cells Dev* **14**, 367–377.
49. Sachs AB, Sarnow P, Hentze MW. (1997) Starting at the beginning, middle, and end: translation initiation in eukaryotes. *Cell* **89**, 831–838.
50. Doubrovin M, Ponomarev V, Beresten T, et al. (2001) Imaging transcriptional regulation of p53-dependent genes with positron emission tomography in vivo. *Proc Natl Acad Sci USA* **V98**, 9300–9305.
51. Yu Y, Annala AJ, Barrio JR, et al. (2000) Quantification of target gene expression by imaging reporter gene expression in living animals. *Nat Med* **6**, 933–937.
52. Ponomarev V, Doubrovin M, Lyddane C, et al. (2001) Imaging TCR-dependent NFAT-mediated T-cell activation with positron emission tomography in vivo. *Neoplasia* **3**, 480–488.

53. Ray P, Wu AM, Gambhir SS. (2003) Optical bioluminescence and positron emission tomography imaging of a novel fusion reporter gene in tumor xenografts of living mice. *Cancer Res* **63**, 1160–1165.
54. Yaghoubi SS, Gambhir SS. (2006) Measuring herpes simplex virus thymidine kinase reporter gene expression in vitro. *Nature Protocols* **1**, 2137–2142.

12

Retroviral Modification of Mesenchymal Stem Cells for Gene Therapy of Hemophilia A

Christopher B. Doering

Summary

Mesenchymal stem cells (MSCs) are a promising target for the delivery of secreted proteins due to their ease of isolation, expansion, and genetic modification. The bleeding disorder hemophilia A results from the deficiency of a secreted blood clotting factor termed factor VIII (fVIII). Hemophilia A could be cured by gene-transfer-based procedures targeting virtually any cell type, including MSCs. Here, we describe methods for retroviral modification of MSCs incorporating a high-expression porcine (HEP)-fVIII transgene and a murine model of hemophilia A. MSCs were isolated from bone marrow of hemophilia A mice, expanded, and transduced ex vivo. Genetically modified MSCs secreted high levels of HEP-fVIII into the conditioned medium. HEP-fVIII was purified from the conditioned medium and demonstrated to have a specific activity, relative electrophoretic mobility, and proteolytic activation pattern similar to HEP-fVIII produced by other commercial cell lines. Collectively, these data support the concept that MSCs can be utilized as a cellular vehicle for successful gene-transfer-based therapy of hemophilia A and other disorders resulting from the deficiency of a secreted protein.

Key Words: Mesenchymal stem cells; marrow-derived stromal cells; gene therapy; retroviral vector; murine stem cell virus; hemophilia A; porcine factor VIII.

1. Introduction

Mesenchymal stem cells (MSCs) are an ideal cell population for the delivery of nucleic acid-based therapeutics. Properties contributing to their attractiveness include the *(1)* ease of isolation, *(2)* robust growth potential ex vivo, and *(3)*

From: *Methods in Molecular Biology, vol. 433: Volume 1: Production and In Vivo Applications of Gene Transfer Vectors*
Edited by: J. M. Le Doux © Humana Press, Totowa, NJ

amenability to genetic modification (for reviews, see **refs. *1*** and ***2***). The original description of clonal, plastic adherent cells derived from bone marrow was by Friedenstein and colleagues in the 1960–70s ***(3–5)***. They demonstrated that these cells have the capacity to not only to form fibroblast colony-forming units (F-CFUs) but also to differentiate in vitro under defined conditions into osteoblasts, chondrocytes, and adipocytes. It is of note that because MSCs have not been demonstrated in vivo, at the single cell level, to regenerate or maintain a tissue compartment and thus do not meet the strict definition of a stem cell, they are sometimes referred to as marrow-derived stromal cells, also abbreviated MSCs. During the last decade, multiple laboratories have confirmed and expanded on the findings of Friedenstein ***(6–9)***. These studies have brought significant attention to MSCs and led to the idea that MSCs can be used therapeutically to treat disorders resulting from genetic defects affecting mesenchymal lineages. MSCs now have been tested in several human clinical trials with some reported success, e.g., allogeneic MSCs cells were used in a clinical trial for the treatment of the genetic collagen disorder, osteogenesis imperfecta ***(10,11)***.

In addition to treating diseases of mesenchymal origin, MSCs also have been studied as a cellular vehicle for gene-transfer applications either designed to replace a missing protein or used in a trojan horse manner to target and eliminate cancer cells. Genetic modification of MSCs has been accomplished using a variety of methods including electroporation in addition to viral transduction using a variety of vector systems such as adeno-associated virus, adenovirus, lentivirus, and oncoretrovirus. Furthermore, several protocols for the delivery of MSCs have been described, and it has been shown that MSCs can take up residence in multiple locations within and outside of the bone marrow niche and can survive for extended periods within artificial biosupport structures following subcutaneous transplantation ***(12–17)***.

Hemophilia A is a monogenic disorder that results from deficiency of a secreted protein termed factor VIII (fVIII) that facilitates blood clotting. Because of the limited amount of fVIII (in the range of 0.01–1 nM) needed in circulation to correct the bleeding phenotype, hemophilia A has been targeted as a prime candidate disease for gene therapy applications. One strategy is to genetically modify autologous or allogeneic MSCs to biosynthesize and secrete fVIII and then transplant the modified MSCs into fVIII-deficient patients. Recently, we demonstrated that murine MSCs can be modified by retroviral transduction to express clinically-significant levels of a high-expression porcine (HEP)-fVIII construct. Here, we describe the methodology used to retrovirally modify MSCs to express therapeutic amounts of HEP-fVIII.

2. Materials

2.1. MSC Isolation, Expansion, and Characterization

1. Exon 16 fVIII-knockout mice *(19)*.
2. RPMI 1640 culture medium containing 10% heat-inactivated fetal bovine serum, 10% horse serum, 10 units/ml penicillin, and 10 μg/ml streptomycin (Gibco, Invitrogen, Carlsbad, CA).
3. Dulbecco's phosphate-buffered saline (DPBS) without magnesium or calcium chloride (Gibco).
4. TrypLE Express Stable Trypsin Replacement Enzyme (Gibco).
5. Six-well and 10-cm tissue culture dishes (Corning, Corning, NY).
6. Small animal surgical tools (scissors, curved and straight tip forceps).
7. Sterile razor blades (Fisher Scientific, Hampton, NH).
8. Three-milliliter syringes with 22-gauge needles attached.
9. Seventy-micrometer sterile filters (BD Biosciences-Clontech, Mountain View, CA).
10. Two-milliliter cryovial tubes (Corning).
11. Hemacytometer (Fisher Scientific).

2.2. Production of Recombinant MSCV-HEP-fVIII Retrovirus

1. Dulbecco's Modified Eagle's Medium (DMEM) containing 10% fetal bovine serum, 10 units/ml penicillin, and 10 μg/ml streptomycin.
2. Ecopack 293 retroviral packaging cells, BD RetroPack™ PT67 packaging cells, and murine stem cell virus (MSCV) vector plasmid (BD Biosciences-Clontech).
3. Biocoat Collagen I-coated 10-cm tissue culture dishes (BD Biosciences).
4. HiSpeed Plasmid Mini Kit (Qiagen, Valencia, CA).
5. Lipofectamine 2000 (Invitrogen).
6. Sterile polypropylene centrifugation tubes with screw-caps (maximum volume, 35 ml) (Fisher Scientific).
7. NIH 3T3 murine fibroblasts (American Type Culture Collection, Manassas, VA).
8. Polybrene (10 mg/ml) (Specialty Media, Phillipsburg, NJ).
9. QIAamp DNA Micro Kit (Qiagen).
10. Power SYBR Green PCR Master Mix and Taqman reverse transcription PCR Kit (Applied Biosystems, Foster City, CA).
11. 7500 sequence detection system (Applied Biosystems).
12. Oligonucleotide primers specific for HEP-fVIII: forward 5′-CGGAGGAAAGC TGATGAAGAGG-3′ and reverse 5′-GCAGAGATGTAGTGCACCCAGG-3′.

2.3. Transduction and Analysis of fVIII Production

1. AIM V serum-free medium (Gibco).
2. Pooled normal human plasma (FACT) and human fVIII-deficient plasma (George King Biomedical, Overland Park, KA).

3. Activated partial thromboplastin reagent (Organon Technika, Durham, NC).
4. STart 4 semi-automated hemostasis analyzer (Diagnostica Stago, Parsippany, NJ).
5. 20 mM calcium chloride.

3. Methods

3.1. MSC Isolation and Expansion

1. Tibias and femurs are isolated from donor exon 16-knockout hemophilia A mice, the associated muscle tissue is removed, and the marrow-containing bones are placed into a sterile 10-cm tissue culture dish containing 5–10 ml of DPBS.
2. Approximately 0.1 cm is removed from the end of each bone using a sterile straight-edged razor blade.
3. A 22-gauge needle attached to a 3-ml syringe containing complete RPMI medium is inserted into one end of each bone and medium is forced through the bone cavity until the marrow is expelled from the other end of the bone.
4. This process is repeated for each tibia and femur and the cells are combined, dissociated by repeated pipetting, passed through a 70-μm sterile filter, and counted using a hemacytometer.
5. Once the cell yield is determined, the cells are sedimented by centrifugation at 600 × *g*, 4 °C for 10 min and resuspended at a concentration of 5×10^6 cells/ml in complete RPMI. Six-well tissue culture plates are seeded with 3.5 ml of the cell suspension/well, i.e., 1.75×10^7 cells/well (*see* **Note 1**).
6. Once plated, the cells are incubated at 37 °C with 5% CO_2 for 72 h.
7. Once the adherent cells occupy ≥90% of the well surface (typically 72 h after plating), the conditioned medium is removed and the cells are washed three times with complete RPMI to remove non-adherent cells (*see* **Note 2**).
8. For the initial passages, the cells are split 1:2 onto new six-well culture plates and eventually 10-cm dishes. **Fig. 1** depicts a confluent culture of MSCs isolated from a C57BL/6 hemophilia A mouse.
9. This process is repeated until four to six passages are completed and the cells display a doubling time of approximately 24 h. At this point, enough cells should be available for the establishment of a master cell bank and preparation of cells for transduction (*see* **Subheading 3.4.**).
10. A master cell bank is established by resuspending approximately 10^7 cells in 10 ml complete RPMI supplemented with 7.5% dimethyl sulfoxide. One milliliter is dispersed into 10 × 2 ml cryovial tubes and slow frozen in an ethanol-insulated freezing chamber to –80°C prior to long-term storage in liquid nitrogen (*see* **Note 3**).

3.2. Production of Recombinant MSCV-HEP-fVIII Retrovirus

1. HEP-fVIII cDNA was subcloned from pfVIII/ReNeo *(20)* into the retroviral expression plasmid pMSCV as described previously *(18)*.

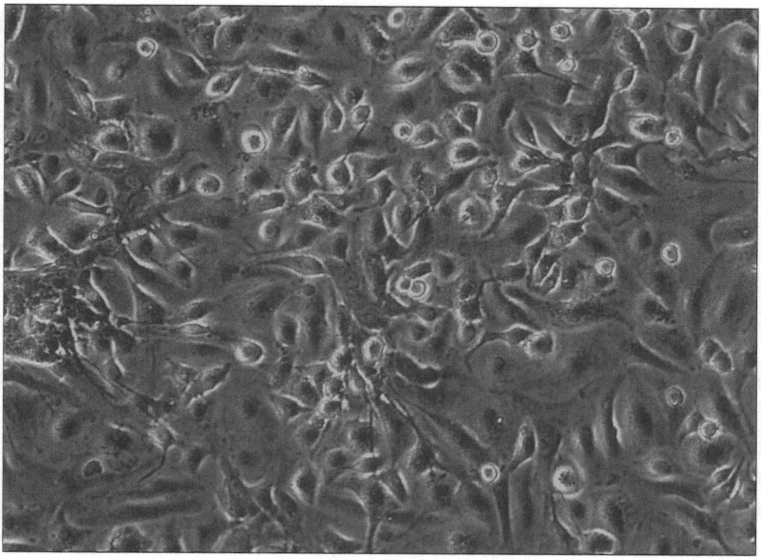

Fig. 1. Microphotographic image of early passage murine mesenchymal stem cells (MSCs).

2. Stable producer cell lines can be generated by transiently transfecting BD RetroPack™ PT67 cells with the MSCV-HEP-fVIII plasmid DNA according to the manufacturers instructions using Lipofectamine 2000 (*see* **Note 4**).
3. Conditioned medium containing MSCV-HEP-fVIII pseudotyped with dualtropic, 10A1 envelope protein is collected daily, 0.45 μm filtered, placed onto the ecotropic packaging cell line, BD EcoPack2™, and plated onto Biocoat Collagen I-coated 10-cm tissue culture dishes. Eight μg/ml polybrene is added to the transduction medium. Typically, we perform five rounds of transduction (8–16 h each). For each transduction, the previous virus-containing medium is removed and fresh viral stocks are thawed and are added to the plated cells (*see* **Note 5**).
4. Following five rounds of transduction, the transduced EcoPack2™ cells are harvested and re-plated onto 5–10 collagen-coated 10-cm culture dishes. Once the cells reach 70–80% confluence, conditioned medium is collected daily, 0.45 μm filtered, and stored at –80°C until concentration. At this time, cells are frozen back to serve as a master cell bank of stable producer cells as described in **Subheading 3.1., step 10** (*see* **Note 6**).
5. Ecotropic recombinant MSCV is concentrated 20- to 50-fold by sedimentation at 9000 × *g* for 16 h at 4°C, resuspended in complete RPMI, and stored at –80°C in 1-ml aliquots (*see* **Note 7**).

3.3. Determination of Viral Titer

The number of functional viral particles (fvp) per ml present in the viral concentrate is determined by transduction of murine 3T3 fibroblasts and subsequent analysis of the number of HEP-fVIII transgenes present per diploid-genome equivalent using quantitative polymerase chain reaction (qPCR).

1. Approximately 10^5 3T3 cells are seeded onto individual wells of a six-well tissue culture plate and cultured overnight at 37°C with 5% CO_2.
2. Sixteen to 24 h after plating, the cell number per well is determined for one of the representative wells by hemacytometer count. If there are greater than 10^5 cells/well (indicating that the cells are dividing), the cells are transduced using varying amounts of concentrated viral stock. Typically, we add 10 or 100 μl of concentrated viral stock to individual wells. The total volume per well is brought up to 1 ml with complete RPMI containing 8 μg/ml polybrene.
3. Forty-eight hours post transduction, genomic DNA is harvested from the wells using a genomic DNA isolation kit (Qiagen) and the DNA concentration of each is determined by absorbance at 260 nm using a spectrophotometer.
4. qPCR is performed on 50 ng of genomic DNA using the (2×) Power SYBR Green PCR Master Mix. The 25 μl reaction contains the following components: 1× SYBR Green Master Mix, 50 mM each primer, 50 ng genomic DNA, and sterile H_2O. The reaction conditions are as follows: a single incubation at 95°C for 10 min, 40 cycles of amplification at 95°C for 15 s then 60°C for 1 min. Post-reaction dissociation analysis is performed to confirm the amplification of a single product.
5. Calculation of the transgene copy number is achieved by interpolation from a standard curve consisting of MSCV-HEP-fVIII plasmid DNA serially diluted into untransduced, control 3T3 genomic DNA. Six pg/diploid genome equivalent is used as a conversion factor to calculate the copy number on a per cell basis.

3.4. Retroviral Transduction of MSCs

1. Prior to transduction, approximately 2–5 × 10^5 MSCs are seeded onto six-well plates and the cells are incubated for 24 h at 37°C with 5% CO_2.
2. One milliliter of concentrated virus is supplemented with 8 μg/ml polybrene and added to each well. The viral titer is typically ~10^6/ml, which results in a multiplicity of infection (MOI) of 2–5.
3. The transduction protocol is repeated two additional times at 8–24 h intervals using an identical amount of viral concentrate, which due to cell division results in a slightly lower MOI with each successive transduction.
4. Post transduction, MSCs are replated onto 5 to 10 × 10-cm tissue culture dishes and grown to 75–90% confluence.
5. Next, approximately 10^7 cells are harvested, resuspended in 10 ml complete RPMI supplemented with 7.5% dimethyl sulfoxide, and 1 ml each is dispersed into 10 × 2 ml cryovial tubes and slow frozen in an ethanol-insulated freezing chamber to –80°C then transferred to liquid nitrogen for long-term storage.
6. The remaining cells are replated and assayed for fVIII production as described in **Subheading 3.5**.

3.5. Analysis of fVIII Production from Genetically-Modified MSCs

The efficiency of fVIII expression from genetically-modified MSCs is determined by analysis of the amount of HEP-fVIII activity produced relative to the steady-state HEP-fVIII mRNA levels and average transgene copy number per cell.

1. For fVIII expression analysis, 10^5 MSCs are seeded onto individual wells of a six-well plate and cultured until each well is approximately 95% confluent.
2. Twenty-four hours prior to measurement of fVIII activity, the cells are rinsed twice with serum-free AIM V medium. Next, 1 ml of fresh medium is added to each well and the cells are returned to the incubator (*see* **Note 8**).
3. FVIII activity measurements are made using an activated partial thromboplastin reagent-based one-stage coagulation assay. Briefly, 5 μl of conditioned medium (sample) or standard (pooled normal human plasma) is added to 50 μl fVIII-deficient plasma and incubated for 4 min at 37°C, followed by addition of 50 μl of 20 mM $CaCl_2$ to initiate the clotting reaction. The time to clot formation is measured viscometrically using a semi-automated hemostasis analyzer. A standard curve is generated by performing the identical assay on serial dilutions of pooled normal human plasma spiked into fVIII-deficient plasma. The results are analyzed by interpolation from a linear regression of the clotting time versus the logarithm of the reciprocal plasma dilution.
4. Integrated HEP-fVIII transgene and mRNA levels are measured by qPCR and reverse transcription (RT)-PCR, respectively, as described previously ***(18)***. The primers and reagents are identical to those described in **Subheadings 2.2** and **3.3 (step 4)** except for the addition of 0.5 μl Multiscribe reverse transcriptase and 0.125 μl RNase inhibitor to each RT-PCR reaction and insertion of an initial 30 min incubation at 48°C to facilitate reverse transcription of total RNA into single-stranded cDNA. Five nanogram of total RNA is used in each RT-PCR reaction. Absolute HEP-fVIII mRNA copy numbers are determined by interpolation from a standard curve generated using serially diluted in vitro transcribed HEP-fVIII RNA that is assayed in the presence of 5 ng control yeast tRNA (*see* **Note 9**).

4. Notes

1. Typically, we recover 2–5 × 10^7 nucleated cells per mouse.
2. MSCs are separated from hematopoietic cells by their selective adherence to tissue culture plastic and survival following serial passage.
3. Following long-term culture, human and murine MSCs have been demonstrated to adopt a transformed phenotype ***(21,22)***. Therefore, it is important to minimize the duration of ex vivo culture prior to in vivo administration of genetically modified MSCs.
4. BD RetroPack™ PT67 cells stably express *gag-pol* and the dualtropic 10A1 envelope protein.

5. We have observed that fVIII production increases with increasing rounds of transduction until three to five rounds of transduction have been performed at which point no further increase in HEP-fVIII production is observed.
6. Because retroviral transduction results in stable integration of the MSCV-HEP-fVIII proviral genome into the EcoPack2™ cells, the producer cells generated using this protocol stably biosynthesize ecotropic-pseudotyped MSCV-HEP-fVIII.
7. Using the protocol described in **Subheading 3.2.**, we routinely generate stable ecotropic retroviral producer cell lines that produce viral stocks containing 10^5 to 10^6 fvp/ml following concentration.
8. The use of serum-free medium yields more precise and likely accurate fVIII activity measurements due to the absence of components present in fetal bovine serum that can affect fVIII activity measurement using a one-stage activated-partial thromboplastin time-based coagulation assay.
9. In a previous study, we determined that genetically modified MSCs generated using the protocol described here produced HEP-fVIII at a rate of 14 units/10^6 cells/24 h, contained mean fVIII mRNA levels of ~700 transcripts/cell, and harbored ~1.5 HEP-fVIII transgenes per diploid genome equivalent. This fVIII expression efficiency is similar to that observed previously from baby hamster kidney-derived cells *(20)*.
10. HEP-fVIII contains a domain structure designated A1-A2-ap-A3-C1-C2. The majority of MSC biosynthesized HEP-fVIII is processed into fVIII heavy (A1-A2) and light (ap-A3-C1-C2) chains and is proteolytically activated by thrombin into an A1/A2/A3-C1-C2 heterotrimer (*see* **Fig. 2**). HEP-fVIII was purified and analyzed by sodium dodecyl sulfate–polyacrylamide gel electrophoresis (SDS–PAGE) as described previously *(20)*.

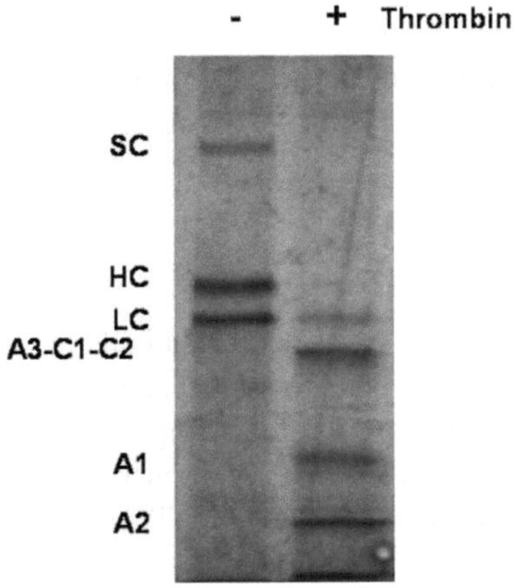

Fig. 2. Sodium dodecyl sulfate–polyacrylamide gel electrophoresis (SDS–PAGE) analysis of purified high-expression porcine (HEP)-FVIII biosynthesized by genetically-modified mesenchymal stem cells (MSCs) (*see* **Note 10**).

Acknowledgments

This work was supported by grants from the Emory University Research Committee, the Woodruff Health Sciences Center Woodruff Fund, the Emory/Egelston Children's Research Center, and the Gene Therapy Initiative at Children's Healthcare of Atlanta.

References

1. Prockop, D.J. Marrow stromal cells as stem cells for nonhematopoietic tissues. *Science* **276**, 71–4 (1997).
2. Devine, S.M. Mesenchymal stem cells: will they have a role in the clinic? *J Cell Biochem Suppl* **38**, 73–9 (2002).
3. Friedenstein, A.J., Chailakhjan, R.K. and Lalykina, K.S. The development of fibroblast colonies in monolayer cultures of guinea-pig bone marrow and spleen cells. *Cell Tissue Kinet* **3**, 393–403 (1970).
4. Friedenstein, A.J., Gorskaja, J.F. and Kulagina, N.N. Fibroblast precursors in normal and irradiated mouse hematopoietic organs. *Exp Hematol* **4**, 267–74 (1976).
5. Friedenstein, A.J., Petrakova, K.V., Kurolesova, A.I. and Frolova, G.P. Heterotopic of bone marrow. Analysis of precursor cells for osteogenic and hematopoietic tissues. *Transplantation* **6**, 230–47 (1968).
6. Pittenger, M.F. et al. Multilineage potential of adult human mesenchymal stem cells. *Science* **284**, 143–7 (1999).
7. Meirelles Lda, S. and Nardi, N.B. Murine marrow-derived mesenchymal stem cell: isolation, in vitro expansion, and characterization. *Br J Haematol* **123**, 702–11 (2003).
8. Jiang, Y. et al. Pluripotency of mesenchymal stem cells derived from adult marrow. *Nature* **418**, 41–9 (2002).
9. Peister, A., Mellad, J.A., Larson, B.L., Hall, B.M., Gibson, L.F. and Prockop, D.J. Adult stem cells from bone marrow (MSCs) isolated from different strains of inbred mice vary in surface epitopes, rates of proliferation, and differentiation potential. *Blood* **103**, 1662–8 (2004).
10. Horwitz, E.M. et al. Transplantability and therapeutic effects of bone marrow-derived mesenchymal cells in children with osteogenesis imperfecta. *Nat Med* **5**, 309–13 (1999).
11. Horwitz, E.M. et al. Clinical responses to bone marrow transplantation in children with severe osteogenesis imperfecta. *Blood* **97**, 1227–31 (2001).
12. Ding, L., Lu, S., Batchu, R., Iii, R.S. and Munshi, N. Bone marrow stromal cells as a vehicle for gene transfer. *Gene Ther* **6**, 1611–6 (1999).
13. Gao, J., Dennis, J.E., Muzic, R.F., Lundberg, M. and Caplan, A.I. The dynamic in vivo distribution of bone marrow-derived mesenchymal stem cells after infusion. *Cells Tissues Organs* **169**, 12–20 (2001).
14. Sato, Y. et al. Human mesenchymal stem cells xenografted directly to rat liver differentiated into human hepatocytes without fusion. *Blood* **106**, 756–63 (2005).
15. Eliopoulos, N., Al-Khaldi, A., Crosato, M., Lachapelle, K. and Galipeau, J. A neovascularized organoid derived from retrovirally engineered bone marrow stroma

leads to prolonged in vivo systemic delivery of erythropoietin in nonmyeloablated, immunocompetent mice. *Gene Ther* **10**, 478–89 (2003).
16. Eliopoulos, N., Gagnon, R.F., Francois, M. and Galipeau, J. Erythropoietin delivery by genetically engineered bone marrow stromal cells for correction of anemia in mice with chronic renal failure. *J Am Soc Nephrol* **17**, 1576–84 (2006).
17. Eliopoulos, N., Lejeune, L., Martineau, D. and Galipeau, J. Human-compatible collagen matrix for prolonged and reversible systemic delivery of erythropoietin in mice from gene-modified marrow stromal cells. *Mol Ther* **10**, 741–8 (2004).
18. Gangadharan, B., Parker, E.T., Ide, L.M., Spencer, H.T. and Doering, C.B. High-level expression of porcine factor VIII from genetically modified bone marrow-derived stem cells. *Blood* **107**, 3859–64 (2006).
19. Bi, L., Lawler, A.M., Antonarakis, S.E., High, K.A., Gearhart, J.D. and Kazazian, H.H. Targeted disruption of the mouse factor VIII gene produces a model of haemophilia A. *Nat Genet* **10**, 119–21 (1995).
20. Doering, C.B., Healey, J.F., Parker, E.T., Barrow, R.T. and Lollar, P. High-level expression of recombinant porcine coagulation factor VIII. *J Biol Chem* **277**, 38345–9 (2002).
21. Miura, M. et al. Accumulated chromosomal instability in murine bone marrow mesenchymal stem cells leads to malignant transformation. *Stem Cells* **24**, 1095–103 (2006).
22. Rubio, D., Garcia-Castro, J., Martín, M.C., de la Fuente, R., Cigudosa, J.C., Lloyd, A.C. and Bernad, A. Spontaneous human adult stem cell transformation. *Cancer Res* **65**, 3035–9 (2005).

13

Transduction of Murine Hematopoietic Stem Cells and In Vivo Selection of Gene-Modified Cells

Lucienne M. Ide, Elisabeth Javazon, and H. Trent Spencer

Summary

Hematopoietic stem cells (HSCs) were among the first targets of genetic manipulation for the purpose of treating human diseases. The translational aspects of the first human clinical trials were based on results obtained using the mouse as an experimental model. Murine studies have shown that the major limitations of HSC gene therapy are similar to those encountered when using non-hematopoietic cells as targets and include *(1)* an inability to genetically modify sufficient numbers of target cells, *(2)* the loss of transgene function over time, and *(3)* potential complications due to vector integration. With continued improvements in transduction protocols, murine HSC transduction and transplantation are now routine with transduction efficiencies >50% easily achievable and even >90% feasible. However, attaining high-level engraftment of gene-modified cells after transplantation is still problematic. Basic transduction conditions entail cytokine stimulation of HSC populations, such as stem cell antigen-1 positive (Sca-1^+) cells isolated from bone marrow, in serum-free media followed by multiple additions of recombinant retrovirus. Analysis of peripheral blood 12 weeks post transplantation of transduced cells into lethally irradiated recipients shows genetic marking in all hematopoietic lineages. Transduction of HSCs is then confirmed by transplanting bone marrow cells harvested from primary transplant recipients into lethally irradiated secondary recipients. Analysis of these mice shows that recombinant retroviruses transduce murine HSCs efficiently and stably and that the genetically modified cells are capable of completely repopulating the hematopoietic system.

Key Words: Hematopoietic stem cell; murine bone marrow transduction; recombinant retrovirus; bone marrow transplant.

From: *Methods in Molecular Biology, vol. 433: Volume 1: Production and In Vivo Applications of Gene Transfer Vectors*
Edited by: J. M. Le Doux © Humana Press, Totowa, NJ

1. Introduction

Hematopoietic stem cells (HSCs) provide a renewable source of hematopoietic cells, including lymphocytes, myelocytes, and reticulocytes. A bone marrow transplant is the process of harvesting bone marrow cells, including HSCs, from a donor and transplanting them into a properly conditioned recipient. The ability of HSCs to repopulate the entire hematopoietic system coupled with their ease of isolation and ex vivo culture makes them ideal targets for genetic manipulation. Transduction of HSCs can result in transgene expression in mature, differentiated cells of all hematopoietic lineages. Thereby, transplantation of genetically modified HSCs allows for production of a novel gene product by many different hematopoietic cell types.

HSCs are concentrated in skeletal long bones, such as femora, tibiae, and humeri, and are mostly quiescent with respect to cell division, meaning they rarely divide. This characteristic of HSCs is problematic for using retroviruses as gene-delivery vehicles as the most frequently used oncogenic retroviruses (such as MSCV-based retroviruses) require cell division in order to successfully transduce cells *(1,2)*. Therefore, cytokine stimulation is used to induce cell division, increasing the efficiency of gene transfer, yet cytokine stimulation can induce HSC differentiation, resulting in a diminished ability of these cells to repopulate the recipient *(3–6)*.

Genetically engineered HSCs have proven beneficial for the treatment of some diseases, most notably immunodeficiencies *(7,8)* [see **refs.** *9* and *10* for a review on the use of gene-modified HSC to treat adenosine deaminase (ADA)- and X-linked severe combined immunodeficiency (SCID) diseases]. Although there have been some successes using genetically modified HSCs in humans, it is now clear that a potential complication of recombinant retroviruses is alteration of endogenous gene function at the vicinity of viral nucleic acid integration *(11,12)*. This disruption is termed insertional mutagenesis. Therefore, the foremost issue facing the field of HSC gene therapy at this time is that the most efficient method for introducing nucleic acid sequences into target cells has the real potential of adversely affecting the phenotype of the genetically altered cell. This was recently evidenced by Dr. Fischer's X-linked SCID trial in which nucleic acid sequences transferred by recombinant retroviruses integrated into oncogenic sites, leading to leukemia in three children. In this trial, gene transfer was accomplished using recombinant oncogenic-based retroviruses. It is now known that these recombinant viruses preferentially deliver genes to areas of active transcription *(13,14)*. For the children who developed leukemia, the transferred gene integrated near the promoter of the LMO-2 proto-oncogene and disrupted LMO-2 expression. Until this study, the issues of insertional mutagenesis were only theoretical and were thought to be highly unlikely, based on the erroneous idea that retroviruses integrated randomly into the genome. Therefore, investigations focusing on insertion-site analysis were limited. It is now clear that under certain situations recombinant viruses can contribute to oncogenic changes in a cell if they insert into the genome in the proximity of a gene that regulates growth.

In the past 2 years, there has been tremendous effort describing the mechanisms involved in HSC transformation following genetic manipulation, and the mechanisms are now well understood. We now know that for retroviral gene transfer of HSCs to be broadly applicable for the treatment of non-lethal diseases, we must more precisely regulate the introduction of nucleic acid sequences into target cells. The mouse will likely be the best model for studying these types of issues. We can begin to refine transduction protocols to provide safer approaches for genetically manipulating target cells. For example, using standard transduction conditions, it is predicted that fewer integration events will occur when using isolated populations of HSCs, as shown in **Fig. 1**. This "debulking" procedure decreases the total number of integration events, decreasing the probability of generating insertion sites that lead to mutagenic transformations while retaining high gene transfer efficiency in HSCs.

Several additional strategies can be utilized to improve viral vector safety and reduce the possibility of insertional mutagenesis including self-inactivating (SIN) vectors, insulator sequences, and choice of viral vector. SIN vectors

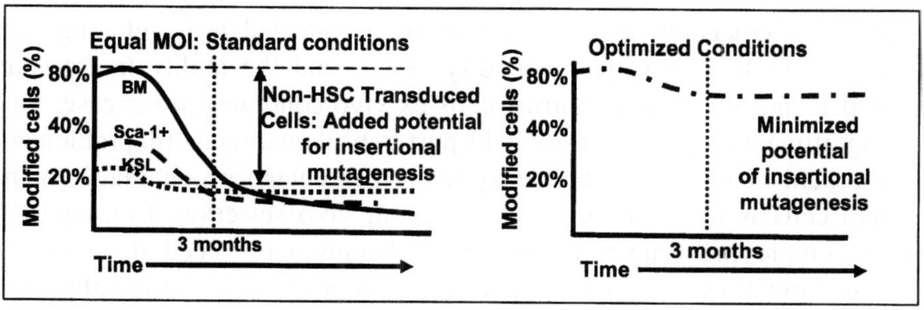

Fig. 1. Effect of optimizing gene transfer into purified hematopoietic stem cells (HSCs). Bone marrow (BM) contains HSCs, progenitor cells, and lineage-committed cells. Isolating Sca-1^+ cells or Kit^+, Sca-1^+, lineage negative (KSL) cells increases the percentage of HSCs by removing the more differentiated populations. By targeting "purer" populations of HSCs, greater virus-to-cell ratios can be achieved. But more importantly, by eliminating the transduction of non-HSCs, the chances of insertional mutagenesis are drastically decreased. As shown in the right-hand panel, as conditions are optimized for isolated populations of HSCs, the percentage of circulating modified cells, i.e., engrafted gene-modified cells, will increase. Therefore, the purpose of using isolated populations of HSCs is twofold *(1)* to "debulk" the number of target cells so increased MOIs can be achieved and *(2)* to decrease the chances of insertional mutagenesis.

decrease the possibility that expression of cellular genes adjacent to the vector integration site will be upregulated by disrupting the promoter activity of the viral long terminal repeats (LTRs) after entry into the cell. This strategy takes advantage of the fact that the U3 regions of both LTRs are generated from the U3 region of the 3′ LTR during reverse transcription. Therefore, a deletion in this region leads to elimination of transcriptional activity from the LTR. This also decreases the possibility of competition between the internal promoter (which is necessary in these vectors) and the LTRs. Insulator sequences are thought to maintain chromatin structure of transcriptionally active regions, isolating them from the effects of transcriptional activity of surrounding chromatin. They can be employed to prevent transcriptional activation of regions downstream of the viral integration site or to protect a transgene from positional effects when using vectors encoding multiple transgenes. Lastly, certain viral vectors such as lentiviruses (e.g., HIV-1 recombinant viruses) are thought to be less "genotoxic" than the oncoretroviral vectors most commonly used in the past. This can be attributed to the tendency of HIV-based vectors to integrate within transcriptionally active genes as opposed to oncoretroviruses, which tend to integrate in the promoter/enhancer region. Although lentiviral vector integration may disrupt cellular gene expression, diploid host cells will still have one intact copy of the gene, and therefore should not suffer from a loss of function mutation (for a more comprehensive review of these issues, see **ref.** ***(15)***).

Although tremendous effort continues to be directed at developing more efficient methods to genetically modify HSCs, and the ability to transduce greater percentages of bone marrow cells ex vivo continues to increase, long-term engraftment of gene-modified cells post transplantation is still problematic. One mechanism of increasing the systemic percentage of circulating gene-modified cells is to incorporate the use of in vivo selection. Consequently, methods employing drug-resistance gene therapy have been developed to overcome limitations of gene transfer into hematopoietic cells. Originally, drug-resistance gene therapy studies were designed to decrease the myelosuppressive side effects encountered during administration of chemotherapy. It was hoped that by genetically modifying blood cells with genes that confer drug resistance, one could lessen the toxic side effects of chemotherapy, and to date, the majority of drug-resistance gene therapy applications have been designed for this purpose. In addition, genes that confer drug resistance can also function as dominant selectable markers for the purpose of enriching genetically modified cells ***(16)***. Hematopoietic cells have been enriched in vivo by gene transfer of the cDNAs that encode P-glycoprotein, dihydrofolate reductase (DHFR), and O^6-alkylguanine DNA alkyltransferase (AGT). This procedure has provided a means to circumvent low transduction efficiencies ***(17–22)***. Currently, in vivo selection systems that use variants of AGT, which are encoded by the methylguanine-DNA-methyltransferase (MGMT) cDNA, are arguably the most efficient model. Retroviral vectors encoding selectable markers such as variants of MGMT and a therapeutic gene can be created so that genetically engineered cells can be preferentially expanded in vivo compared with non-modified cells, thereby increasing the percentage of cells expressing the therapeutic protein.

MGMT encodes the DNA repair protein AGT, which removes alkyl lesions from the O^6 position of guanine of DNA. Methylating and chloroethylating agents, such as 1,3-bis(2-chloroethyl)-nitrosourea (BCNU) and temozolomide, result in DNA alkylation and multiple DNA strand breaks and are thus cytotoxic. However, cells expressing MGMT are resistant to these agents *(23)*. Many cells endogenously express AGT, and tumor cells have been shown to resist alkylating agents by increasing expression of AGT. HSCs appear to express low levels of AGT making them more sensitive to the toxicity of alkylating agents compared to other tissues. Therefore, hematopoietic toxicity is the dose-limiting side effect for many anti-neoplastic alkylating agents. O^6-benzylguanine (6-BG) is an inhibitor of AGT and has been used to sensitize cells to some methylating agents, including BCNU. It was predicted that if HSCs could be genetically modified to express 6-BG-resistant variants of AGT, then gene-modified cells should be resistant to the combination treatment of 6-BG/BCNU, and non-modified cells should be selectively killed. Indeed, it has been shown, both in vitro and in vivo, that the percentage of cells modified with cDNA sequences encoding drug-resistant variants of AGT are more resistant to alkylating agents compared with non-modified cells *(23–25)*, and MGMT-gene-modified cells can be preferentially expanded following 6-BG and BCNU treatment. Typically, the first round of selection with 6-BG and BCNU results in a modest increase in transduced cells; however, nearly 100% transduction has been achieved with just two to three rounds of selection *(21,25,26)*. For example, T cell cultures and murine hematopoietic progenitor cells with only 3% transduction efficiency were selected to approximately 100% and 70%, respectively, following treatment with 6-BG and BCNU *(25)*. Successful expansion of gene-modified cells has been achieved in human, murine, and canine cells demonstrating the versatility and potential for this gene therapy strategy.

2. Materials

2.1. Viral Production

1. EcoPack-293 cell line (BD Biosciences Clontech, Mountain View, CA).
2. Dulbecco's Modified Eagle's Medium (DMEM; Cellgro/Mediatech, Herndon, VA) supplemented with 10% fetal bovine serum (FBS; Cellgro/Mediatech) and 100 IU/ml penicillin/100 μg/ml streptomycin (Cellgro/Mediatech).

3. Stempro-34 Serum-Free Expansion Media (Invitrogen, Carlsbad, CA).
4. 0.45-μm syringe filter (Pall Life Sciences, Ann Arbor, MI), 60-ml syringe (BD Biosciences, Bedford, MA).
5. Sterile centrifuge tubes.
6. Centrifuge capable of spinning appropriate tubes at 13,800 × *g*.
7. Biocoat collagen-coated cell culture dishes, 100 × 20 mm (BD Biosciences).

2.2. Stem Cell Isolation

1. Buffer: phosphate-buffered saline (PBS) with 1 mM EDTA, calcium and magnesium free (Invitrogen) supplemented with 2% FBS (Cellgro/Mediatech) and 100 IU/ml penicillin/100 μg/ml streptomycin (Cellgro/Mediatech). Keep cold (4–8°C).
2. Stempro-34 Serum-Free Expansion Media reconstituted with provided nutrient supplement (Invitrogen) and L-Glutamine (200 mM, Invitrogen).
3. Human interleukin (IL)-11, mouse stem cell factor (mSCF), mouse IL-3, mouse Flt3 Ligand (R and D Systems, Minneapolis, MN).
4. 40-μm nylon cell strainer (BD Biosciences).
5. Anti-Sca-1 microbeads (Miltenyi Biotec, Auburn, CA).
6. Miltenyi MS+ columns (Miltenyi Biotec).
7. Magnet configured to accept Miltenyi columns.
8. 23-gauge needle (Becton Dickenson, Franklin Lakes, NJ).
9. 3-ml syringe (Becton Dickenson).

2.3. Transduction

1. Retronectin (Takara Bio, Otsu, Shiga, Japan).
2. 2% Bovine Serum Albumin (BSA), sterile (Sigma-Aldrich, St. Louis, MO).
3. Hanks balanced salt solution (Sigma-Aldrich), with 2.5% 1 M HEPES (Cellgro/Mediatech).
4. Non-tissue culture-treated six-well plates (Beckton Dickinson, Franklin Lakes, NJ).

2.4. Transplantation

1. PBS with 1 mM EDTA, Ca^{2+} and Mg^{2+} free (Invitrogen).
2. Isoflourane for inhalation (Hospira Inc., Lake Forest, IL).
3. 1-ml insulin syringe (Beckton Dickinson).
4. Sterile disposable cell scraper (Fisher Scientific, Hampton, NH).
5. Either total body irradiation (TBI) or chemotherapy can be used for pretransplantation conditioning. Eleven Gy TBI from a cesium-137 gamma radiation source, given as a split dose 4 h apart, constitutes lethal irradiation and should allow for >90% engraftment of donor cells whereas 5.5 cGy permits 50–60% engraftment of donor cells. There are several chemotherapy alternatives to irradiation, busulfan being one of the most commonly used drugs.

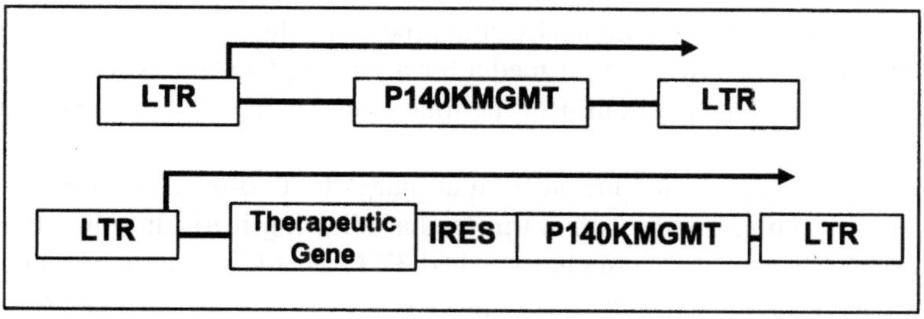

Fig. 2. Schematic of recombinant retroviral vectors that can be used for protecting gene-modified hematopoietic stem cells (HSCs) and for subsequent in vivo selection. The top construct can be used to increase the percentage of circulating gene-modified cells, e.g., to enrich for hematopoietic chimerism post transplant. The bottom construct can be used to increase the percentage of gene-modified cells expressing a therapeutic cDNA sequence.

2.5. In Vivo Selection

Depending on variables such as virus titer, cytokine cocktails, stem cell source, and preconditioning regimens, the engraftment of transduced cells may vary from only a few percent to as many as the majority of cells being gene-modified. In some cases, it may be advantageous to increase the percentage of gene-modified cells. As mentioned above, variants of MGMT are currently the best choices for in vivo selection of gene-modified cells. In addition, dual expression vectors, such as shown in **Fig. 2**, can be used to increase the percentage of cells expressing a desired therapeutic construct. The following reagents are needed to achieve in vivo selection using the MGMT selection system.

1. All necessary supplies for transduction/transplantation.
2. 6-BG (Sigma-Aldrich).
3. BCNU (Carmustine; Sigma-Aldrich).
4. 40% Polyethylene glycol (PEG)-400 in PBS (preheated at 37°C).
5. Ethanol (EtOH), absolute, 200 proof for molecular biology (Sigma-Aldrich).

3. Methods

3.1. Viral Production and Concentration

For transduction of murine HSCs, viral supernatant from Ecopack stable producer cells, which produce recombinant virus pseudotyped with the Ecotropic envelope, achieves excellent transduction efficiency at a multiplicity of infection (MOI) between 1 and 5.

1. Change the media conditioned by viral producer cells when they reach ~80–90% confluency. Use only 7 ml of media for a 10-cm plate (*see* **Note 1**).
2. Cells should be nearly confluent the following day at which time you can begin viral collection.
3. Very gently remove the media from all plates to a 50-ml conical tube(s).
4. Carefully replace with 7 ml warm media, allowing it to run down the side of the dish. Be sure the media has reached 37°C as cold media may cause cells to peel off of the plate. Immediately place dishes back in the incubator.
5. Filter collected media into a new 50-ml tube(s) using a 60-ml syringe and a 0.45-μm filter (*see* **Note 2**).
6. If virus will not be used or concentrated immediately, freeze at –80°C (*see* **Note 3**).
7. Media can be collected twice per day (separated by >7 h) for 4–5 days or until cells begin to peel off of the plate.
8. At this point, cells should either be split 1:4 to new plates for further viral collection, frozen in liquid nitrogen for later use, or discarded.
9. To increase viral titer, supernatant can be concentrated by overnight centrifugation. Thaw viral supernatants briefly in 37°C water bath. Pool virus and divide among sterile centrifuge tubes. Tube size depends on total volume of supernatant to be concentrated and determines the rotor to be used.
10. Centrifuge viral media at 9,000 × *g* overnight (minimum 16 h) at 4°C.
11. A small viral pellet should be visible (*see* **Note 4**). Carefully remove media from tube and resuspend the pellet in StemPro34 media at 1/20th to 1/40th the original volume (*see* **Note 5**).
12. Store in 1-ml aliquots at –80°C (*see* **Note 6**).
13. Virus should be titered on NIH 3T3 cells or a similarly appropriate cell line that is easily cultured and efficiently transduced. Plate the cells at a density of 100,000 cells per well of a six-well plate the afternoon prior to titering, allowing at least one extra well for counting cells the following day. The following morning, count one well to ensure that cells have at least doubled to 200,000 cells per well (*see* **Note 7**). Add virus to fresh media to a total volume of 1 ml with 8 μg/ml polybrene. Allow well to grow to confluency and analyze by flow cytometry or real-time PCR for percent modified cells or copy number integration, respectively (for example of viral titer calculation, *see* **Note 8**).

3.2. Stem Cell Isolation

Keep cells and solutions cold to prevent non-specific labeling. Volumes for magnetic labeling are stated per 10^7 labeled cells. For fewer cells, use the volumes given below; for higher cell numbers, scale up the reagents accordingly. MS separation columns can handle 1×10^7 positive cells or 2×10^8 total cells each.

1. In a sterile hood, flush bone marrow from the long bones of the hind legs of a mouse/mice in buffer (PBS supplemented with 2% FBS and 100 IU/ml

penicillin/100 μg/ml streptomycin) using a 23-gauge needle on a 3-ml syringe into a 15-ml conical tube. Keep on ice. One mouse typically yields 5–8 × 10^7 cells. [Some investigators use 5-fluorouracil (5-FU) primed bone marrow to increase the yield of HSCs going through cell division—*see* **Note 9**.]

2. Using a syringe, mix cells well to achieve a single cell suspension and pass through a 40-μm nylon filter to remove larger debris. Wet filter with buffer before use.
3. Add buffer to a total volume of 12 ml and centrifuge at 300 × *g* for 10 min at 4°C, then remove and discard the supernatant.
4. Resuspend cell pellet in 1 ml of buffer per donor mouse. Remove an aliquot for counting and calculating the yield of bone marrow cells (*see* **Note 10**). Cells for pre-sort flow cytometry should also be set aside at this point.
5. Centrifuge at 300 × *g* for 10 min. Pipette off supernatant completely.
6. Resuspend pellet in 80 μl of buffer per 10^7 cells.
7. Add 20 μl of anti-Sca-1 microbeads per 10^7 cells.
8. Mix well and incubate for 15 min at 4°C. (Working on ice may increase incubation times.)
9. Wash cells by adding 1 ml of buffer per 10^7 cells and centrifuge at 300 × *g* for 10 min. Remove the supernatant completely and resuspend the cell pellet in 500 μl buffer (for up to 10^8 cells).
10. Place separation column in separation unit. Prepare column by rinsing with 500 μl of buffer. Discard effluent and change collection tube.
11. Apply cell suspension onto column and collect the effluent. Wash with 500 μl of buffer three times and collect total effluent as the negative fraction. Count cells in negative fraction and take a sample for flow cytometry.
12. Remove the column from the separator and place on a new collection tube.
13. Apply 1 ml of buffer to the column and firmly flush out cells using the plunger supplied. Count Sca-1^+ cells and take a sample for flow cytometry. (This can be repeated over a new column to increase purity—*see* **Note 11**.)
14. Prepare Stempro-34 SFEM with nutrient supplement and L-glutamine. Ten milliliters of media requires 250 μl of 40× supplement and 100 μl of 100× L-glutamine (200 mM) to a final concentration of 2 mM.
15. Centrifuge the cells and resuspend the cell pellet in 1 ml Stemspan serum-free media with mSCF (100 ng/ml), mIL-3 (20 ng/ml), hIL-11 (100 ng/ml), and hFL (100 ng/ml). Culture for 48–72 h at 37°C.

3.3. Transduction

1. Retronectin should be resuspended at 1 mg/ml, aseptically filtered through a 0.22-μm membrane, and stored in aliquots at –80°C until time of use. Dilute stock solution to 20–100 μg/ml for coating plates. Prepare Retronectin-coated plates by adding 2 ml of sterile Retronectin to each well of a non-tissue-culture-treated six-well plate. Allow to stand for 2 h at room temperature.
2. Remove Retronectin solution and add sterile 2% BSA for blocking. Allow to stand for 30 min at room temperature.

3. Wash once with sterile 3 ml Hank's balanced salt solution + 2.5% 1 M HEPES. Plates are now ready for transduction.
4. Count cells after prestimulation. Suspend 1×10^6 cells in 1 ml of virus containing Stempro-34 serum-free expansion media with mSCF (100 ng/ml), mIL-3 (20 ng/ml), hIL-11 (100 ng/ml), hFL (100 ng/ml), nutrient supplement, and L-glutamine. Add the cell suspension to a six-well plate at 1 ml/well and centrifuge at $1000 \times g$ for 20 min at 25°C. Plates can be sealed with parafilm to ensure sterility during centrifugation. Incubate at 37°C. (See **Note 12** for calculation of MOI.)
5. Twelve to 24 hours later, add an additional 1 ml of virus in Stempro-34 serum-free expansion media with mSCF (100 ng/ml), mIL-3 (20 ng/ml), hIL-11 (100 ng/ml), hFL (100 ng/ml), nutrient supplement, and L-glutamine. Centrifuge plate at $1000 \times g$ for 20 min. Incubate at 37°C.

3.4. Transplantation

1. Six to 24 hours after the second transduction, collect and pool the cells. (Note: As cells often adhere to the retronectin coating, it may be necessary to use a cell scraper to retrieve as many cells as possible from the plate.) Centrifuge in a 14-ml conical tube at $300 \times g$ for 10 min.
2. Resuspend cells in 300 μl of sterile PBS per 1×10^6 cells originally plated. Take an aliquot for counting and flow cytometry if applicable. Cells can also be plated in methylcellulose at this point for a progenitor assay (*see* **Note 13**). Representative flow cytometry data are shown in **Fig. 3** for Sca-1$^+$ cells transduced with virus encoding green fluorescent protein (GFP); 75% of the c-kit$^+$, sca-1$^+$, lineage negative cells are GFP positive on the day of transplantation.

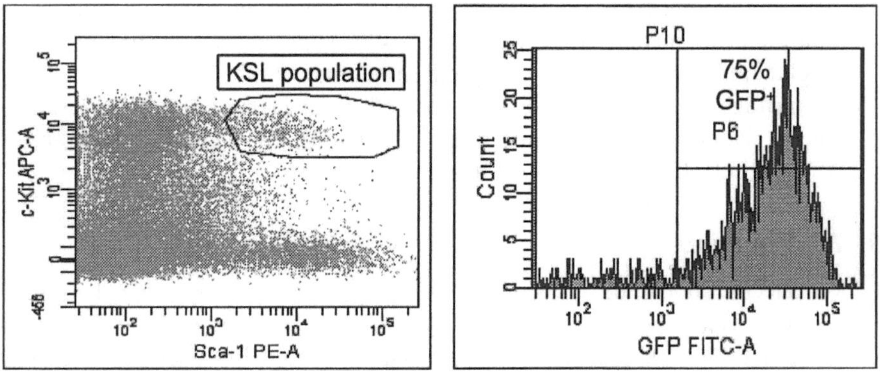

Fig. 3. Flow cytometric analysis of c-Kit$^+$/Sca-1$^+$/lineage$^-$ (KSL) cells. The KSL population of cells contains a high percentage of repopulating HSCs. The right panel shows that 75% of KSL cells express the transgene on the day of transplantation.

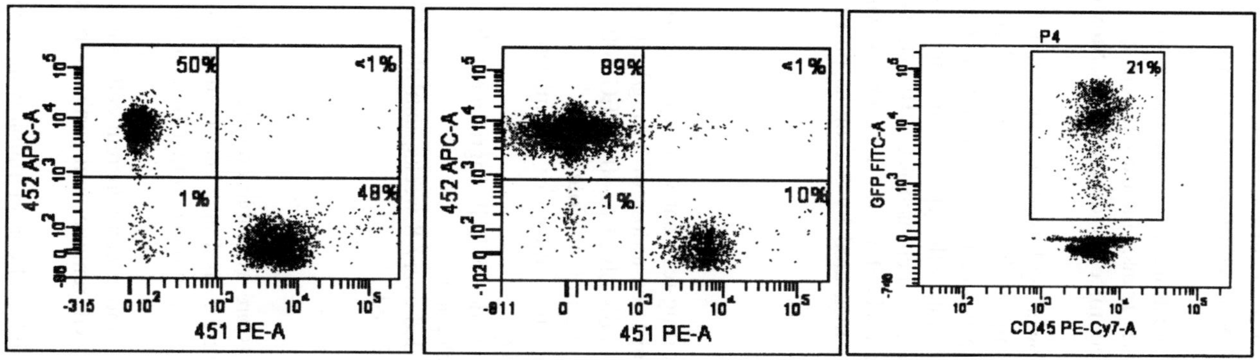

Fig. 4. Engraftment of gene-modified cells in transplanted mice. The left and middle panels show engraftment of CD45.2 cells transplanted into CD45.1 mice 12 weeks post transplantation after sublethal irradiation (5.5 Gy, left panel) and lethal irradiation (11 Gy, middle panel). The right panel shows engraftment of GFP-transgenic C57BL/6 cells into 45.2 wild-type C57BL/6 mice just 6 weeks post transplantation following reduced-intensity conditioning.

3. Adjust concentration by adding PBS for injection of 200 μl per mouse at desired cell number. About 300,000–600,000 Sca-1$^+$ cells reliably reconstitute a mouse conditioned with myeloablative or reduced-intensity conditioning. If you have poor yield of cells, you can repeat centrifugation and resuspend cells in desired volume.
4. Mix cell suspension well by pipetting. Load individual syringes with 200 μl of cell suspension for each mouse. Transplant via tail vein or retroorbital injection into appropriately conditioned mice as described in **Subheading 2.4.5** (*see* **Note 14**).
5. Engraftment can often be followed by flow cytometry, taking advantage of minor differences in cell-surface antigens between donor and recipient mice. GFP-transgenic C57BL/6 mice make excellent donors (as long as GFP is not one of your transgene(s)) for transplantation into wild-type C56BL/6 mice. Another alternative is using mice with minor HLA mismatch such as C57BL/6 CD45.1 versus CD45.2. Representative flow data for each of these cases are shown in **Fig. 4**. Stable engraftment numbers can be measured by 3 months post-transplantation (*see* **Notes 14** and **15**).

3.5. *In Vivo* Selection

1. If using vectors containing a drug-resistant variant of MGMT, 4–8 weeks post-transplant mice can be administered a dose of 30 mg/kg 6-BG via an intraperitoneal (i.p.) injection. 6-BG is dissolved in 40% PEG-400 in PBS preheated to 37°C at a concentration of 3 mg/ml (*see* **Note 16**).
2. One hour after the 6-BG injection, mice are administered an i.p. injection of 15 mg/kg BCNU (*see* **Note 17**).
3. Analyze peripheral blood for donor cell engraftment by flow cytometry a minimum of 4 weeks after selection.
4. Repeat drug selection as needed every 8 weeks, until the desired percentage of donor cell engraftment is achieved.

4. Notes

1. Ecopack virus producer cells are available commercially. They do not adhere well to standard tissue culture dishes and may come off of the plates during repeated viral collection. It is recommended that Biocoat collagen-coated tissue culture plates be used.
2. A 0.45-μm filter is used to remove non-adherent cells and cellular debris from dead cells. A 0.20-μm filter will remove viral particles from the culture media, greatly reducing viral titer.
3. If viral titer is sufficiently high and concentration is not required, freeze supernatant for later use in individual aliquots to avoid repeated freeze thaw cycles that reduce viral titer.

4. If no pellet is visible after centrifugation, place centrifuge tubes on ice for 10–15 min. A translucent white pellet may become visible at the colder temperature.
5. In order to calculate the concentration factor, first determine desired MOI. An MOI of 5 per transduction should be sufficient to achieve 50–70% genetic modification of cells. Cells are transduced as 1×10^6 cells per well of a six-well plate. Therefore, a viral titer of $1–2 \times 10^6$ will provide a minimum amount of virus.
6. When freezing aliquots of concentrated virus for later use, freeze one aliquot separately to be used for viral titering if the titer is not yet known. This avoids thawing and refreezing concentrated stocks.
7. The cells used for titering must be actively growing at the time of viral addition in order to accurately calculate the viral titer. If the cells have not doubled by the following morning, allow them to grow for an additional 12–24 h and repeat the cell count.
8. Viral titer may be determined by flow cytometry or real-time PCR; the choice of methodology depends mostly on the transgene. Transgenes such as GFP which encode a fluorescent protein lend themselves to titering by flow cytometry, but for many transgenes no appropriate antibody is available or the encoded protein may be secreted in which case flow cytometry would not be appropriate unless cells were treated with an agent to prevent protein secretion. When transduction efficiencies are kept below approximately 30%, the two methods yield similar results. Above that level of transduction, multiple integration events may be occurring in individual cells which would not be reflected accurately by flow cytometry but would be accounted for by real-time PCR. Following is an example of viral titer calculation by flow cytometry; for this example, we will assume the virus encodes GFP. Assume that there were 250,000 cells/well at the time of viral addition and that 100 µl of virus was added to 900 µl of media. If 30% of cells are GFP positive, the titer would be calculated as follows:

$$250{,}000\ (\mathit{cells\ per\ well}) \times 0.3\ (\mathit{fraction\ of\ GFP^+\ cells}) = 7.5 \times 10^4\ \mathit{functional\ viral\ particles}/100\ \mu l = 7.5 \times 10^5\ \mathit{functional\ viral\ particles}/ml.$$

Viral titer can also be calculated by real-time PCR. The output in this case is copy number (number of total gene integration events) per genome copy. Making the same assumption as above and using 50 ng of genomic DNA, which is equivalent to 8666 genomic copies of DNA, assume that real-time results show 500 copies of proviral sequence. The calculation would be:

$$250{,}000\ (\mathit{cells\ per\ well}) \times 500\ \mathit{copies}/50\ ng \times 50\ ng/8666\ \mathit{copies\ of\ genomic\ DNA} = 1.6 \times 10^5\ fvp/100\ \mu l = 1.6 \times 10^6\ fvp/100\ \mu l$$

For the most accurate titering, it is recommended to use multiple dilutions of viral supernatant such as undiluted, 1:10, and 1:100, especially in the case of titering by flow cytometry. Once transduction efficiency reaches 30%, it is very likely that some viral particles are infecting the same cells and therefore viral titer will be underestimated.

9. Some investigators use 5-FU "primed" bone marrow. 5-FU (150 mg/kg given i.p.) is administered to mice that will be used as bone marrow donors. At this dose, 5-FU kills cells of committed lineage, which induces HSC to cycle. The cycling cells are more readily transduced with recombinant retroviruses. Using the cytokine stimulation methods described in this chapter, we have found that 5-FU primed marrow is not necessary. If 5-FU is used, make a stock solution of 10 mg/kg in sterile PBS or water. It is difficult to dissolve 5-FU at higher concentrations.
10. Freshly isolated whole bone marrow cells should be counted in 2% acetic acid, which is used to lyse red blood cells (this is not necessary once Sca-1 isolation is performed). All cell counts should be done with Trypan blue or a similar reagent that allows for live versus dead cell determination.
11. Typical yield is 5–10% of starting cell number. Purity of 80–90% Sca-1^+ cells is typically achieved with a single pass through the column.
12. Our working definition for MOI is determined by dividing functional viral particles by the number of cells and multiplying by total number of viral exposures. For example, if 1 ml of 1×10^6 fvp/ml virus is added to 1×10^6 cells twice the MOI is 2.
13. Excellent methylcellulose preparations are available from companies such as StemCell Technologies, providing an in vitro method for assessing progenitor cell viability and transduction efficiency when using a transgene encoding a visible protein product such as GFP.
14. Either retroorbital or intravenous tail vein injection is a suitable method of HSC transplantation, but based on our non-published competitive repopulation studies, we think retroorbital and tail vein injections are equivalent with respect to repopulation of the hematopoietic system.
15. Until stable engraftment of donor cells is achieved, engraftment and therapeutic gene expression levels can be quite erratic during this exordium phase. This may be due to clonal proliferation and then demise of short-term repopulating cells before progeny of long-term repopulating cells come to dominate the donor cell pool.
16. Prepare 6-BG on the day of injection. If using a fragile strain of mice (such as transgenic sickle mice), we have found that 40% PEG is a thick solution and can result in a high mortality rate following IP injection. In this case 6-BG can be dissolved in 70% EtOH at 15 mg/ml. Incubate 6-BG at 37°C in shaker until dissolved (usually requires about 1 h to completely dissolve), then filter through a 0.22-μm filter. This high concentration of EtOH is required to dissolve 6-BG. However, high concentrations of EtOH are also lethal to mice. Therefore, first dissolve 6-BG in 70% EtOH and then dilute in 10% EtOH to 1.5 mg/ml (if 6-BG is diluted in water or PBS 6-BG will precipitate). Mice are then administered an i.p. dose of 30 mg/kg.

17. Resuspend BCNU at 30 mg/ml in 100% EtOH and freeze stocks at –80°C. Immediately prior to injection, thaw stock on ice, dilute with dH_2O to 1.5 mg/ml final concentration, and immediately administer 15 mg/kg (i.e., 200 μl/20 g mouse). Because BCNU is not stable in water, it is important that the diluted stock is injected immediately.

References

1. Attar EC, Scadden DT. Regulation of hematopoietic stem cell growth. Leukemia. 2004.
2. Lewis PF, Emerman M. Passage through mitosis is required for oncoretroviruses but not for the human immunodeficiency virus. J Virol. 1994;68:510–516.
3. Mazurier F, Gan OI, McKenzie JL, Doedens M, Dick JE. Lentivector-mediated clonal tracking reveals intrinsic heterogeneity in the human hematopoietic stem cell compartment and culture-induced stem cell impairment. Blood. 2004;103:545–552.
4. McNiece I, Briddell R. Ex vivo expansion of hematopoietic progenitor cells and mature cells. Exp Hematol. 2001;29:3–11.
5. McNiece IK, Almeida-Porada G, Shpall EJ, Zanjani E. Ex vivo expanded cord blood cells provide rapid engraftment in fetal sheep but lack long-term engrafting potential. Exp Hematol. 2002;30:612–616.
6. Verfaillie CM. Hematopoietic stem cells for transplantation. Nat Immunol. 2002;3:314–317.
7. Becker PS. The current status of gene therapy in autologous transplantation. Acta Haematol. 2005;114:188–197.
8. Ott MG, Schmidt M, Schwarzwaelder K, et al. Correction of X-linked chronic granulomatous disease by gene therapy, augmented by insertional activation of MDS1-EVI1, PRDM16 or SETBP1. Nat Med. 2006;12:401–409.
9. Buckley RH. Molecular defects in human severe combined immunodeficiency and approaches to immune reconstitution. Annu Rev Immunol. 2004;22:625–655.
10. Nienhuis AW, Dunbar CE, Sorrentino BP. Genotoxicity of retroviral integration in hematopoietic cells. Mol Ther. 2006;13:1031–1049.
11. Cavazzana-Calvo M, Hacein-Bey S, de Saint Basile G, et al. Gene therapy of human severe combined immunodeficiency (SCID)-X1 disease. Science. 2000;288: 669–672.
12. Cavazzana-Calvo M, Hacein-Bey-Abina S, Fischer A. Gene therapy of X-linked severe combined immunodeficiency. Curr Opin Allergy Clin Immunol. 2002;2: 507–509.
13. Wu X, Burgess SM. Integration target site selection for retroviruses and transposable elements. Cell Mol Life Sci. 2004;61:2588–2596.
14. VandenDriessche T, Collen D, Chuah MK. Biosafety of onco-retroviral vectors. Curr Gene Ther. 2003;3:501–515.
15. Baum C, Kustikova O, Modlich U, Li Z, Fehse B. Mutagenesis and oncogenesis by chromosomal insertion of gene transfer vectors. Hum Gene Ther. 2006;17: 253–263.

16. Neff T, Beard BC, Kiem HP. Survival of the fittest: in vivo selection and stem cell gene therapy. Blood. 2006;107:1751–1760.
17. Sorrentino BP. Gene therapy to protect haematopoietic cells from cytotoxic cancer drugs. Nat Rev Cancer. 2002;2:431–441.
18. Licht T, Goldenberg SK, Vieira WD, Gottesman MM, Pastan I. Drug selection of MDR1-transduced hematopoietic cells ex vivo increases transgene expression and chemoresistance in reconstituted bone marrow in mice. Gene Ther. 2000;7: 348–358.
19. Spencer HT, Sleep SE, Rehg JE, Blakley RL, Sorrentino BP. A gene transfer strategy for making bone marrow cells resistant to trimetrexate. Blood. 1996;87:2579–2587.
20. Allay JA, Persons DA, Galipeau J, et al. In vivo selection of retrovirally transduced hematopoietic stem cells. Nat Med. 1998;4:1136–1143.
21. Davis BM, Koc ON, Gerson SL. Limiting numbers of G156A O(6)-methylguanine-DNA methyltransferase-transduced marrow progenitors repopulate nonmyeloablated mice after drug selection. Blood. 2000;95:3078–3084.
22. Warlick CA, Diers MD, Wagner JE, McIvor RS. In vivo selection of antifolate-resistant transgenic hematopoietic stem cells in a murine bone marrow transplant model. J Pharmacol Exp Ther. 2002;300:50–56.
23. Koc ON. Gene Therapy of Cancer; 1999.
24. Chinnasamy D, Fairbairn LJ, Neuenfeldt J, et al. Lentivirus-mediated expression of mutant MGMTP140K protects human CD34+ cells against the combined toxicity of O6-benzylguanine and 1,3-bis(2-chloroethyl)-nitrosourea or temozolomide. Hum Gene Ther. 2004;15:758–769.
25. Davis BM, Humeau L, Dropulic B. In vivo selection for human and murine hematopoietic cells transduced with a therapeutic MGMT lentiviral vector that inhibits HIV replication. Mol Ther. 2004;9:160–172.
26. Sawai N, Zhou S, Vanin EF, Houghton P, Brent TP, Sorrentino BP. Protection and in vivo selection of hematopoietic stem cells using temozolomide, O6-benzylguanine, and an alkyltransferase-expressing retroviral vector. Mol Ther. 2001;3:78–87.

14

Assessment of CFTR Function after Gene Transfer In Vitro and In Vivo

Uta Griesenbach, Felix M. Munkonge, Stephanie Sumner-Jones, Emma Holder, Stephen N. Smith, A. Christopher Boyd, Deborah R. Gill, Stephen C. Hyde, David Porteous, and Eric W. F. W. Alton; on behalf of the UK Cystic Fibrosis Gene Therapy Consortium

Summary

Cystic fibrosis (CF) a monogenic lethal disease and, therefore, ideally suited for the development of gene therapy. The first clinical trials were carried out shortly after cloning the CF gene in 1989. Since then, 25 trials have been carried out. Proof of principle for low-level airway gene transfer was established in most, but not all, trials. It is currently unclear whether current gene transfer efficiency will lead to improvements in clinically relevant endpoints such as inflammation or infection. In addition to addressing this important question, we and others are further improving airway gene transfer, by modifying existing and developing new gene transfer agents. Here, we describe pre-clinical methods related to assessing correction of the CF chloride transport defect.

Key Words: Cystic fibrosis; gene therapy; gene transfer; pre-clinical assays; CFTR.

1. Introduction

Gene therapy is particularly attractive for diseases that currently do not have satisfactory treatment options and is probably easier for monogenic disorders than for more complex polygenic diseases. Cystic fibrosis (CF) fulfills these criteria and is, therefore, a good candidate for gene therapy-based treatment. CF affects various organs but chronic lung infection and subsequent inflammation followed by organ failure are the major cause of death. The cloning of the

From: *Methods in Molecular Biology, vol. 433: Volume 1: Production and In Vivo Applications of Gene Transfer Vectors*
Edited by: J. M. Le Doux © Humana Press, Totowa, NJ

CF gene, the *cystic fibrosis transmembrane conductance regulator* (*CFTR*) in 1989 *(1)*, opened the door for CF gene therapy. Proof of principle for *CFTR* gene transfer was quickly established in vitro and in animal models *(2–4)*. The first clinical trials in CF patients were carried out in 1993, and to date, 25 trials have been completed incorporating approximately 450 CF patients. These included viral and non-viral gene transfer to both the nasal and bronchial airway epithelium (for the most recent viral and non-viral trial, see **refs.** ***5*** and ***6***). In general, gene transfer was well tolerated and proof of principle for *CFTR* gene transfer (as measured by vector-specific mRNA or CFTR-mediated chloride transport) has been established in some, but not all, studies.

Overall, the CFTR protein is not expressed abundantly in the lungs, but high expression is seen in serous cells in the submucosal glands and isolated epithelial cells in the airways *(7)*. It is currently unclear which of these cell types should be the main targets for CF gene therapy, but considering that CF (at least in the early stages) presents as a small airway disease, we are working on the basis that airway epithelial cells (AECs) are highly relevant targets for CF gene therapy. Despite the ease of non-invasive access to the lung, gene transfer into AECs has been comparatively inefficient. It is apparent these cells have evolved effective barriers (reviewed in **ref.** ***8***) to prevent uptake of foreign particles, including gene transfer agents (GTAs).

In general, non-viral vectors are less efficient than viral vectors. Importantly, however, they are less likely to induce an immune response, thereby more likely allowing for repeat administration, crucial for the treatment of chronic diseases such as CF. Considering the obvious advantages of non-viral gene transfer for the life-long treatment of CF patients, the UK CF Gene Therapy Consortium (http://www.cfgenetherapy.org.uk) is focusing on the further optimization of non-viral gene transfer to AECs.

We, and others, have developed a wide range of in vitro assays relevant for CF gene therapy, as well as methods aimed at assessing airway gene transfer in small (mice) and large (sheep) animal models. Here, we will focus on describing methods relevant for assessing CFTR function after gene transfer in vitro and in vivo.

2. Materials

2.1. CFTR Transfection and Assessment of CFTR Function In Vitro

1. Human embryonic kidney cells (HEK293T) and human colon carcinoma cells (T84) [American Type Culture Collection (ATCC), LGC Promochem, Teddington, Middlesex, UK].
2. Pre-coated poly-L-lysine six-well plates (BD Biosciences, Bedford, MA, USA).
3. Lipofectamine™ 2000 (Invitrogen Ltd, Paisley, UK).

4. Pre-coated poly-L-lysine six-well plates (35-mm diameter) (Becton-Dickinson, Oxford, UK).
5. OPTI-MEM® I Reduced Serum Medium *modification of MEM (Eagle's)* (Invitrogen Ltd).
6. Fetal bovine serum (FBS) and phosphate-buffered saline (PBS).
7. $I^{125}I$ (Amersham, Little Chalfont, UK).
8. Efflux buffer: 140 mM NaCl, 5 mM KCl, 1.3 mM $CaCl_2$, 0.5 mM $MgCl_2$, 5 mM glucose, and 10 mM HEPES.
9. 0.2 N NaOH to lyse cells.
10. Forskolin and 3-isobutyl-1-methylxanthine (IBMX) (Sigma-Aldrich Poole Doised, UK).
11. Dulbecco's Modified Eagle's Medium (DMEM) (Invitrogen Ltd), supplemented with 10% FBS, 1 mM glucose (Sigma-Aldrich), 5 mM L-glutamine (Sigma-Aldrich), 100 IU/ml penicillin (Sigma-Aldrich), and 100 μg/ml streptomycin (Sigma-Aldrich).
12. β-counter (Packard Biosciences, Pangbourne, UK).

2.2. Transfection of Nasal Respiratory Epithelial Cells In Vivo

1. Ketaset, Domitor, and Antisedan (National Veterinary Service, Stoke on Trent, UK).
2. Ointment (Lacrilube, Boots, UK).
3. Polyethylene catheters (inner diameter 2 mm, outer diameter 3 mm; Jencons PLS, Leighton Buzzard, UK).
4. Syringe pump (Cole-Palmer, IL, USA).

2.3. Quantification of Luciferase Expression in Nasal Tissue

1. Reporter Lysis Buffer (RLB) and Luciferase Assay System T281 (Promega, Madison, WI).
2. QiaShredder column (Qiagen, Crawley, UK).
3. BioRad protein assay kit (BioRad Laboratories, Hercules, CA).

2.4. Selective Isolation of Respiratory Epithelial Cells and Preparation of RNA

1. Interdental brush (CPS 07, Curaprox UK Ltd, Kimbolton, UK).
2. RLT buffer (Qiagen Ltd).
3. β-Mercaptoethanol (Sigma-Aldrich).
4. RNeasy Mini kit (for total RNA isolation from animal cells) and RNeasy Mini protocol for RNA clean-up (Qiagen Ltd).
5. Ambion DNA-free™-kit (Ambion LTD, Huntingdon, UK).

2.5. Assessment of Ion Transport in the Nose of CF-Knockout Mice

1. HEPES–Krebs (HK) Buffer: 10 mM HEPES, 140 mM NaCl, 6 mM KCl, 1 mM $MgCl_2$, 2 mM $CaCl_2$, and 10 mM glucose.

2. HEPES–Krebs–Amiloride (HKA) solution: 10 mM HEPES, 140 mM NaCl, 6 mM KCl, 1 mM $MgCl_2$, 2 mM $CaCl_2$, 10 mM glucose, and 0.1 mM amiloride.
3. Low Chloride (LC) solution: 10 mM HEPES, 1 mM $MgCl_2$, 2 mM $CaCl_2$, 10 mM glucose, 0.1 mM amiloride, 140 mM Na glutamate, and 6 mM K glutamate.
4. PD equipment (Logan Research LTD, Rochester, UK).

3. Methods

Although assessment of reporter gene expression is useful, the ultimate aim of CF gene therapy is the transfer of a functional *CFTR* gene into relevant cells in the airways in vivo. Here, we describe pre-clinical in vitro and in vivo methods addressing this aim.

3.1. CFTR Transfection and Assessment of CFTR Function In Vitro

The radiotracer halide efflux assay (commonly known as the "sample and replace assay"), first introduced by Venglarik et al. ***(9)***, is probably the most utilized method available for the assessment of CFTR protein function in vitro. The assay is based on pre-loading cells with a radiotracer ion, whose efflux out of the cells, via cyclic-AMP-dependent opening of CFTR, is quantified in the surrounding medium. I^{125} (half-life 60 days, emitting γ-particles with an average energy of 35 keV) is used as alternative to ^{36}Cl (half-life 300,000 years, emitting β-particles with an average energy of 0.252 keV) in order to minimize the radiation dose to personnel and to disposal costs. In addition, cyclic-AMP-induced chloride channels are permeable to iodide whereas many other chloride transport processes are not ***(10)***, further making I^{125} a good surrogate for chloride.

3.1.1. Transfection of Cells

1. Seed HEK293T cells at a density of 1×10^5 cells/well in purchased pre-coated poly-L-lysine six-well plates 24 h before transfection.
2. Complex 1 μg of a eukaryotic expression plasmid carrying the CFTR cDNA (*see* **Note 1**) or an appropriate negative control plasmid to 2.5 μl Lipofectamine™ 2000 (1 mg/ml) in serum-free Opti-MEM medium (total volume of 400 μl per well), and incubate for 20 min at room temperature (*see* **Note 2**).
3. Wash cells with PBS, add the liposome/DNA complex, and incubate in a humidified atmosphere at 37°C with 5% CO_2. Discard the transfection medium after 6 h and replace with 1 ml Opti-MEM media supplemented with 10% FBS. Replace the medium after 18 h and incubate cells for an additional 24 h at 37°C with 5% CO_2.
4. Use T84 cells expressing high levels of endogenous CFTR as positive controls. Maintain these cells in DMEM and culture them in a humidified atmosphere

at 37°C with 5% CO_2. Carry out experiments with cells at passage 5–15. For the I^{125} efflux assay, transfer sub-cultured cells into six-well tissue culture plates at an initial density 1×10^5cells/well. Carry out the efflux assay 72 h after seeding.

3.1.2. Radiotracer Efflux Assay

1. Assay CFTR chloride channel activity by measuring the rate of I^{125} efflux, essentially as described by Becq and colleagues *(11)*. Perform all experiments at room temperature.
2. Wash cells twice with 2 ml efflux buffer.
3. Incubate cells with efflux buffer containing 0.2 μCi I^{125}/ml for 30 min at 37°C to permit the I^{125} to enter the cells.
4. Wash cells again with efflux buffer. Remove the buffer after 1 min and quickly replace with 1 ml of fresh buffer. Repeat this procedure every minute for 11 min. Collect samples for the next 3 min and use these to establish a stable baseline.
5. Use efflux buffer containing the cAMP-agonist forskolin (20 μM) and the phosphodiesterase inhibitor IBMX (100 μM) to replace and collect the remaining aliquots for another 6–10 min.
6. At the end of the incubation, remove all media and solubilize cells in 1 ml 0.2 N NaOH.
7. Determine radioactivity in all buffer samples and in the lysed cells using a β-counter.

3.1.3. Data Analysis

The fraction of intracellular I^{125} lost during each time point is determined, and time-dependent rates of I^{125} efflux are calculated.

1. Derive I^{125} efflux rates using the equation $\ln(I^{125}{}_{t1}/I^{125}{}_{t2})/(t_1 - t_2)$, where In is the efflux rate, $I^{125}{}_{t}$ is the intracellular I^{125} at time *t*, and t_1 and t_2 are successive time points. Normalize the I^{125} efflux rates to efflux at the time of Forskolin/IBMX addition (time t_0).
2. Construct curves by plotting rates of I^{125} efflux versus time (with exclusion of the points used to establish the baseline) (*see* **Fig. 1**).
3. To calculate cumulative levels of I^{125}efflux following agonist-stimulation, on areas under the time–I^{125} efflux curves. Calculate the area-under-curve (*see* **Fig. 2**) by the trapezium rule (essentially place adjustable, moveable geometric shapes under the curve for 0–5 min and sum the area contained in the shapes) *(12)*.

3.2. Transfection of Nasal Respiratory Epithelial Cells In Vivo

There is no natural animal model of CF, but a large number of CF-knockout mouse strains have been generated. CF-knockout mice have CF-like gut disease,

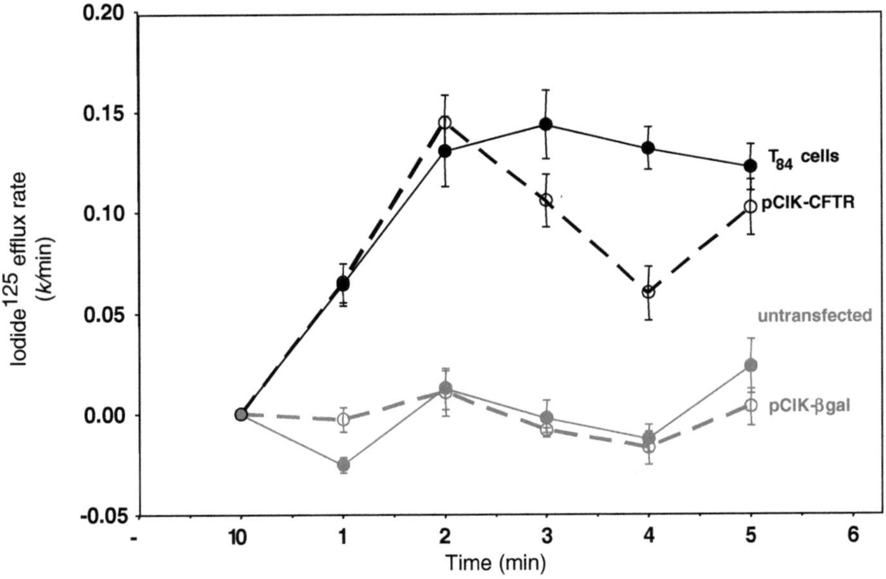

Fig. 1. Forskolin/3-isobutyl-1-methylxanthine (IBMX)-induced I^{125}efflux after *CFTR* gene transfection. HEK293T cells were transfected with a plasmid carrying the *CFTR* cDNA (pCIk-CFTR) or an appropriate control plasmid (pCIK-βgal) or left untransfected. T84 cells, expressing endogenous levels of CFTR, are included as positive control. Following addition of the cAMP agonist forskolin and the phosphodiesterase inhibitor IBMX, CFTR-mediated chloride channel activity is increased in T84 and CFTR cDNA-transfected HEK293T cells.

but unfortunately do not recapitulate CF lung disease ***(13)***. The nasal epithelium, however, shows the characteristic CF ion transport defects: (a) reduced chloride secretion due to absent CFTR-mediated chloride transport, (b) increased sodium absorption due to absent CFTR-mediated inhibition of the epithelial sodium channel (ENaC). This tissue is, therefore, suitable to assess if gene therapy is able to correct CFTR-specific endpoints.

In general, CF mice are generated from heterozygote breeding pairs (i.e., on average only one in four pups is CF), are runted, and die before or soon after weaning from intestinal blockage. We generally recommend the use of gut-corrected CF-knockout mice generated in Jeff Whitsett's laboratory ***(14)***. These mice express human CFTR under the control of the fatty-acid-binding protein promoter (FABp), which is active in the intestine and, therefore, corrects the CF gut disease, but does not alter the CF phenotype in the nasal epithelium. FABp-CF knockout mice can be bred from homozygote-knockout breeding pairs (all pups are CF) and have near-normal life expectancy.

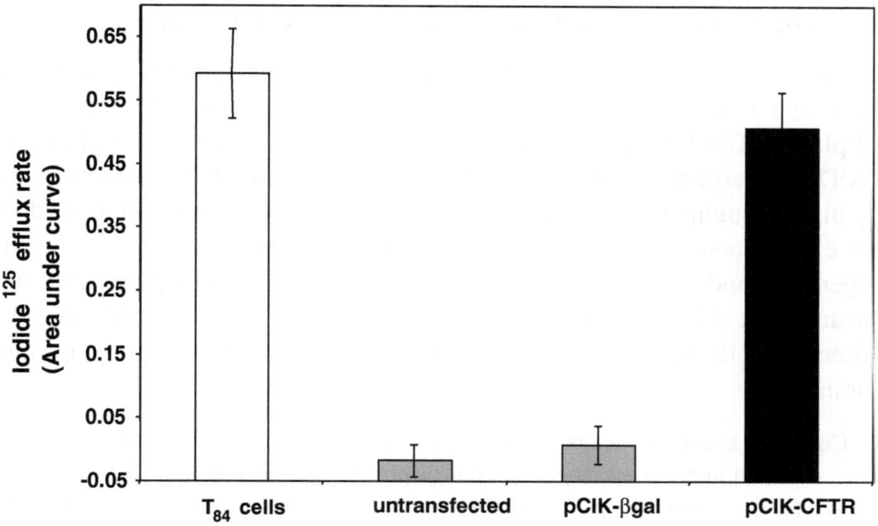

Fig. 2. Forskolin/3-isobutyl-1-methylxanthine (IBMX)-induced I^{125} efflux expressed as area under the time-efflux curve. The data from **Fig. 1** were used to calculate the area under the time-efflux curve (0–5 min). $^{***}p < 0.001$ when compared with untransfected controls.

3.2.1. In Vivo Gene Transfer Via Nasal Perfusion

1. Anesthetize mice with Ketaset/Domitor (76 mg/kg Ketaset, 1 mg/kg Domitor) and place on a heating board (*see* **Note 3**).
2. Make a paper-pad by folding a blue towel a few times and place under the mouse's head so that the fold of the paper is near the tip of the nose. Use the paper to absorb any excess liquid coming out of the nose during the perfusion. This prevents the mice from choking.
3. Tape whiskers and back legs to the heating board using Micropore tape. Apply Lacrilube ointment to the eyes to prevent drying out during the perfusion. Carefully place the tongue to one side with forceps.
4. Prime polyethylene catheters (<0.5 mm outer diameter, *see* **Note 4**) to the tip with GTA and insert into the nose (approximately 2.5 mm deep, *see* **Note 5**) and tape to the board, which is then tilted to approximately 45 degrees with the mouse's head facing down to avoid drowning.
5. Catheters can be inserted into one or both nostrils.
6. Using a syringe pump, perfuse GTAs (100–500 μl) carrying either a reporter gene or the *CFTR* gene onto the nasal epithelium (7 μl/min) for 15–75 min (*see* **Note 6**).
7. Reverse the anesthesia of mice by intraperitoneal (i.p.) injection of the antidote antisedan (1 mg/kg) and place in a recovery box at 30°C for 2 h.

3.3. Quantification of Luciferase Expression in Nasal Tissue

A large number of non-viral GTAs have been assessed for airway gene transfer. Please refer to the individual formulations for details on how to generate the final plasmid DNA-containing formulations. In general, careful optimization of GTA:DNA charge ratios and DNA concentration is required. To ensure a moderately high throughput during the initial optimization phase, we recommend doing these experiments using a luciferase reporter gene. Luciferase assays are fast, inexpensive, and sensitive and allow rapid identification of optimal formulations and gene delivery conditions for a given GTA. This information can then be used to facilitate functional assays of *CFTR* gene transfer which are more labor intensive.

1. Complex a eukaryotic expression plasmid carrying the firefly luciferase (lux) gene or an appropriate negative control plasmid to the non-viral GTA of choice and perfuse onto the nasal epithelium of mice as described above (*see* **Note 7**).
2. Cull mice at appropriate time-points (*see* **Note 8**) after transfection. Remove snout, skin, and nasal plate using a scalpel. Remove the septum, rinse in PBS, and place in 200 μl 1× RLB (*see* **Note 9**).
3. Homogenize the nasal tissue using spring scissors for 30 s, followed by three freeze–thaw cycles alternating at –80°C and room temperature (*see* **Note 10**).
4. Centrifuge samples for 5 min at 13,000 × g. Remove the supernatant and transfer to a QiaShredder column and centrifuge again (5 min, 13,000 × g_{av}). Store the supernatant at –80°C until detection of lux activity.
5. Thaw the tissue extracts at room temperature. Mix appropriate volumes, to ensure working in the dynamic range of the assay, with Luciferase Assay Reagent and assay for luciferase activity (relative light units—RLU) in a luminometer (*see* **Note 11**).
6. Use the BioRad protein assay kit to determine the total protein content of each sample to allow normalization of lux expression to total protein.
7. Place appropriate volumes, to ensure working in the dynamic range of the assay, in triplicate in a 96-well plate alongside a set of BSA protein standards (ranging from 0 to 1.5 mg/ml protein) diluted in water.
8. Add reagents to the extracts as described by the manufacturer (BioRad Laboratories) and incubate the plate at room temperature for 15–60 min.
9. Quantify the total protein concentration spectrophotometrically (750 nm) and express all data as luciferase activity in RLU/mg protein.

3.4. Selective Isolation of Respiratory Epithelial Cells and Preparation of RNA

The mouse nasal cavity is lined by squamous, respiratory, and olfactory epithelial cells. Only the respiratory epithelium, however, is a relevant target for CF gene therapy. Luciferase assay carried out on nasal homogenates are fast

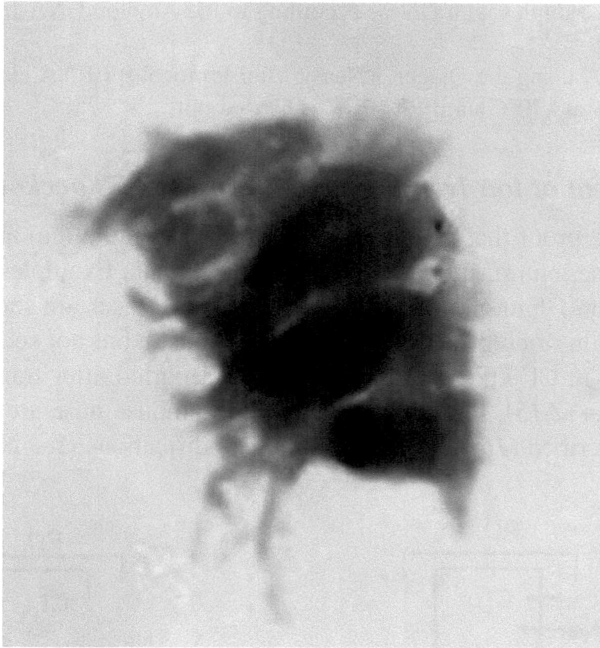

Fig. 3. Mouse nasal respiratory epithelial cells. Nasal cells were harvested by nasal brushing and stained with hematoxylin and eosin. The vast majority of these cells (approximately 80%) are ciliated respiratory epithelial cells.

but do not indicate which cell types have been transfected. We have, therefore, developed a technique that enabled us to isolate nasal respiratory epithelial cells selectively (Emma Holder, manuscript in preparation) (*see* **Fig. 3**). Total RNA suitable for analysis of plasmid-specific reverse transcription (RT)–PCR can be prepared from these cells. RT–PCR conditions (such as primer design and temperatures) are dependent on the plasmid used and will not be described here.

1. Cull mice by cervical dislocation and remove skin, snout, and cartilage to expose the nasal cavity.
2. Insert an interdental brush into the right nasal cavity and roll it along the septum (*see* **Note 12**).
3. Place the brush in a 1.5-ml Eppendorf tube containing 350 µl of RLT buffer (plus 1% β-mercaptoethanol) and resuspend cells by vigorous shaking for 30 s (*see* **Note 13**).
4. Incubate samples for 5 min at 37°C, and add to a QIAshredder spin column and centrifuge (2 min,10,000 × g_{av} at room temperature).
5. Add 350 µl of 70% ethanol and mix gently by pipetting and use the Qiagen RNeasy Mini protocol for total RNA isolation from animal cells.

6. Treat RNA samples with DNase according to the Ambion DNA-free™ treatment protocol.
7. Purify RNA using the Qiagen RNeasy Mini Protocol for RNA clean-up.
8. Store RNA at –70°C until RT–PCR is carried out.

3.5. Assessment of Ion Transport in the Nose of CF-Knockout Mice

The ultimate proof that gene transfer into airway epithelium has occurred is correction of the ion transport defect in mice or human. Examples of characteristic CF and non-CF nasal potential difference (PD) traces are shown in **Fig. 4**. Partial correction of chloride transport (LC response), but not sodium transport (baseline PD), in CF knockout mice has been reported after transfection with non-viral GTAs *(2,15)*. PD measurements in the mouse nose are performed as previously described *(16)* with a few minor modifications (*see* **Note 14**).

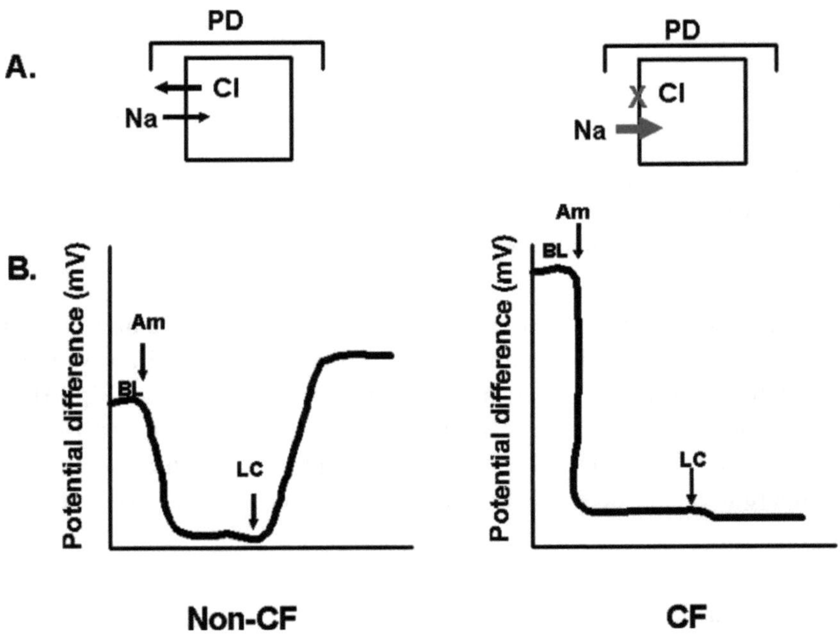

Fig. 4. Schematic presentation of ion transport and potential difference (PD) measurements. (**A**) Sodium (Na^+) and chloride (Cl^{--}) transport across the apical membrane of CF (right) and non-CF (left) epithelial cells. (**B**) Nasal PD measurements in CF and non-CF mice. BL: baseline PD; Am: exposure of nasal epithelium to amiloride, which inhibits sodium channels and leads to a reduction in PD; LC: exposure of nasal epithelium to low chloride solution, which generates a driving force for chloride transport and opens CFTR. This leads to an increase in PD in non-CF, but not in CF, subjects.

1. Perfuse mice with a plasmid carrying the *CFTR* cDNA complexed to a GTA as described above.
2. At appropriate time-points (*see* **Note 8**), after transfections anesthetize animals with Ketaset (76 mg/kg) and Domitor (1 mg/kg) and place on a heating board (*see* **Note 3**).
3. Make a paper-pad by folding a blue towel a few times and place under the mouse's head so that the fold of the paper is near the tip of the nose. The paper is used to absorb any excess liquid coming out of the nose during the perfusion. This prevents the mice from choking.
4. Tape whiskers and back legs to the board using Micropore tape and apply ointment (Lacrilube) to the eyes to prevent drying out during the perfusion. Carefully pull the tongue to one side with forceps. Insert a pointed paper wick into the mouth to collect excess fluid.
5. Fill a venflon catheter (20 gauge) filled with HEPES–Krebs buffer and insert intraperitoneally as a reference electrode. Once in place, withdraw the needle, connect the reference line, and tape to the heating board (*see* **Note 15**).
6. Fill a catheter with HEPES–Krebs buffer and insert into the mouth to measure the buccal PD. This should be around 10–20 mV and indicates that the set-up is working correctly.
7. Use a double-lumen catheter (outer diameter 0.5 mm) to simultaneously perfuse solutions into the nasal cavity and record nasal PD. One lumen contains the perfusate solution, and the other lumen is filled with HEPES–Krebs buffer throughout the procedure and serves as the measuring electrode. Insert the double-lumen catheter into the nose (2.5 mm), and tape to the board which is tilted to approximately 45 degrees with the mouse's head facing down to avoid inhalation of liquids. Record PD using specialized PD equipment.
8. Measure baseline PD using a double-lumen catheter primed with HKA solution to the tip (no perfusion). PD generally stabilizes over a few minutes. Read baseline PD after PD has stabilized for at least 30 s.
9. Perfuse HKA solution perfused (25 μl/min) with a syringe pump until a stable plateau (= amiloride response) has been reached (after approximately 5 min), followed by perfusion with LC HKA for 8 min.
10. Read LC PD (mV) 7 min after starting the LC perfusion (*see* **Note 16**). Measure the LC response [Delta PD (mV)] by calculating the difference (in mV) between the post-amiloride PD and the peak of the LC PD (delta mV).
11. Reverse the anesthesia of the mouse by i.p. injection of Antisedan (1 mg/kg) and place in a recovery box at 30°C for 2 h.

4. Notes

1. The *CFTR* cDNA can be obtained from Dr. Johanna Rommens, Hospital for Sick Children, Toronto, Canada.
2. Transfection conditions (amount of lipid, amount of DNA, and transfection time) are cell type-dependent and have to be carefully optimized.

3. Take care not to warm the mice. This can be achieved by placing a few paper towels between the heating board and the animal.
4. Catheters are briefly held into a flame and quickly pulled to the required thickness.
5. Mark catheter with a pen at approximately 2.5 mm.
6. Initial optimization experiments with a luciferase reporter can also be carried out in wild-type mice, if availability of CF-knockout mice is a rate-limiting factor.
7. Charge ratios (DNA:non-viral GTA), DNA concentration, and perfusion volume and time are GTA-dependent and need to be carefully optimized. We have perfused mice for a maximum time of 6 h using a perfusion volume of 6–7 μl/min.
8. The optimal time-point for assessing reporter or *CFTR* gene expression will depend on the promoter used to regulate reporter gene expression.
9. Detection of luciferase expression in tissue homogenate will not provide information about which cells have been transfected.
10. If required, nasal tissue can be frozen in liquid nitrogen and stored at –80°C before homogenization in RLB-buffer.
11. A variety of luminometers can be used such as the TD20/20 luminometer (Turner BioSystems, Sunnyvale, CA) or the Lucy 1 luminometer (Anthos, Eugendorf, Australia).
12. A black suture thread is tied approximately 2 mm from the tip of the brush to mark insertion depth. At 2–2.5 mm into the nose, the majority of cells lining the septum are respiratory epithelial cells.
13. If required, samples can be stored at –70°C in RLT buffer.
14. Power calculations indicate that correction of CF LC responses to 10, 20, and 30% of wild-type values requires 100, 26, and 12 mice, respectively to be studied in each of the CFTR-transfected and the control groups.
15. It is very important that air bubbles have been completely removed.
16. The start of perfusion of HKA and the LC solution have to be clearly marked on tracing sheet. In general, the LC solution reaches the epithelium 2 min after starting the perfusion, although the exact time varies slightly between catheters, due to subtle differences in length and diameter.

Acknowledgments

The work was funded by the CF Trust and a Senior CF Trust Fellowship (U.G.). We thank Lucinda Somerton for help with preparing the manuscript.

References

1. Riordan, J. R., Rommens, J. M., Kerem, B., Alon, N., Rozmahel, R., Grzelczak, Z., Zielenski, J., Lok, S., Plavsic, N., and Chou, J. L. (1989) Identification of the cystic fibrosis gene: cloning and characterization of complementary DNA. *Science* **245**, 1066–1073.

2. Alton, E. W., Middleton, P. G., Caplen, N. J., Smith, S. N., Steel, D. M., Munkonge, F. M., Jeffery, P. K., Geddes, D. M., Hart, S. L., and Williamson, R. (1993) Non-invasive liposome-mediated gene delivery can correct the ion transport defect in cystic fibrosis mutant mice. *Nat. Genet.* **5**, 135–142.
3. Drumm, M. L., Pope, H. A., Cliff, W. H., Rommens, J. M., Marvin, S. A., Tsui, L. C., Collins, F. S., Frizzell, R. A., and Wilson, J. M. (1990) Correction of the cystic fibrosis defect in vitro by retrovirus-mediated gene transfer. *Cell* **62**, 1227–1233.
4. Hyde, S. C., Gill, D. R., Higgins, C. F., Trezise, A. E., MacVinish, L. J., Cuthbert, A. W., Ratcliff, R., Evans, M. J., and Colledge, W. H. (1993) Correction of the ion transport defect in cystic fibrosis transgenic mice by gene therapy. *Nature* **362**, 250–255.
5. Konstan, M. W., Davis, P. B., Wagener, J. S., Hilliard, K. A., Stern, R. C., Milgram, L. J., Kowalczyk, T. H., Hyatt, S. L., Fink, T. L., Gedeon, C. R., Oette, S. M., Payne, J. M., Muhammad, O., Ziady, A. G., Moen, R. C., and Cooper, M. J. (2004) Compacted DNA nanoparticles administered to the nasal mucosa of cystic fibrosis subjects are safe and demonstrate partial to complete cystic fibrosis transmembrane regulator reconstitution. *Hum. Gene Ther.* **15**, 1255–1269.
6. Moss, R. B., Rodman, D., Spencer, L. T., Aitken, M. L., Zeitlin, P. L., Waltz, D., Milla, C., Brody, A. S., Clancy, J. P., Ramsey, B., Hamblett, N., and Heald, A. E. (2004) Repeated adeno-associated virus serotype 2 aerosol-mediated cystic fibrosis transmembrane regulator gene transfer to the lungs of patients with cystic fibrosis: a multicenter, double-blind, placebo-controlled trial. *Chest* **125**, 509–521.
7. Engelhardt, J. F., Yankaskas, J. R., Ernst, S. A., Yang, Y., Marino, C. R., Boucher, R. C., Cohn, J. A., and Wilson, J. M. (1992) Submucosal glands are the predominant site of CFTR expression in the human bronchus. *Nat. Genet.* **2**, 240–248.
8. Ferrari, S., Griesenbach, U., Geddes, D. M., and Alton, E. (2003) Immunological hurdles to lung gene therapy. *Clin. Exp. Immunol.* **132**, 1–8.
9. Venglarik, C. J., Bridges, R. J., and Frizzell, R. A. (1990) A simple assay for agonist-regulated Cl and K conductances in salt-secreting epithelial cells. *Am. J. Physiol.* **259**, C358–C364.
10. Wright, E. M. and Diamond, J. M. (1977) Anion selectivity in biological systems. *Physiol. Rev.* **57**, 109–156.
11. Derand, R., Bulteau-Pignoux, L., and Becq, F. (2003) Comparative pharmacology of the activity of wild-type and G551D mutated CFTR chloride channel: effect of the benzimidazolone derivative NS004. *J. Membr. Biol.* **194**, 109–117.
12. Baggot, D. (1978) Pharmacokinetics of kanamycin in the dog. *J. Vet. Pharmacol. Ther.* **1**, 163–170s.
13. Dorin, J. R. (2006) *Animal Models in Progress in Respiratory Research: Cystic Fibrosis in the 21st Century* (eds: A. Bush, E. Alton, J Davies, U Griesenbach, A Jaffe), Karger, London, 84–92.
14. Zhou, L., Dey, C. R., Wert, S. E., DuVall, M. D., Frizzell, R. A., and Whitsett, J. A. (1994) Correction of lethal intestinal defect in a mouse model of cystic fibrosis by human CFTR. *Science* **266**, 1705–1708.

15. Jiang, C., O'Connor, S. P., Fang, S. L., Wang, K. X., Marshall, J., Williams, J. L., Wilburn, B., Echelard, Y., and Cheng, S. H. (1998) Efficiency of cationic lipid-mediated transfection of polarized and differentiated airway epithelial cells in vitro and in vivo. *Hum. Gene Ther.* **9**, 1531–1542.
16. Smith, S. N., Middleton, P. G., Chadwick, S., Jaffe, A., Bush, K. A., Rolleston, S., Farley, R., Delaney, S. J., Wainwright, B., Geddes, D. M., and Alton, E. W. (1999) The in vivo effects of milrinone on the airways of cystic fibrosis mice and human subjects. *Am. J. Respir. Cell Mol. Biol.* **20**, 129–134.

15

Oncolytic Adenoviruses for Cancer Gene Therapy

Ta-Chiang Liu, Stephen H. Thorne, and David H. Kirn

Summary

The use of replication-competent oncolytic viruses has largely advanced cancer gene therapy. Oncolytic virus not only possesses unique mechanisms of action that are distinct from other treatment modalities, its self-perpetuating nature provides an ideal platform for therapeutic transgene insertion. Tumor selectivity can be achieved by deleting viral genes that are critical for growth in normal cells but dispensable in tumor cells, transcriptional control under tumor-specific promoters, fiber modification targeting tumor-specific cellular receptors, or the use of inherent tumor-specific viruses. Transgene products can be amplified along with viral replication, thus maximizing therapeutic effect. Using adenovirus as a template, this chapter describes common assays used for the study of oncolytic viruses, with special emphasis on in vitro and in vivo viral replication determination.

Key Words: Oncolytic; adenovirus; cancer; virotherapy; particle count; inactivation; titration; bioluminescence imaging; luciferase.

1. Introduction

Replication-selective oncolytic viruses have emerged as a new platform for cancer treatment. The history of using live viruses for cancer patients dated back a century ago, and the advances of molecular biology and virology have fostered the development of genetically engineered viruses in the last decade. Virotherapy agents can be typically categorized into a few groups based on their mechanisms:

From: *Methods in Molecular Biology, vol. 433: Volume 1: Production and In Vivo Applications of Gene Transfer Vectors*
Edited by: J. M. Le Doux © Humana Press, Totowa, NJ

1. Inherently tumor-selective: Newcastle disease virus (NDV), measles virus, vesicular stomatitis virus (VSV), etc. These viruses are mostly RNA viruses that, upon infection, induce interferon release from the host immune system. Normal cells are protected from further viral replication and spread under interferon through apoptosis induction, while viral replication can still proceed in tumor cells as they are resistant to interferon-mediated apoptosis.
2. Gene deletion approach: First described in herpes simplex virus (HSV), this approach deletes the viral genes critical for growth in normal cells but which are expendable in tumor cells. Adenovirus E1B-55K-deleted mutant (*dl*1520; Onyx-015) was the first to be used in clinical trials and showed success with chemotherapy. Other adenoviral mutants with different viral gene deletions (e.g., E1A-CR2, E1B-19K, E3-gp19K, VAI RNA) have been subsequently developed. This approach has also been utilized in other virus species, for example, vaccinia virus mutants (e.g., VGF-, TK-deleted mutants), and HSV mutants (ICP6-/γ34.5-deleted G207) have been developed and tested in clinical trials.
3. Promoter- or fiber-engineering: This limits the viral replication to specific cell types that activate specific tumor-specific promoters (transcriptional control), or limits the infection of the viruses to desired targets (fiber modification). For example, adenovirus CG7870 replicates in prostate-specific antigen (PSA) -overexpressing prostate cancer cells, and modified adenovirus fibers have been shown to successfully limit the infection to certain tumor types.

Apart from tumor selectivity, the most important characteristic of oncolytic virus is its self-perpetuating nature, which allows amplification of the virus several orders of magnitude at the end of replication cycle. Therefore, assays determining virus replication in vitro and in vivo will be the main focus of this chapter. Readers should refer to other chapters in this book for vector engineering.

2. Materials

2.1. Viral Particle Count Determination

1. Virus lysis buffer: 0.5% sodium dodecyl sulfate (SDS), 0.02 M Tris–HCl (pH 7.4).
2. Spectrophotometer capable of reading at optical density at 260 nm (OD_{260}), and cuvette.
3. 2-ml Eppendorf tubes.
4. Heating blocks.

2.2. Virus Inactivation

1. Ultraviolet (UV) light source (Stratagene Linker, La Jolla, CA, USA).
2. 8-Methoxypsoralen (Sigma, Saint Louis, MO, USA).
3. Sephadex G-50 (Sx-G-50; 10×1 cm) columns (Pharmacia, Kirkland, Quebac, Canada).

4. Human Embryonic Kidney (HEK) 293 cells [American Type Culture Collection (ATCC, Manassas, VA, USA)], cultured in Dulbecco's Modified Eagle's Medium (DMEM; Gibco, Carlsbad, CA, USA) supplemented with 10% Fetal Calf Serum (FCS; Hyclone, Logan, UT, USA).

2.3. Virus Titration by Limiting Dilution

1. A549, KB, or HEK 293 cells (all from ATCC), cultured in DMEM (Gibco) supplemented with 10% FCS (Hyclone).
2. 12-well multichannel pipettes holding 20 and 200 μl.

2.4. Virus Titration by Plaque Assay

1. Overlay medium for plaque assay (per 100 ml): 50 ml 2×DMEM, 2 ml FCS, 15 ml sterile water, and 33 ml 3% SeaPlaque agarose (BMA, Rockland, ME, USA). For scoring overlay medium, add 0.3% Neutral red solution.
2. Pasteur pipettes and pipette tips (Ranin, Woburn, MA, USA).
3. Pipetman and aspiration pipettes (Falcon, Franklin Lakes, NJ, USA).
4. Sterile 250-ml Erlenmeyer flasks (Corning, Lowell, PA, USA).
5. Microwave.

2.5. Ex Vivo Infection of Patient Tissue

1. Surgical tools: scalpels, scissors, forceps.
2. DMEM.
3. Tissue culture dishes and 12-well plates (Falcon).
4. Paraformaldehyde (EMS, Hatfield, PA, USA).

2.6. Bioluminescence Imaging

1. Plate reading luminometer/fluorometer.
2. D-Luciferin, 30 mg/ml in PBS, filter sterilized (Biosynth, Naperville, IL, USA).
3. Coelentrazine 5 mg/ml in methanol (Millipore Chemicon, Temecula, CA, USA).
4. Mice, normally immunodeficient (nu/nu or SCID) for studies with oncolytic adenoviruses, typically obtained form commercial vendors (e.g., Jackson laboratories, Bar Harbor, ME, USA or Charles River, Wilmington, MA, USA).
5. IVIS imaging system or equivalent whole animal imaging system (Xenogen Corp., Alameda, CA) with heated (37 °C) stage and appropriate image analysis software (e.g., Living Image, Xenogen Corp.).
6. 70% Ethanol.
7. Anesthetic (ideally inhaled, such as isoflurane, but may be injectable, such as avertin).
8. Black walled, clear bottom 96-well plates.
9. Luciferase assay system (Promega, Madison, WI, USA).

3. Methods

3.1. Viral Particle Count Determination

Particle counts are being used more often than plaque-forming units (pfu) as quantification of stock virus or input dose for oncolytic adenoviruses. This protocol is to quantitate viral particles (pt) in purified virus preparations by measurement of absorbance at 260 nm (*see* **Note 1**).

1. Label tubes (tubes 1 and 2) for two viral dilutions and (tubes 3 and 4) for two blank tubes containing the corresponding dilutions of dialysis buffer alone.
2. Add 100 μl and 200 μl viral stock to tubes 1 and 2, respectively.
3. Add 100 μl and 200 μl of dialysis buffer to tubes 3 and 4, respectively.
4. Add 100 μl of dialysis buffer to tubes 1 and 3.
5. Finally add 200 μl of the 2×Lysis buffer to all tubes, close the tops, vortex, and place tubes in heating block. Incubate at 56 °C for 10 min.
6. Remove tubes and vortex 3 s at maximum speed, let cool at room temperature for 5–10 min, and vortex.
7. Switch on spectrophotometer, use the 260/280-program, and let the UV-lamp warm up for at least 10 min before reading samples.
8. Add 0.6 ml sterile H_2O to make final volume 1 ml in all tubes. Vortex.
9. Place clean quartz cuvette with sterile H_2O in spectrophotometer and read blank. This will set the baseline value to zero. Check that spectrophotometer is stable by reading absorbance at 260 nm one to two times. If stable ($OD_{260} = 0$), measure samples.
10. Replace H_2O in the cuvette with the first blank solution (tube 3) and measure OD_{260}. The blank reading should be in the range 0.002–0.012. If higher, check buffers.
11. Replace blank solution with sample 1 and measure OD_{260}.
12. Repeat for blank 4 and sample 2. Rinse cuvette with H_2O (empty excess liquid) between solutions to minimize carry-over.
13. Acceptance criteria: The highest sample concentration has to be above 0.05 OD units. The absorbance of sample dilution 2 should always result in twice the absorbance of sample dilution 1, for example, dilution 1 = 0.050 and dilution 2 = 0.090–0.110. If sample dilution 1 is below 0.02 OD units, it is necessary to repeat the absorbance measurement using more virus (for example, 150 μl and 300 μl virus stock).
14. To calculate the number of viral particles per ml in the stock solution, use the following formula: $OD_{260} \times \text{dilution factor} \times 1.12 \times 10^{12} = \text{pt/ml}$.

3.2. Virus Inactivation

Sometimes it is important to determine whether the numbers of virus capsids or particles alone, and not infective viral DNA, have an impact in certain circumstances (e.g., induction of Cytopathic effect (CPE) or immune reaction). It is therefore necessary to inactivate the virus so that replication is limited.

1. Mix the virus with 8-methoxypsoralen to a final concentration of 0.33 μg/μl and expose the mixture to UV light 365 nm on ice for 3×10 min.
2. Desalt the reaction mixture using a Sx-G-50 (10×1 cm) column.
3. Verify the inactivation of virus by plaque and cytopathic effect on HEK293 cells; the virus should be more than 1000 times less potent than the original viral preparation.

3.3. Virus Titration by Limiting Dilution

One basic assay for studying oncolytic viruses is the titration assay, which is used to determine the amount of replication-competent virus in stocks and to quantify the extent of replication within virus-treated cells. The number of infectious virions can be quantified using the limiting dilution titration assay $TCID_{50}$ (tissue culture inhibitory dose, 50%) ***(1,2)*** (*see* **Note 2**), which is derived from the method described first by Reed and Muench ***(3)***. The traditional plaque assay is described later.

1. Seed HEK 293 cells into 96-well plates at 10,000 cells/well in 200 μl DMEM + 10% FCS.
2. Dilute the test samples (virus stock or sample obtained for replication assessment) based on the expected titer of the adenovirus vector 24 h later. Add 20 μl volume of diluted sample to each well in the top row (Row A) across the plate of HEK293 cells. Use a multichannel pipette to do 10-fold serial dilutions down the 96-well plate by removing 22 μl from each well in a row and adding it to the 200 μl of volume in the row below. Repeat this for each row through Row G so that each row represents 12 cultures at each dilution. Leave Row H uninfected as a control. Incubate the infected cells at 37 °C and 5% CO_2 for 10–14 days.
3. Examine the titer plates of HEK293 cells visually for CPE caused by productive virus infections between days 10 and 14. A well is scored positive for CPE if any of the cells have rounded up or if a plaque has formed in any part of the well. Record the number of wells in a row that are positive for CPE. Examples of the morphology of the HEK293 cells at day 14 following application of the sample are shown in **Fig. 1**.
4. Enter the number of wells in a row that are positive for CPE into the $TCID_{50}$ calculation. A detailed description of the mathematics behind the $TCID_{50}$ calculation is given in **ref. *1***. A Microsoft Excel spreadsheet is shown in **Table 1** for the calculation of the viral titer (adapted from **ref. *1***).
5. If CPE is observed in ≤6 of the wells at the lowest dilution, no titer can be returned by the $TCID_{50}$ calculation and the titer is expressed as being "less than the limit of detection." The limit of detection is calculated by entering a dilution

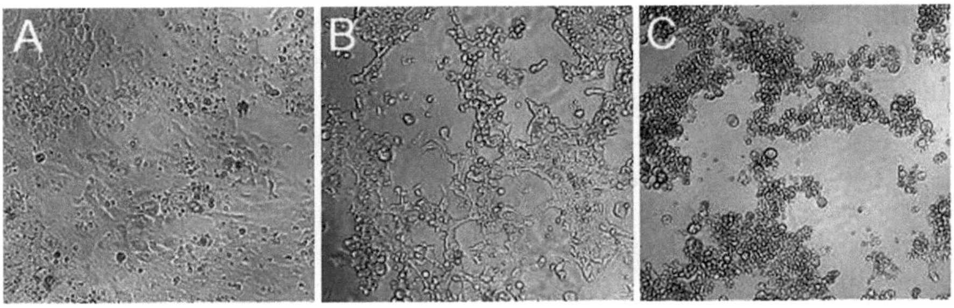

Fig. 1. Cytopathic effect (cpe). Typical cytopathic effect visualized 14 days after titration. **(A)** Uninfected cells. **(B)** Example of a well showing partial cpe and was thus scored as positive. **(C)** Example of a well showing full cpe.

10-fold lower than the lowest dilution used in column A, and entering "12" as the number of infected wells in column B of the $TCID_{50}$ formula (*see* **Table 1**). The pfu/ml value returned by the equation is the limit of detection.

3.4. Virus Titration by Plaque Assay

This is carried out using the HEK293 cell line in a six-well-plate format.

1. Plate out HEK293 cells in six-well plates (approximately 10,000 cells/well). Adjust the seeding density so that the cells reach 75–80% confluency in 24–48 h.
2. Set up the dilutions across the following range: 10^8, 10^9, 10^{10}, 10^{11}, and 10^{12}. Each well will be infected with virus in a total of 1 ml of medium. Set up triplicate plates for each sample.
3. Defrost the viruses and place on ice. Set up the dilutions as described, using DMEM medium (without serum), in sterile tubes. Ensure that each tube is mixed thoroughly after each dilution.
4. To infect the six-well plates, aspirate the medium from each of the wells, and add 1 ml of the virus dilution to the cells; gently agitate the plate after addition to the wells to ensure even exposure of the cells. Return the plates to the incubator after addition of the virus.
5. Incubate the plates for approximately 1.5 h. Then aspirate the supernatant, and replace with 2 ml overlay medium per well.
6. Make the liquid medium components up in a sterile Erlenmeyer flask. Aliquot 33 ml into a 50-ml sterile tube. Heat the SeaPlaque agarose (3% in sterile distilled water, autoclaved) in a microwave until fully melted and bubbling slightly. Have the medium distributed in the tubes and the agarose melted.
7. Because 50 ml of overlay will go onto four six-well plates (using 2 ml of overlay per well), remove four plates from the incubator. Aspirate the medium form each well, leaving a very slight film of medium on each well to prevent drying. Add the melted agarose to the 33 ml of medium up to the 50 ml mark on the

Table 1
Microsoft Excel Spreadsheet for $TCID_{50}$ Calculation

	A	B	C	D	E	F	G	H	I	J
1	Dilutions	Number of infected wells	Number of uninfected wells	Total number of infected	Total number of uninfected	% total infected				
2	*1E-02*	*12*	=$B$9 – B2	=SUM (B2:$B$8)	=SUM ($C$2:C2)	=D2/ (D2+E2)	=IF(F2>0.5, "TRUE", "FALSE")	=IF(AND (G2="TRUE", G3="FALSE"), F2,0)	=IF(AND (G10="TRUE", G2="FALSE"), F2,0)	=IF(H2>0, LOG(A2),0)
3	*1E-03*	*7*	=$B$9 – B3	=SUM (B3:$B$8)	=SUM ($C$2:C3)	=D3/ (D3+E3)	=IF(F3>0.5, "TRUE", "FALSE")	=IF(AND (G3="TRUE", G4="FALSE"), F3,0)	=IF(AND (G2="TRUE", G3="FALSE"), F3,0)	=IF(H3>0, LOG(A3),0)
4	*1E-04*	*3*	=$B$9 – B4	=SUM (B4:$B$8)	=SUM ($C$2:C4)	=D4/ (D4+E4)	=IF(F4>0.5, "TRUE", "FALSE")	=IF(AND (G4="TRUE", G5="FALSE"), F4,0)	=IF(AND (G3="TRUE", G4="FALSE"), F4,0)	=IF(H4>0, LOG(A4),0)
5	*1E-05*	*1*	=$B$9 – B5	=SUM (B5:$B$8)	=SUM ($C$2:C5)	=D5/ (D5+E5)	=IF(F5>0.5, "TRUE", "FALSE")	=IF(AND (G5="TRUE", G6="FALSE"), F5,0)	=IF(AND (G4="TRUE", G5="FALSE"), F5,0)	=IF(H5>0, LOG(A5),0)
6	*1E-06*	*0*	=$B$9 – B6	=SUM (B6:$B$8)	=SUM ($C$2:C6)	=D6/ (D6+E6)	=IF(F6>0.5, "TRUE", "FALSE")	=IF(AND (G6="TRUE", G7="FALSE"), F6,0)	=IF(AND (G5="TRUE", G6="FALSE"), F6,0)	=IF(H6>0, LOG(A6),0)

(Continued)

Table 1
(Continued)

	A	B	C	D	E	F	G	H	I	J
7	*1E-07*	*0*	=$B$9 – B7	=SUM (B7:$B$8)	=SUM ($C$2:C7)	=D7/ (D7+E7)	=IF(F7>0.5, "TRUE", "FALSE")	=IF(AND (G7="TRUE", G8="FALSE"), F7,0)	=IF(AND (G6="TRUE", G7="FALSE"), F7,0)	=IF(H7>0, LOG(A7),0)
8	*1E-08*	*0*	=$B$9 – B8	=SUM (B8:$B$8)	=SUM ($C$2:C8)	=D8/ (D8+E8)	=IF(F8>0.5, "TRUE", "FALSE")	=IF(AND (G8="TRUE", G9="FALSE"), F8,0)	=IF(AND (G7="TRUE", G8="FALSE"), F8,0)	=IF(H8>0, LOG(A8),0)
9	Number of of wells	*12*								
10	ml/well	0.02			0.5	=SUM (H2:H8)	=SUM (I2:I8)	=SUM (J2:J8)		
11	Prop. Dist.	=(H10 – 0.5)/ (H10 – I10)								
12	Log TCID50	=J10 – B11								
13	TCID50	=10B̂12								
14	1/TCID50	=1/B13								
15	TCID50/ml	=B14/B10								
16	pfu/ml	=B15*0.69								

Note: Data in italics are entered by the investigator on a per experiment basis. Adapted from **ref. *1***.

tube. Replace cap and mix quickly. Hold tube in hand to ensure that it is not too hot to add to the cells; it should be warm, about the same temperature as a baby bottle. Carefully layer 2 ml of overlay to each well by letting it glide down the side of each well. Do not drop directly onto the cell layer. Let solidify in the hood (approximately 15 min) and then return to the incubator.

8. Check daily to ensure that agarose is not drying out. Look at cells under microscope to make sure they are still alive and well and to monitor the progression of plaque formation.
9. At day 6, add an additional 1 ml volume of scoring overlay medium to each well. This will feed the cells and the dye makes the plaques easier to visualize.
10. Count the plaques at days 7, 10, and 14. Use data that give about 10–75 plaques per well. Too few will result in large standard errors and too many are too difficult to count because plaques will merge/overlap. Calculate the pfu/ml as follows: average plaque number for 3 wells$\times$dilution = pfu/ml. For example, if there are 3 wells with 60, 50, and 55 plaques each on a dilution of 10^{-10}, the final titer is 5.5×10^{11} pfu/ml.

3.5. Ex Vivo Infection of Patient Tissue

This protocol describes assessing viral infection/replication in primary human tissues obtained from surgical specimens as a direct measure of virus activity *(4)*.

1. Obtain surgically removed tissue from operating room and process for infection within 30 min.
2. Cut tumor tissue or normal tissue into slices measuring approximately 2 mm square. Cut enough sections to harvest a sample every 24 h for 3 days.
3. Place each chunk into a well of a 12-well plate. Each well contains 0.5 ml of DMEM plus 5×10^9 pfu of adenovirus or no virus (*see* **Note 3**). Incubate the samples in a 37 °C incubator with 5% CO_2.
4. Harvest tumor and normal tissue samples every 24 h by removing the sample from the medium and fixing in formalin.
5. Embed the samples in paraffin, section, and stain for general histopathology or immunohistochemistry. The detail of these procedures is beyond the scope of this book and can be found in **ref.** *5*.

3.6. Labeling Cells or Viruses for Bioluminescence Imaging

Non-invasive whole animal imaging provides several advantages over traditional animal models. The ability to repeatedly image the same animals means that they become their own internal controls and also dramatically reduces the number of animals required to run experiments. In addition, unexpected and expected signals detected during whole animal imaging can be used to define time points and tissues selected for ex vivo analysis, and so better define the routes of infection, timings of therapeutic responses, and any off-target toxicities caused by infection of non-tumor tissues. The design of experiments will be determined by what parameters are being investigated (e.g., viral biodistribution, therapeutic effect), such that labeling of tumors, virus, or immune cells may be incorporated into the imaging approaches. The first decision must therefore be the choice of agent to be imaged and the reporter used to label them.

1. Labeling adenoviral strains: The limited cloning capacity of replication-competent adenovirus vectors means that only a single reporter gene is likely to be expressed, either a luciferase enzyme or a fluorescent protein. However, fluorescent fusion proteins may be used to save packaging space in the vector, for example, luciferase–green fluorescent protein (GFP) fusion proteins are of small size and viral protein–GFP fusions have been described *(6,7)*. Although in vivo fluorescence imaging is possible, it is likely that bioluminescence imaging will be primarily used for in vivo analysis and fluorescence labeling used for ex vivo analysis of viral gene expression (e.g., FACS or immunofluorescence).
2. The choice of luciferase enzyme incorporated for viral labeling is important. In order to achieve maximum light output from within a living organism, a significant part of the spectrum of the light produced from any luciferase should be greater than 600 nm, as light below this wavelength will be adsorbed by hemoglobin. The most commonly used luciferase is that of the firefly (Promega), and this produces light with a spectral peak of 560 nm; however, at 37 °C this shifts to 590 nm making it suitable for in vivo use *(8)*. Other insect luciferases, such as those of the click beetle or railroad worm, produce a variety of spectra, and although green luciferases are often brighter in vitro, the massive adsorption at these wavelengths in vivo makes them less useful. Several luciferases have also been cloned from marine organisms, such as Renilla and Aequorea, but these produce blue-green light (peak of 475 nm). However, the advantage of Renilla luciferase is that it utilizes a different substrate (coelentrazine) to the insect luciferases (luciferin) and so dual bioluminescence imaging is possible (see below and **Note 4**).
3. Another consideration is the choice of promoter to use. Typically if viral biodistribution is being examined, a strong constitutive promoter (e.g., CMV) would be used to drive luciferase expression. If a tissue-specific or inducible promoter system is to be used, consideration should be given to the strength of the promoter as it is likely that weak promoters will not be sufficient for in vivo analysis, but two-step promoter systems may overcome these issues *(9)*. Cre-lox systems may also be incorporated, with weakly or transiently virally-expressed Cre mediating recombination of cellular loxP sites and so inducing luciferase expression in infected cells.
4. Dual reporter systems: In the study of oncolytic viruses, it would be preferable to label tumor cells, virus, and even immune cells and so study the progression of disease, therapeutic delivery, response, and viral clearance, all in parallel in vivo in the same animal. If this is attempted, careful consideration should be given to what imaging modalities are used to study different parts of this system. Although

PET/SPECT, MRI, CT, US, or X-ray may all be incorporated to image one or more of the components of this therapy, this protocol concentrates on optical techniques. It is likely that insect luciferases will provide the strongest signal, and the least background, and so should be used to image the most important or least prevalent component (*see* **Note 5**). Renilla luciferase will produce a weaker signal and more background, and so is best utilized to image large tumors or to track primary sites of viral infection and gene expression. Finally, fluorescent labels can also be incorporated. As with choice of luciferase, an emission spectra producing some light greater than 600 nm is required. This means that red or red-shifted fluorescent proteins will work most efficiently. Alternatively, it may be possible to covalently label or load virus or cells with near-infrared fluorescent probes, such as AlexaFluor-647 or Cy5.5, but these cannot be used to examine replication. Also, because light is required to both enter (excitation) and exit (emission) the organism and because auto-fluorescence is often high, producing high background, detection of fluorescent probes can be limited, and so should only be used for strong signals (or for ex vivo analysis).

5. Animal models: Because of the adsorption and diffraction limitations of optical imaging, tissues greater than about 1 cm deep cannot be imaged. This means that the ideal animal model to use will be the mouse. Although larger animals (such as rats or even rabbits) have been used, typically this is for focused imaging of one target organ or tissue that is close to the surface of the animal. Choice of mouse strain may be limited by the availability of cell lines, but in general white mice are preferable (as the pigmentation in the skin of black or brown mice further adsorbs emitted light). Some normally dark mice (e.g., C57B/6) have been bred as albinos. Animal fur also diffracts light, and so shaving or use of depilatory cream may enhance weak signal. Alternatively, athymic nu/nu mice may be used (and this strain is essential when fluorescence imaging is to be used).

3.7. In Vivo Imaging

1. Assay the in vitro properties of any new constructs prior to in vivo analysis (*see* **Note 6**).
2. Run pilot experiments in vivo to characterize the model (*see* **Note 7**).
3. Deliver luciferin substrate to mice. Imaging of insect luciferase will typically require injection of 200 μl of 30 mg/ml luciferin intraperitoneally 5–20 min prior to imaging; coelentrazine for Renilla luciferase will typically be injected intravenously (tail vein or retino-orbital injection), with 100 μl of 5 mg/ml injected no longer than 5 min prior to imaging.
4. Anesthetize mice. The exact order of delivery of substrate and anesthesia will depend on the system being used. The choice of anesthetic may also be important; some injectable anesthetics may interfere with certain metabolic pathways that may be integral to the study (*see* **Note 7**). In addition, mice take longer to recover and so fewer imaging time points can be used. Inhaled anesthetics (such as isoflurane, 3%) are therefore typically preferable; however, there is some evidence

that these may produce a low level of background luminescence, and so may not be suitable for detection of weak signal, especially from the lungs.

5. Arrange mice in the imaging chamber so that the expected source of signal is closest to the top of the animal (e.g., ventral view for imaging major organs, dorsal for imaging spine, left side for imaging spleen). Wipe imaging stage with 70% ethanol prior to use, and place a piece of black paper on the stage to protect it. Use dividers between the animals (if multiple animals are to be imaged at once), to prevent strong signal from one animal reflecting off its neighbor. Remove all non-essential materials from the imaging chamber. Heat the stage to 37 °C, and check the animals for level of anesthesia for 30 s prior to closing the chamber (*see* **Note 8**).
6. Take image. Several variables may be adjusted to improve the quality of the image, including height of stage, focus, length of exposure, binning and aperture setting (*see* **Note 9**). In order to help locate the source of a signal, reposition and re-image the mice from different angles. Recent advances in software available for image analysis may include 3D reconstructions, whereby animals are imaged consecutively at different wavelengths, and the relative levels of blue and red light emitted can be used to determine the depth of the signal. An example is shown in **Fig. 2**.

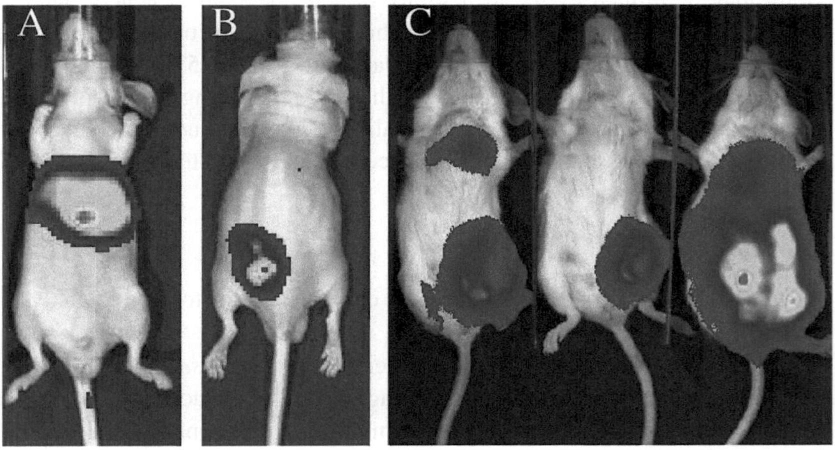

Fig. 2. Examples of Bioluminescence imaging of adenoviral vectors (**A** and **B**) and tumor cell lines (**C**). (**A**) Athymic nu/nu mouse 24 h following intravenous (tail vein) injection with 1×10^{10} pfu of Ad-luc. (**B**) Athymic nu/nu mouse bearing unlabeled subcutaneous HCT 116 tumor, 24 h following intratumoral injection of 1×10^{10} pfu of Ad-luc. (**C**) BALB/c mice bearing 4T1-luc tumors; mammary fat pad with metastasis to the lungs (left); mammary fat pad without metastasis (center); intraperitoneal (right) (Note: The fat pad tumors have begun necrosis in the core, as seen by the donut-shaped bioluminescent signal).

7. Recover mice. Once imaging is complete, recover the mice on a heated stage, until capable of self-righting.
8. Image analysis: A variety of software packages exist for image analysis, including quantification, such as Living Image (Xenogen Corp.).
9. Verification of imaging results: It is important to verify patterns of gene expression and the source of any signals ex vivo in a subset of animals. Bioluminescence signal from firefly luciferase will remain strong for a period of 45 min postmortem, and so individual organs can be imaged to verify the tissue of origin of any signal (although the signal no longer remains quantifiable). In addition it should be remembered that bioluminescent signal represents biodistribution of gene expression and not viral delivery. Luciferase expression can be verified ex vivo using luciferase assay systems (on frozen, ground tissues, and normalized to protein level), and viral genomes can be quantified by Q-PCR (or viral infectious units quantified by plaque assay) to determine the proportional differences between viral delivery and gene expression in different tissues.

4. Notes

1. To verify correct virus concentration in each virus stock, at least two people should determine OD_{260} by preparing the samples separately. It is also recommended that particle counts be determined before using a new stock for a major experiment.
2. The limiting dilution assay can be used to measure titers of replication-competent adenovirus in any cell type of interest. The actual optimal dilutions will, however, be different from sample to sample. It is therefore recommended that a trial assay be done on samples from each untested cell line/virus before embarking on a large-scale titration process.
3. To control for differences in transduction when comparing the vector production in different cell types for an oncolytic vector, one must relate vector production relative to a wild-type control such as Ad*dl*327 or Ad5.
4. Insect luciferases require ATP and oxygen to convert their substrate (luciferin) and produce light; the ATP requirement is typically satisfied by the intracellular environment, while the oxygen requirements are low and are typically similar to the levels required to maintain a cell. Therefore these luciferases are expected to function within hypoxic, but not necrotic regions of tumors. Marine luciferases (e.g., Renilla), do not require ATP, and so may also be able to function extracellularly. The major limitations of Renilla luciferase are that the majority of the light produced will be adsorbed (especially from within deeper tissues) and that the coelentrazine substrate is unstable. This means a much higher background signal is produced, and the substrate only remains within an animal for a short period of time. However, this can be an advantage when using dual imaging of both firefly and Renilla luciferases in a single animal, as it is possible to image the Renilla first following addition of coelentrazine, and then to image the firefly luciferase (as soon as 1 h later) following addition of luciferin. Both substrates

are non-toxic, are rapidly distributed throughout the body, and can cross most membranes, including cellular membranes and the blood–brain barrier.

5. Source of luciferase-labeled cell types: Immune cells can be obtained from a universal transgenic donor mouse, or may be transiently or stably labeled ex vivo, while tumor cells should be stably transfected.
6. Cell lines or viruses constructed to express reporter genes should always be checked extensively in vitro to verify there has been no unexpected alteration to their properties. This may include verification of proliferation rate, morphology, any cellular characteristics important to a particular study for a cell line, replication rates and levels, and cytopathic effect for viruses. In addition, the photon/cell light output should be determined for cell lines (in black walled 96-well plates), and the light produced following infection at different multiplicities of infection and different times after infection should be examined for viral constructs.
7. An initial pilot experiment is always recommended. Verify tumor growth patterns and measure rates for different injected doses of cancer cell lines; this may be particularly relevant for imaging studies, as early tumor development (prior to the development of overt signs of disease following injection of low doses of tumor cells) can be followed. In addition, dosing with limiting numbers of cells injected into the expected location of the tumor (e.g., subcutaneous, intraperitoneal) or injecting limiting numbers of cells infected with a labeled virus, followed immediately by imaging, can be used to predict limits of detection, or to help quantify signals. Finally, sequences of images should be taken at different times post addition of luciferin substrate in order to verify the optimal timing of imaging. Anesthetic to be used should also be tested to ensure the minimum required dose is used, and if possible, to verify that the anesthetic used does not interfere with the study. As most injectable anesthetics are metabolized by the subject, they may affect metabolic pathways, stress responses, or oxygen utilization. In addition, protocols for all planned experiments should first be approved by institutional review boards.
8. Many plastics and other materials (including animal bedding, chow, and dander) will produce phosphorescence, and so careful selection of any item used in the imaging chamber is important. In particular, anesthesia nose cones, paper used to protect the stage, plastic dividers, and animal shields should always be carefully tested, and all bedding, dander, and chow should be cleaned from the stage before imaging.
9. The height of the stage should be adjusted so that the animals to be imaged best fit the field of view with minimal empty space. It is recommended that initially a 60-s image be tried with medium binning if the strength of the signal is not known. If this saturates the camera, then quantification is not possible, and so shorter exposure times should be tried and smaller binning used. If at 1 s, minimum binning image still saturates the CCD camera, then the aperture may be reduced. If the original image produces weak or no signal, then increased exposure times (up to 5 min) and increased binning may be tried. Images longer than 5 min will likely not improve the signal, as the background will also increase.

If imaging of a weak signal is required, behind a stronger signal (e.g., when studying metastasis), it may be possible to place a shield over the source of the bright signal, so revealing a weaker signal elsewhere in the animal.

References

1. O'Reilly, D. R., Muller, L. K. and Luckow, V. A. (1994) *Baculovirus Expression Vectors: A Laboratory Manual*, Oxford University Press, pp. 132–34.
2. Liu, T. C., Hallden, G., Wang, Y., Brooks, G., Francis, J., Lemoine, N. and Kirn, D. (2004) An E1B-19 kDa gene deletion mutant adenovirus demonstrates tumor necrosis factor-enhanced cancer selectivity and enhanced oncolytic potency. *Mol Ther* **9**, 786–803.
3. Reed, L. and Muench, H. (1938) A simple method for estimating fifty percent endpoints. *Am J Hyg* **27**, 858–63.
4. Wang, Y., Thorne, S., Hannock, J., Francis, J., Au, T., Reid, T., Lemoine, N., Kirn, D. and Hallden, G. (2005) A novel assay to assess primary human cancer infectibility by replication-selective oncolytic adenoviruses. *Clin Cancer Res* **11**, 351–60.
5. Bancroft, J.D. and Gamble M. (Ed). (2002) *Theory and Practice of Histological Techniques* (5th Edition), Churchill Livingstone, pp. 63–139.
6. Edinger, M., Sweeney, T. J., Tucker, A. A., Olomu, A. B., Negrin, R. S. and Contag, C. H. (1999) Noninvasive assessment of tumor cell proliferation in animal models. *Neoplasia* **1**, 303–10.
7. Le, L. P., Li, J., Ternovoi, V. V., Siegal, G. P. and Curiel, D. T. (2005) Fluorescently tagged canine adenovirus via modification with protein IX-enhanced green fluorescent protein. *J Gen Virol* **86**, 3201–8.
8. Zhao, H., Doyle, T. C., Coquoz, O., Kalish, F., Rice, B. W. and Contag, C. H. (2005) Emission spectra of bioluminescent reporters and interaction with mammalian tissue determine the sensitivity of detection in vivo. *J Biomed Opt* **10**, 41210.
9. Iyer, M., Salazar, F. B., Lewis, X., Zhang, L., Wu, L., Carey, M. and Gambhir, S. S. (2005) Non-invasive imaging of a transgenic mouse model using a prostate-specific two-step transcriptional amplification strategy. *Transgenic Res* **14**, 47–55.

16

Design of Trans-Splicing Adeno-Associated Viral Vectors for Duchenne Muscular Dystrophy Gene Therapy

Yi Lai, Dejia Li, Yongping Yue, and Dongsheng Duan

Summary

The development of *trans*-splicing vectors opens the door for delivering a large therapeutic gene with adeno-associated viral vectors (AAV). One potential application is to deliver the 6 kb mini-dystrophin gene for Duchenne muscular dystrophy (DMD) gene therapy. However, early attempts have been very disappointing because of low transduction efficiency. We have recently identified mRNA accumulation as a critical barrier for the *trans*-splicing AAV vectors. This barrier can be overcome by rational selection of the gene splitting site. Here we outline a detailed RNase protection assay-based strategy to determine the optimal gene splitting site for the mini-dystrophin gene. We also provide methods to evaluate transduction efficiency of the mini-dystrophin *trans*-splicing vectors in *mdx* mouse, a model for DMD.

Key Words: AAV; adeno-associated virus; Duchenne muscular dystrophy; dystrophin; mini-dystrophin; *trans*-splicing; gene therapy; muscle; *mdx*.

1. Introduction

Duchenne muscular dystrophy (DMD) is the most common lethal childhood muscle-wasting disease affecting all body muscles. It is caused by mutations in the dystrophin gene. Current treatment can only alleviate certain symptoms but cannot cure the disease. An ultimate solution to DMD requires replacing the mutated dystrophin gene with a functional gene. To achieve this goal, one needs to have a vehicle that can efficiently deliver a therapeutic gene to all muscles in the body. Adeno-associated viral vector (AAV) seems an ideal tool

From: *Methods in Molecular Biology, vol. 433: Volume 1: Production and In Vivo Applications of Gene Transfer Vectors*
Edited by: J. M. Le Doux © Humana Press, Totowa, NJ

because it readily transduces muscle fibers either by direct muscle injection or through systemic delivery. The challenge with AAV-mediated DMD gene therapy lies in the small viral packaging capacity.

Wild type AAV carries a ~4.7 kb single-stranded DNA genome. It contains two flanking inverted terminal repeats (ITRs) at the ends and two internal open reading frames encoding viral capsid (cap gene) and replication proteins (rep gene), respectively. In a recombinant vector, cap and rep genes can be replaced by a reporter and/or therapeutic expression cassette. This dictates a $\leq$5 kb vector genome size for an AAV vector. The full-length dystrophin protein is 427 kD and contains the N-terminal, central rod, cysteine-rich, and C-terminal domains. It is encoded in an 11.6 kb coding sequence. Apparently, this exceeds the packaging capacity of a single AAV virion. To solve this problem, investigators have developed the mini- and micro-dystrophin genes (*see* **Fig. 1A**) *(1,2)*. The microgenes are less than 4 kb and do not carry the C-terminal domain. They

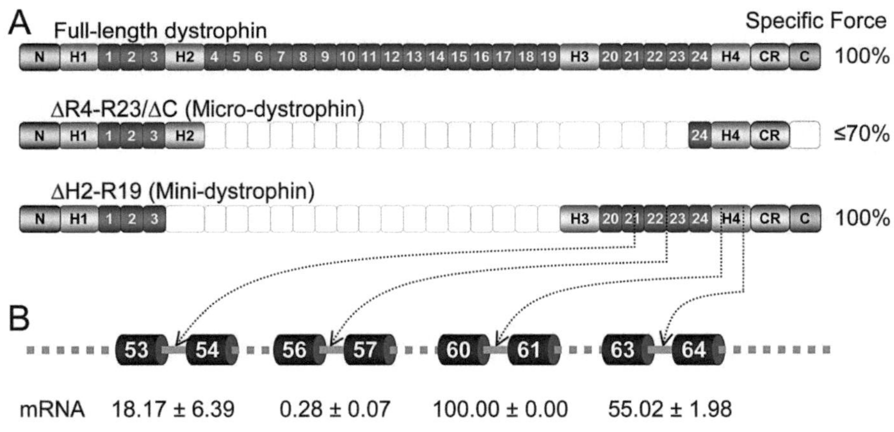

Fig. 1. Dystrophin isoform structure and the effect of the gene splitting site on mRNA accumulation in mini-dystrophin. (**A**) Schematic outline of the full-length, micro- and mini-dystrophin proteins. Four functional domains in the full-length protein are the N-terminal domain (N), the rod domain (including 24 spectrin-like repeats and four hinges, H1 to H4), the cysteine-rich domain (CR) and the C-terminal domain. Regions that are deleted in micro- and mini-dystrophins are shown as white boxes. Specific force data are derived from Harper et al. (2002) *Nat Med* (2) (**B**) Relative mRNA levels according to RPA results from the mini-cassette depicted in **Fig. 2B**. The data are derived from Lai et al. (2005) *Nat Biotech* (7). The cartoon on the top of the panel B depicts the exon/intron/exon junctions where the minigene is split in each pair of the *trans*-splicing vectors. The dotted lines refer to the corresponding locations in the rod domain.

can fit into a single AAV vector but cannot restore specific force to the normal level *(2)*. The 6 kb minigene is an optimized version of a truncated, but fairly functional human dystrophin gene *(3)*. It contains all the functional domains of the full-length protein. The original gene is capable of supporting daily activity in a 60-year-old patient *(3)*. The optimized gene performed even better than the original gene, and it resulted in a muscle-specific force indistinguishable from that of wildtype mice *(2,4)*. Efficient delivering of the 6 kb minigene with AAV vector may lead to a cure to DMD.

The *trans*-splicing dual vector approach effectively doubles the AAV packaging capacity and raises the hope of delivering the 6 kb minigene with AAV vector *(5–7)*. In this context *trans* -splicing is defined as the reconstruction of an intact transcript by splicing in "*trans*" between two covalently linked vector genomes each carrying a part of the transgene (*see* **Fig. 2**). This approach stems from a unique feature in AAV transduction biology. After entering cells, viral genomes form head-to-tail concatamers through ITR-mediated recombination. By taking advantage of cellular splicing machinery, one can engineer splicing signals at the viral genome junction to remove the noncoding component at the junction and restore a functional expression cassette (*see* **Fig. 2**). The *trans*-splicing system is composed of two vectors.

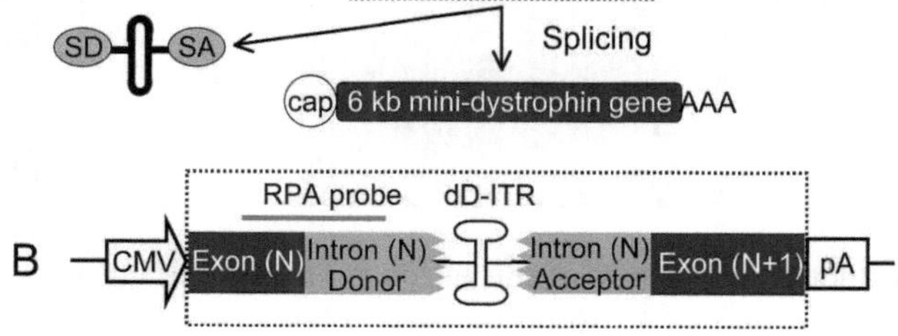

Fig. 2. *Trans*-splicing AAV-mediated expression of the mini-dystrophin gene. (**A**) Schematic outline of the transgene reconstitution in mini-dystrophin *trans*-splicing vectors. (**B**) The mini-cassette used in RPA to screen for the best gene splitting site. The dotted boxes in panels A and B represent the same region. SD, splicing donor; SA, splicing acceptor; ITR, AAV inverted terminal repeat; dD-ITR, the double D ITR structure formed in recombination.

An AV.Donor vector carries the promoter, the 5′ transgene, and the splicing donor signal. An AV.Acceptor vector carries the splicing acceptor signal, the 3′ transgene, and the polyA sequence.

Successful transduction of the *trans*-splicing vectors depends on efficient co-infection of AV.Donor and AV.Acceptor, efficient viral genome recombination, and mRNA production from the reconstituted genome. We have previously demonstrated that co-infection is not a barrier. However, mRNA accumulation represents a predominant rate-limiting step *(8)*. On the basis of this observation, we have developed an RNase-protection assay (RPA)-based screening strategy to rationally design the *trans*-splicing vectors for the 6 kb minigene. Using the conserved splicing value as a guide, we have analyzed a series of potential gene splitting sites in the mini-dystrophin gene. Our screening has yielded a pair of extremely efficient minigene *trans*-splicing vectors. This pair of vectors is based on splitting the gene at endogenous intron 60. Their transduction efficiency has reached that of a single AAV vector (*see* **Fig. 3**).

In this protocol, we outline the procedures used in screening the gene splitting site, generating the mini-dystrophin *trans*-splicing vectors, and confirming mini-dystrophin expression in the *mdx* mouse model of DMD.

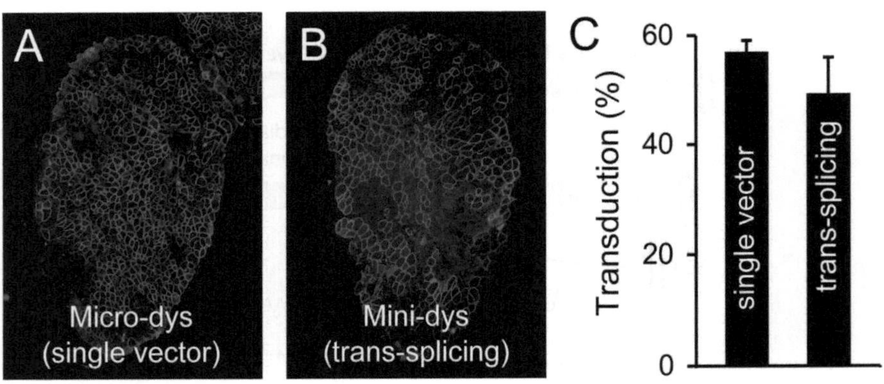

Fig. 3. Transduction efficiency of the intron 60 *trans*-splicing vectors reaches that of the single intact vector. (**A**) Representative photomicrographs of the *mdx* EDL muscles infected by AV.Micro-dys (single intact vector). (**B**) Representative photomicrographs of the *mdx* EDL muscles co-infected by AV.Mini-dys.Donor and AV.Mini-dys.Acceptor (the intron 60 *trans*-splicing vectors). (**C**) Quantitative comparison of the transduction efficiency of the single intact vector and the *trans*-splicing vectors. Data are means ± standard error of mean. $N = 8$ for AV.Micro-dys infection; $N = 15$ for *trans*-splicing vector infection; $p = 0.43$.

2. Materials

2.1. Generating the Mini-dystrophin Trans-Splicing Vectors

2.1.1. Selection of the Potential Gene Splitting Sites

There is no special material needed besides the dystrophin gene sequence itself.

2.1.2. RPA-Mediated Screening of the Potential Gene Splitting Site

1. 293 cells (ATCC). This is an adenovirus E1 gene transformed human kidney cell line. They are used as a packaging cell line for recombinant AAV production. These cells are propagated in high-glucose Dulbecco's modified Eagle's medium (DMEM) containing 10% fetal bovine serum (FBS) and 1% penicillin G/streptomycin (see below).
2. Cell culture medium: DMEM, high glucose with L-glutamine (Gibco-BRL, Grand Island, NY, USA). Store at 4 °C.
3. FBS (Hyclone). Store at –20 °C.
4. Penicillin G-Streptomycin: Penicillin 100 U/ml DMEM culture medium and Streptomycin 100 μg/ml DMEM culture medium (Gibco-BRL). Store at –20 °C.
5. 1× Trypsin–EDTA: 0.25% Trypsin, 1 mM EDTA/4Na (Gibco-BRL). Store at 4 °C.
6. Lipofectamine and Plus reagent (Invitrogen, Carlsbad, CA, USA).
7. Mini-dystrophin gene plasmid pΔH2-R19. This plasmid is obtained from Dr. Jeffrey Chamberlain (University of Washington, Seattle, WA, USA).
8. pcDNA3.1(+) (Invitrogen).
9. pGL3-control (Promega, Madison, WI, USA).
10. pGEM3Z (Promega).
11. pTRI-β-actin-human antisense control template (Ambion, Austin, TX, USA).
12. Guanidine isothiocyanate (Amresco, Solon, OH, USA).
13. Cesium chloride (Research Products International Corp, Mount. Prospect, IL, USA).
14. SP6 RNA polymerase (Invitrogen).
15. RNasin (RNase inhibitor, Roche Applied Science, Indianapolis, IN, USA).
16. 5 mM rNTP (Amersham Pharmacia, Piscataway, NJ, USA).
17. αP^{32} rUTP (Perkin Elmer, Boston, MA, USA).
18. Vanadyl Ribonucleoside complex (VRC) (Gibco-BRL), an RNase inhibitor used during anti-sense RNA probe purification.
19. RQ1 DNase (Promega) to remove the DNA template after RNA probe is generated from in vitro transcription.
20. tRNA (Sigma).
21. RNase A (Roche Applied Science).
22. RNase T1 (Roche Applied Science).

2.1.3. Generating the cis-Plasmids for Recombinant AAV Packaging

1. Vent DNA Polymerase (New England Biolabs): This is a high-fidelity thermophilic DNA polymerase. It has a low polymerase chain reaction (PCR)-related mutation rate.
2. Electroporation-competent *Escherichia coli* SURE cells (Stratagene, LA Jolla, CA, USA): AAV ITRs can be rearranged or deleted by endogenous DNA repair systems when they are propagated in bacterial cells. SURE cells are defective in DNA repair and recombination pathways (such as uvrC, umuC, SbcC, RecJ, recB, and recJ). Store at –80 °C.
3. *E. coli* Pulser (Bio-Rad).
4. Gene Pulser cuvets, 0.1 cm gap (Bio-Rad).
5. 14-ml polypropylene round-bottomed tubes (Becton Dickinson Labware, Franklin Lakes, NJ, USA).
6. S.O.C. medium (Gibco-BRL). Store at room temperature.
7. Amp selection LB agar plates (100 µg/ml, Ampicillin). Store at 4 °C.
8. Standard materials for large-scale plasmid preparation. To achieve the best transfection efficiency for AAV production, we recommend preparing the plasmids by $CsCl_2$/ethidium bromide equilibrium centrifugation.

2.1.4. Recombinant AAV-6 Production

1. Adenoviral helper plasmid: (pHelper, Stratagene).
2. Helper plasmid for AAV-2 rep gene (pMT-Rep2): This plasmid provides the Rep gene required for recombinant AAV replication and packaging ***(9)***. AAV-2 rep proteins can fully support AAV-6 production. This plasmid is obtained from Dr. A. Dusty Miller (Fred Hutchinson Cancer Research Center, Seattle, WA, USA).
3. Helper plasmid for AAV-6 capsids (pCMVCap6): This plasmid provides the Cap gene required for rAAV-6 packaging ***(9)***. This plasmid is obtained from Dr. A. Dusty Miller (Fred Hutchinson Cancer Research Center).
4. 2.5 M $CaCl_2$. Sterilize by filtration and store at –20 °C.
5. 2× HBS buffer: 0.3 M NaCl, 1.5 mM Na_2HPO_4, and 40 mM HEPES, pH 7.05 ± 0.05. Sterilize by filtration and store at –20 °C. As pH affects transduction efficiency, it is highly suggested to double check pH before each usage.
6. DNase I (Sigma, 11 mg protein/vial, total 33 K [kuniz] units).
7. 0.25% Trypsin (Gibco-BRL). Store at 4 °C.
8. 10% Sodium deoxycholate. Store at room temperature.
9. Misonic Cell Disruptor S3000 (Misonix, NY).
10. Cell lifter (Corning Incorporated, Corning, NY, USA).
11. HEPES AAV dialysis buffer: 20 mM HEPES, 150 mM NaCl, pH 7.8. Filter-sterilize and store at 4 °C.
12. Dialysis tubing: 12,000 MW cutoffs (Gibco-BRL). Store at 4 °C.
13. AAV digestion buffer: 0.4 M NaOH, 20 mM EDTA. Freshly made before use.

14. Slot blot hybridization solution [5× SSC, 5× Denhardts's solution, 1% sodium dodecyl sulfate (SDS), and 50 % formamide, add 100 μg/ml denatured salmon sperm DNA just before use].
15. HiTrap heparin column (Amersham).

2.2. In Vivo Evaluation of the Mini-Dystrophin Trans-Splicing Vectors

2.2.1. Local Muscle Injection

Two-month-old dystrophin-null *mdx* mice (C57BL/10ScSn-*Dmdmdx*/J; The Jackson Laboratory). Dystrophin expression is aborted in this inbreed strain because of a nonsense mutation in exon 23 ***(10,11)***. (see **Note 1**).

1. Anesthetic reagents and instruments for local muscle injection: 4 μl/g body weight anesthetic cocktail (ketamine 25 mg/ml, xylazine 2.5mg/ml, acepromazine 0.5mg/ml in physiological saline). Sterile forceps and scissors (World Precision Instruments, Inc., Sarasota, FL, USA), needle holders (Accurate Surgical & Scientific Instruments Corp., Westbury, NY, USA) and Guthrie double hook retractor (Fine Science Tools, Inc., Foster City, CA, USA). 5-0 sofsilk suture (Auto Suture Company, Norwalk, CT, USA). 33G gas-tight Hamilton syringe and needle (Hamilton Company Reno, NV, USA).
2. Polyethylene mouse Elizabethan collar (E-collar, Harvard Apparatus).
3. Banamine (also called Flunixin, 1.5 mg/kg) (Schering-Plough Animal Health Corp., Union, NJ, USA)

2.2.2. Quantifying Mini-Dystrophin Expression by Western Blot

1. Muscle homogenization solution: 20 mM $Na_4P_2O_4$, 20 mM $NaHPO_4$, 1 mM $MgCl_2$, 0.5 mM EDTA, and 303 mM sucrose.
2. Muscle microsome preparation wash buffer: 20 mM Tris–HCl (pH 7.0), 60 mM KCl, 303 mM sucrose.
3. Muscle microsome storage solution: 20 mM Tris–HCl (pH 7.0) and 303 mM sucrose.
4. Cheese-cloth (Fisher Scientific, Pittsburgh, PA, USA).
5. Tissue-Tek OCT compound (Sakura Finetek Inc., Torrance, CA, USA).

2.2.3. Quantifying Mini-Dystrophin Expression by Immunostaining

1. Mouse monoclonal antibodies for different regions of the dystrophin protein: Dys-3 (clone Dy10/12B2, IgG2a), an antibody that specifically recognizes the hinge 1 region of human dystrophin (Novocastra, Newcastle, UK). Dys-2 (clone Dy8/6C5, IgG1), an antibody that reacts with the dystrophin C-terminal domain (Novocastra). Mandys-8 (clone 8H11, IgG2b), an antibody that specifically recognizes spectrin-like repeat 11 (exon 32) (Sigma) (*see* **Note 2**).
2. Polyclonal antibodies for different regions of the dystrophin protein: rabbit anti-dystrophin N-terminal antibody, an affinity-purified antibody obtained from

Dr. Jeffrey Chamberlain's laboratory (University of Washington, Seattle, WA, USA). Goat anti-dystrophin C-terminal antibody (Santa Cruz). Rabbit anti-dystrophin spectrin-like repeat 4–6 antibody (Santa Cruz).

3. 2-methylbutane (Sigma).
4. Tissue-Tek OCT (Sakura Finetek Inc., USA).
5. Rabbit anti-mouse IgG (Sigma, final concentration 0.46 mg/ml).
6. Papain (Sigma, final concentration 45 μg/ml).
7. L-cystein (Sigma, final concentration 22 mM).
8. Iodoacetic acid (Sigma, final concentration 10 mM).
9. 20% rabbit serum (Jackson ImmunoResearch Laboratories, Inc.).
10. Alex 488 conjugated rabbit anti-mouse antibodies (Molecular Probe).
11. Alex 594 conjugated rabbit anti-mouse antibodies (Molecular Probe).
12. KPBS: 356 μM KH_2PO_4, 1.64 mM K_2HPO_4, 160 mM NaCl.
13. Gelatin (Sigma).
14. M.O.M kit (Vector Laboratories, Inc.).

3. Methods

3.1. Generating the Mini-dystrophin Trans-Splicing Vectors

3.1.1. Selection of the Potential Gene Splitting Sites

There are several considerations when deciding on the potential sites to split the mini-dystrophin gene. These include the size of the individual vector, the gene structure, and the relative efficiency of transcription and splicing. Specifically, both AV.Donor and AV.Acceptor have to fit into a single AAV virion themselves, respectively; the minigene should be split at a natural exon/exon junction rather than within an exon; and the site should yield the highest level of mRNA in the context of a reconstituted/recombined viral genome.

1. Preliminary selection of the gene splitting site according to the size constraint of a single AAV virion. There are a total of 79 exons in the full-length dystrophin gene. However, the minigene only contains the coding sequence from 44 exons. The 5′ half of the minigene is ~2.2 kb (exon 1 to the beginning of exon 17) spanning the region from the N-terminus to repeat 3. The 3′ half of the minigene is ~3.8 kb (the end of exon 49 to the beginning of exon 79) covering the region from hinge 3 to the C-terminus. Considering the size of the ITR (two ITRs, total ~0.3 kb), the CMV promoter (~0.6 kb), the pA signal (~0.4 kb), and the splicing signals (~ 0.3 kb), any exon boundary between exons 51 and 65 would be appropriate to divide the minigene.
2. Preliminary screening of the gene splitting site according to the conserved splicing value in the wildtype dystrophin gene. Sironi et al. ***(12)*** has systematically calculated the conserved splicing value (CV) and U1 small nuclear RNA binding energy for all the exon/intron/exon junctions in the dystrophin gene (*see* **Note 3**). The mean CVs for the 5′ and 3′ splicing sites are (0.83 ± 0.07) and (0.87 ± 0.09),

respectively. The mean binding energy is –7.76 ± 2.21 kcal/mole *(12)*. We selected three sites that have the highest CV and lowest binding energy as potential sites for gene splitting. These are 56/56/57 (exon/intron/exon), 60/60/61, and 63/63/64 junctions. Their respective CVs and binding energy are (in the order of 5′ splicing site CV, 3′ splicing site CV, and binding energy) 0.84. 0.97, and –10.4 kcal/mole (for the 56/56/57 junction), 0.94, 0.98, and –11.7 kcal/mole (for the 60/60/61 junction), and 0.95, 0.90, and –10.6 kcal/mole (for the 63/63/64 junction). As a control, we also selected the 53/53/54 junction, which has an average binding energy (–7.7 kcal/mole) and moderate CVs (0.90 and 0.60 for the 5′ and 3′ splicing sites, respectively).

3.1.2. RPA-Mediated Screening of the Potential Gene Splitting Site

Formation of the double-D ITR structure is a unique feature in the recombined AAV genome *(13)*. Importantly, the double-D ITR structure in intron down-regulates mRNA accumulation *(8)*. An ideal gene splitting site should yield the highest level of mRNA in the presence of the double-D ITR structure. To identify the best site, we constructed a series of mini-cassettes to mold the reconstituted vector genome. An RPA assay was then used to quantify the relative level of mRNA accumulation (*see* **Fig. 2**).

1. Cloning the endogenous dystrophin intron splicing signal from a human cell line. Use 293 cell genomic DNA as the PCR template. Design the primers to amplify the exon/intron junctions. The primer sequences for each junction can be found in the online supplementary material in Lai et al. (http://www.nature.com/nbt/journal/v23/n11/suppinfo/nbt1153_S1.html) *(7)*. To facilitate subsequent cloning, include restriction site overhangs in the primers as needed (*see* **Note 4**).
2. Reconstituting the mini-cassette in the backbone of pcDNA3.1(+). First, introduce the respective exon/intron junctions in pcDNA3.1(+). For the 53/53/54 junction, insert the splicing donor between the Kpn I/BamH I sites and insert the splicing acceptor between the Xho I/Xba I sites. For the 56/56/57 junction, insert the splicing donor between the Hind III/Kpn I sites and insert the splicing acceptor between the EcoR V/Xba I site. For the 60/60/63 junction, insert the splicing donor between the Kpn I/BamH I sites and insert the splicing acceptor between the Xho I/Xba I sites. For the 63/63/64 junction, insert the splicing donor between the Hind III/Kpn I sites and insert the splicing acceptor between the Xho I/Xba I sites. After inserting splicing signals, clone the double-D ITR structure in the intron region between the EcoR I/EcoR V sites (*see* **Fig. 2B**).
3. Generating the RPA probe. To quantify both spliced and unspliced RNA products, generate an anti-sense RNA probe for each potential site to hybridize with the respective 5′ exon and the intron donor sequence (*see* **Fig. 2B**). Then insert the PCR products into the EcoR I/BamH I sites in pGEM3Z accordingly. After linearization at the EcoR I site, synthesize the RPA probes by in vitro transcription with SP6 RNA polymerase and [α-^{32}P]-rUTP as the substrate *(7,8)*.

4. RPA screening. Co-transfect the plasmids carrying the mini-cassette (4 μg) and a transfection control plasmid (0.5 μg; we used a luciferase plasmid pDD12. In this plasmid, the entire expression cassette from pGL3-control (Promega) was cloned into an AAV packaging plasmid) into 70–80% confluent 293 cells using Lipofectamine and Plus reagent (Invitrogen). About 48 h later, extract protein from one-fifth cells for luciferase assay to normalize transfection efficiency. Lyse remaining cells in 4 M guanidine isothiocyanate and extract total RNA in 5.7 M $CsCl_2$ by ultra-centrifugation at the speed of 116,200 × *g* for at least 12 h at 20 °C. In each RPA reaction, use10 μg total RNA to hybridize with excessive amount of ^{32}P-labeled probe and digest unprotected RNA by RNaseT1 (0.923 μg/ml) and RNaseA (at 18.46 μg/ml). Elecctrophorese the reaction products in 8% denaturing polyacrylamide gel and quantify the bands with the Molecular Imager FX and Quantity One (version 4.2.2 image software; Bio-Rad, Heracules, CA). Normalize the RPA reaction with endogenous human β-actin using an independent probe generated from pTRI-β-actin-human antisense. Normalize RNA signal intensity by transfection efficiency, RPA reaction and the number of ^{32}P-labeled uridine in each protected band. The 60/60/61 junction produces the highest level of mRNA and the 63/63/64 junction is the second best (~50% of that from the 60/60/61 junction) (*see* **Fig. 1B**).

3.1.3. Generating the cis -Plasmids for Recombinant AAV Packaging

Generate two pairs of the *trans*-splicing plasmids representing gene splitting sites at the 60/60/61 and the 63/63/64 junctions, respectively. Each pair includes a *cis*-donor plasmid and a *cis*-acceptor plasmid. To facilitate comparison, both pairs are under transcriptional regulation of the CMV promoter and the SV40 polyA. Use standard molecular cloning methods to generate these constructs. The detailed step-by-step cloning procedure can be found in the online supplementary material in Lai et al. (http://www.nature.com/nbt/journal/v23/n11/suppinfo/nbt1153_S1.html) *(7)*.

3.1.4. Recombinant AAV-6 Production see Notes 5 and 6

.

1. Two days before viral production, split 293 cells at 1:6 to 150 mm culture plates. Seed a total of 15 plates for one production. Change to fresh culture media about 1–2 h before transfection.
2. Preparing DNA-calcium-phosphate precipitate. Co-transfect four plasmids to make AAV-6 vector. These include the *cis*-plasmid, pMT-Rep2, pCMVCap6, and pHelper. For each vector preparation (15 × 150 mm plates), use 187.5 μg of the *cis*-plasmid, 187.5 μg pMT-Rep2, 562.5 μg pCMVCap6, and 562.5 μg pHelper (at a ratio of 1:1:3:3). Mix all plasmids thoroughly in 15.2 ml H_2O. Add 1.68 ml of 2.5 M $CaCl_2$ to a final concentration of 250 mM. Generate DNA-calcium-phosphate precipitate by slowly dropping the DNA/$CaCl_2$ mixture to 16.8 ml of

2× HBS. In general, it takes 15–30 min room temperature incubation to form a high quality precipitate (*see* **Note 7**).
3. Gently apply the DNA-calcium-phosphate precipitate to 293 cells drop-by-drop while swirling the culture plate.
4. At 72 h after transfection, collect cell lysate with a cell lifter. After a 20 min spin in a bench-top centrifuge (1,800 × g at 4 °C), resuspend cell pellet in 9 ml of 10 mM Tris–HCl (pH 8.0). Freeze/thaw cell lysate 8–10 times using dry ice/ethanol and a 40 °C water bath. Sonicate cell lysate at the power output of 5.5 for 10 min (on ice). Digest cell lysate with DNase I at 37 °C for 45 min (we normally use half vial of Sigma DNase I for each 15 plates viral preparation). Sonicate cell lysate again at the power output of 5.5 for 7 min (on ice). Digest lysate with one-tenth volume of 0.05% trypsin and 10% sodium deoxycholate for 30 min at 37 °C. Clear cell lysate by spinning at 3,200 × g for 30 min at 4 °C. Carefully transfer supernatant to a new tube.
5. Adjust the volume to 29 ml with 10 mM Tris–HCl (pH 8.0) and add 18.2 g $CsCl_2$ (this is equal to 0.613 g/ml). The final volume should be about 32.5ml. Incubate for 30 min at 37 °C to dissolve $CsCl_2$. Spin at 3200 × *g* for 30 min at 4 °C. Carefully load the supernatant into six 5-ml Beckman ultracentrifugation tubes in an SW55Ti rotor. Spin at 200,000 × *g* for 40 h at 4 °C.
6. Collect fractions from the bottom of the tube with a 20-G needle. Identify the viral containing fractions by slot blot (see below).
7. Combine fractions with the highest viral titer and centrifuge again at 200,000 × *g* for 40 h at 4 °C. Collect fractions as described in **step 6** and identify highest viral fraction by slot blot (see below).
8. Dialyze viral stock in HEPES buffer (4 °C for 2 × 24 h).
9. Slot blot viral titer determination. Use duplicated sets of viral stock aliquots (1, 5, and 10 μl) and plasmid copy number controls (10^7, 10^8, 10^9, 10^{10}, 10^{11} molecules/μl) in slot blot. Denature samples in 50 μl AAV digestion buffer at 100 °C for 10 min. Then immediately chill samples on ice and bring up volume to 400 μl with digestion buffer. Load samples onto Hybond-N plus membrane with a Bio-Dot SF manifold microfiltration apparatus. After blotting, crosslink DNA to the membrane with UV irradiation. Prehybridize the membrane. Then hybridize the membrane with a ^{32}P-labeled transgene-specific probe in the slot blot hybridization solution. Determine the viral particle titer by comparing the intensity of the viral stock band to that of the plasmid standards.

3.2. *In Vivo* Evaluation of the Mini-Dystrophin Trans-Splicing Vectors

3.2.1. Local Muscle Injection

For all animal experiments, get approval from the institute Animal Care and Use Committee and follow NIH guidelines. Two muscles have been used to evaluate AAV transduction efficiency including the tibialis anterior (TA) muscle and the extensor digitorum longus (EDL) muscle. In *trans*-splicing vector transduction studies, mix equal amount of AV.Donor and AV.Acceptor particles before muscle injection. Then anesthetize experimental animal.

1. Gene delivery to the TA muscle. Expose the proximal end of the TA muscle with a 2~3 mm incision. Insert a 33-G Hamilton needle into the middle belly of the TA muscle. Slowly inject 30 μl *trans*-splicing AAV vectors into the muscle while slowly backing out the injection needle. Suture the wound and monitor the animal until it recovers.
2. Gene delivery to the EDL muscle. Make a 0.5 × 5 mm incision along the longitudinal axis in the lateral surface of the distal hind limb. Separate the EDL tendon from the TA tendon and then gently pull the TA muscle aside with a Guthrie double hook retractor to expose the EDL muscle. Perform two injections from the proximal and the distal ends respectively with a 33G Hamilton syringe. In each injection, directly inject 5 μl virus into the muscle body (a total of 10 μl AAV for each EDL muscle). Suture the wound and monitor the animal until it recovers (*see* **Note 8**).

3.2.2. Quantifying Mini-Dystrophin Expression by Western Blot

1. Muscle microsome preparation. Pulverize muscle in liquid nitrogen by hand grinding. Resuspend the muscle powder in 6 ml muscle homogenization solution. Transfer muscle lysate to a clean Oakridge tube. Spin at 250,000 × *g* for 15 min at 4 °C in a Sorval RC-5B supercentrifuge. Filter supernatant through six layers of cheese cloth. Load the filtered supernatant (~5 ml) to a 5.2-ml SW55 centrifuge tube. Spin at 110,000 × *g* for 30 min at 4 °C. Discard supernatant. Gently resuspend the pellet in 5 ml wash buffer. Spin again at 110,000 × *g* for 30 min at 4 °C. Resuspend the pellet (muscle microsome) in 200 μl storage solution and store at –80 °C.
2. Western blot. Electrophorese 50 μg of muscle microsome preparation in a 6% SDS-polyacrylamide gel. Detect mini-dystrophin with either Dys-2 (1:100 dilution) or Dys-3 (1:20 dilution) antibodies.

3.2.3. Quantifying Mini-Dystrophin Expression by Immunostaining

1. Snap freeze freshly isolated muscle sample in liquid nitrogen-cooled 2-methylbutane in Tissue-Tek OCT.
2. Immunostaining with monoclonal antibody. To use mouse monoclonal antibody in murine tissue, one must first block the binding of the secondary antibody to endogenous mouse immunoglobulins (Igs) ***(14)***. We used papain digested rabbit anti-mouse IgG (including Fab and Fc fragments) for blocking (*see* **Note 9**). Briefly, digest rabbit anti-mouse IgG with Papain in the presence of 1 mM EDTA and 22 mM L-cystein at 37 °C for 16 h. Stop digestion reaction with iodoacetic-acid. For immunostaining, first block 8 μm air-dried cryosections in anti-mouse IgG blocking solution at room temperature for 60 min. After washing in PBS, block the cryosections again with 20% rabbit serum at room temperature for 30

min. After washing in PBS, apply the primary antibody (diluted in 1% rabbit serum, such as Dys-3, Dys-2, or Mandy-8) at a dilution of 1:100 and incubate at 4 °C overnight. After washing in PBS, detect signals with Alex 488 or Alex 594 conjugated rabbit anti-mouse antibodies, respectively.

3. Immunostaining with polyclonal antibody. Briefly rinse 8 μm air-dried cryosections with KPBS. Block with 1% goat serum KPBS at room temperature for 15 min. Wash with 0.2% gelatin KPBS. Apply primary antibody at 4 °C overnight (1:600 for Chamberlain's affinity purified N-terminal antibody; 1:200 for Santa Cruz's C-terminal antibody, and 1:400 for Santa Cruz's repeat 4–6 antibody). Wash with 0.2% gelatin KPBS. Detect signal with Alex 488 or Alex 594 conjugated goat anti-rabbit antibody (for the N-terminal and repeat 4–6 primary antibodies) or rabbit anti-goat antibody (for the C-terminal primary antibody) (*see* **Notes 10** and **11**).

4. Notes

1. Despite the nonsense mutation, occasionally dystrophin positive revertant fibers can be detected in *mdx* muscle ***(15)***. These are called revertant fibers. In revertant fibers, the mutated exon 23 is skipped due to alternative splicing. Revertant fibers can be distinguished from AAV-mediated mini-dystrophin expression by the following criteria. First, revertant fibers can often be detected with antibodies against the missing regions in the mini-dystrophin gene such as Mandy-8 (Sigma, recognizes repeat 11) and Repeats 4–6 polyclonal antibody (Santa Cruz). Second, revertant fibers cannot be recognized by human dystrophin-specific antibody Dys-3.
2. In addition to the commercially available monoclonal antibodies mentioned here, more than 60 monoclonal antibodies have been published. These antibodies react with the different regions of the dystrophin protein. A collection of these antibodies can be found in the website http://www.dmd.nl/antibody.html. A large number of epitope-specific antibodies can be obtained from Prof. Glenn E. Morris (Centre for Inherited Neuromuscular Disease, Leopold Muller ARC Building, RJAH Orthopaedic Hospital, OSWESTRY, Shropshire, SY10 7AG, UK).
3. The conserved splicing value for a particular exon/intron/exon junction can be calculated according to the following formula ***(16)***. For the 5′ splicing site, CV = (X – 47)/548. X = the sum of all the reference values for the positions X and x (*see* **Table 1**). For the 3′ splicing site, CV = (Y – 57)/640 + (Z – 11)/624. Y = the sum of the eight highest values for all the y positions (see **Table** 1). Z = the sum of all the reference values for the positions z and Z (*see* **Table** 1).
4. As mutations in intron may alter splicing profile and mutations in exon may change amino acid sequence, we strongly recommend using high fidelity Taq polymerase for all PCR-mediated cloning. It is also essential to confirm the cloned PCR product by sequencing.

Table 1
Reference Value for CV Calculation in a Primate Gene[a]

	X[b]	X	x	x	x	x	x	x	y	y	y	y	y	y	y	y	y	y	o[c]	z	z	z	Z
A	58	10	0	0	57	71	5	16	9	9	7	7	10	10	7	9	6	6	n/a	3	100	0	28
C	13	4	0	0	2	8	6	15	31	33	31	35	35	35	43	41	39	40	n/a	74	0	0	13
G	15	78	100	0	39	12	84	22	15	13	11	7	7	11	7	8	6	8	n/a	1	0	100	49
T	15	8	0	100	2	9	5	47	45	45	51	51	47	44	42	42	48	46	n/a	22	0	0	10

[a] Adapted from Shapiro and Senapathy 1987 *Nucleic Acid Res* 15(17):7155.
[b] Capital letters represent nucleotides in exon. Small letters represent nucleotides in intron.
[c] Values at this nucleotide position is not used in calculation. n/a, not applicable.

5. AAV-6 has been shown as the best serotype for muscle transduction following direct local injection ***(17,18)***. However, for systemic delivery one may also consider other newly developed AAV serotypes such as AAV-8 and AAV-9.
6. In this protocol, we described a $CsCl_2$ ultracentrifugation-based protocol for AAV-6 purification. However, AAV-6 can also be purified with a HiTrap heparin column using HPLC or FPLC ***(19)***.
7. A high-quality precipitate is essential to high viral yield. We recommend routinely monitoring calcium phosphate precipitate on a coverslip using a phase contrast microscope. Alternatively, the quality of the transfection reagents (such as $CaCl_2$ and 2× HBS) can be pre-tested with reporter plasmids such as GFP, LacZ, or alkaline phosphatase plasmids.
8. To prevent gnawing on the suture and improve wound healing, we recommend placing a polyethylene mouse Elizabethan collar around the mouse neck. This device blocks the head from access to the rest of the body, but still enables eating, drinking, and comfortable movement. To alleviate pain and discomfort at the operation site, we suggest subcutaneous injection of a non-steroid analgesic drug Banamine.
9. For immunostaining with monoclonal antibody, one can also use M.O.M kit.
10. To confirm *trans*-splicing AAV vector-mediated mini-dystrophin expression, it is essential to perform immunostaining with at least three antibodies. Mini-dystrophin should be lighted up with antibodies against the N-terminal and the C-terminal domains, respectively, but it will not be detected with antibody against the regions that are deleted in mini-dystrophin (such as repeat 4–6).
11. The advantage of the polyclonal antibody is its low background staining in mouse muscle. However, polyclonal antibody cannot distinguish between endogenously mouse dystrophin in revertant myofibers and human mini-dystrophin expressed from AAV vectors. Only Dys-3 monoclonal antibody can definitively diagnose human originated mini-dystrophin.

Acknowledgments

We thank Dr. Jeffrey Chamberlain for providing the pΔH2-R19 mini-dystrophin plasmid. We also thank Dr. A. Dusty Miller for the AAV-6 packaging plasmids (pMT-Rep2 and pCMVCap6). This work is supported by grants from the National Institutes of Health (AR-49419, DD) and the Muscular Dystrophy Association (DD). We thank Mr. Arka Ghosh for optimizing AAV-6 preparation protocol. We thank Ms. Chun Long for animal care and Mr. Brian Bostick for helpful discussion.

References

1. Wang, B., Li, J., and Xiao, X. (2000) Adeno-associated virus vector carrying human minidystrophin genes effectively ameliorates muscular dystrophy in mdx mouse model. *Proc Natl Acad Sci USA* **97**, 13714–9.

2. Harper, S. Q., Hauser, M. A., DelloRusso, C., Duan, D., Crawford, R. W., Phelps, S. F., Harper, H. A., Robinson, A. S., Engelhardt, J. F., Brooks, S. V., and Chamberlain, J. S. (2002) Modular flexibility of dystrophin: implications for gene therapy of Duchenne muscular dystrophy. *Nat Med* **8**, 253–61.
3. England, S. B., Nicholson, L. V., Johnson, M. A., Forrest, S. M., Love, D. R., Zubrzycka-Gaarn, E. E., Bulman, D. E., Harris, J. B., and Davies, K. E. (1990) Very mild muscular dystrophy associated with the deletion of 46% of dystrophin. *Nature* **343**, 180–2.
4. Phelps, S. F., Hauser, M. A., Cole, N. M., Rafael, J. A., Hinkle, R. T., Faulkner, J. A., and Chamberlain, J. S. (1995) Expression of full-length and truncated dystrophin mini-genes in transgenic mdx mice. *Hum Mol Genet* **4**, 1251–8.
5. Duan, D., Yue, Y., and Engelhardt, J. F. (2001) Expanding AAV packaging capacity with trans-splicing or overlapping vectors: a quantitative comparison. *Mol Ther* **4**, 383–91.
6. Duan, D., Yue, Y., Yan, Z., and Engelhardt, J. F. (2003) Trans-splicing vectors expand the packaging limits of adeno-associated virus for gene therapy applications. *Methods Mol Med* **76**, 287–307.
7. Lai, Y., Yue, Y., Liu, M., Ghosh, A., Engelhardt, J. F., Chamberlain, J. S., and Duan, D. (2005) Efficient in vivo gene expression by trans-splicing adeno-associated viral vectors. *Nat Biotechnol* **23**, 1435–9.
8. Xu, Z., Yue, Y., Lai, Y., Ye, C., Qiu, J., Pintel, D. J., and Duan, D. (2004) Trans-splicing adeno-associated viral vector-mediated gene therapy is limited by the accumulation of spliced mRNA but not by dual vector coinfection efficiency. *Hum Gene Ther* **15**, 896–905.
9. Halbert, C. L., Allen, J. M., and Miller, A. D. (2001) Adeno-associated virus type 6 (AAV6) vectors mediate efficient transduction of airway epithelial cells in mouse lungs compared to that of AAV2 vectors. *J Virol* **75**, 6615–24.
10. Bulfield, G., Siller, W. G., Wight, P. A., and Moore, K. J. (1984) X chromosome-linked muscular dystrophy (mdx) in the mouse. *Proc Natl Acad Sci USA* **81**, 1189–92.
11. Sicinski, P., Geng, Y., Ryder-Cook, A. S., Barnard, E. A., Darlison, M. G., and Barnard, P. J. (1989) The molecular basis of muscular dystrophy in the mdx mouse: a point mutation. *Science* **244**, 1578–80.
12. Sironi, M., Pozzoli, U., Cagliani, R., Comi, G. P., Bardoni, A., and Bresolin, N. (2001) Analysis of splicing parameters in the dystrophin gene: relevance for physiological and pathogenetic splicing mechanisms. *Hum Genet* **109**, 73–84.
13. Duan, D., Yan, Z., Yue, Y., and Engelhardt, J. F. (1999) Structural analysis of adeno-associated virus transduction intermediates. *Virology* **261**, 8–14.
14. Lu, Q. L., and Partridge, T. A. (1998) A new blocking method for application of murine monoclonal antibody to mouse tissue sections. *J Histochem Cytochem* **46**, 977–84.
15. Lu, Q. L., Morris, G. E., Wilton, S. D., Ly, T., Artem'yeva, O. V., Strong, P., and Partridge, T. A. (2000) Massive idiosyncratic exon skipping corrects the nonsense

mutation in dystrophic mouse muscle and produces functional revertant fibers by clonal expansion. *J Cell Biol* **148**, 985–96.

16. Shapiro, M. B., and Senapathy, P. (1987) RNA splice junctions of different classes of eukaryotes: sequence statistics and functional implications in gene expression. *Nucleic Acids Res* **15**, 7155–74.
17. Wang, Z., Zhu, T., Qiao, C., Zhou, L., Wang, B., Zhang, J., Chen, C., Li, J., and Xiao, X. (2005) Adeno-associated virus serotype 8 efficiently delivers genes to muscle and heart. *Nat Biotechnol* **23**, 321–8.
18. Ghosh, A., Yue, Y., and Duan, D. (2006) Viral serotype and the transgene sequence influence overlapping adeno-associated viral (AAV) vector-mediated gene transfer in skeletal muscle. *J Gene Med* **8**, 298–305.
19. Blankinship, M. J., Gregorevic, P., Allen, J. M., Harper, S. Q., Harper, H., Halbert, C. L., Miller, D. A., and Chamberlain, J. S. (2004) Efficient transduction of skeletal muscle using vectors based on adeno-associated virus serotype 6. *Mol Ther* **10**, 671–8.

17

Prevention of Type 1 Diabetes in NOD Mice by Genetic Engineering of Hematopoietic Stem Cells

Jessamyn Bagley, Chaorui Tian, and John Iacomini

Summary

Type 1 diabetes is caused by autoimmune destruction of insulin-producing cells in the pancreas. Type 1 diabetes could potentially be treated by islet transplantation; however, the recurrence of autoimmunity leads to the destruction of islet grafts in a relatively short time frame. Therefore, a major goal of diabetes research is the induction of tolerance in diabetic patients to prevent recurrence of diabetes. Diabetes is a polygenic disease, and not all the determinants responsible for disease susceptibility have been identified. However, in both humans and mouse models of this disease, one of the principle determining genetic factors in diabetes incidence is the inheritance of mutant MHC class II alleles that are associated with increased occurrence of disease. We have shown that in the NOD mouse model, the introduction of protective MHC class II alleles through retroviral gene therapy can prevent the onset of autoimmune diabetes. Prevention of diabetes appears to be mediated, at least in part, by the deletion of autoreactive T cells in the presence of protective MHC class II. Here, we outline the procedures involved in the modification of murine hematopoietic cells through retroviral transduction, the reconstitution of recipients with modified bone marrow, and the monitoring of gene therapy recipients after reconstitution.

Key Words: Tolerance; T cells; bone marrow transplantation; MHC class II; type 1 diabetes.

1. Introduction

Type 1 diabetes is caused by T cell–mediated autoimmune destruction of insulin-producing ß cells in the pancreas. Although patients can be kept alive by the administration of exogenous insulin, there is presently no cure. Several

From: *Methods in Molecular Biology, vol. 433: Volume 1: Production and In Vivo Applications of Gene Transfer Vectors*
Edited by: J. M. Le Doux © Humana Press, Totowa, NJ

genes contribute to the development of type 1 diabetes, but the principal determining genetic component in humans is the inheritance of mutant MHC alleles associated with susceptibility to disease (at-risk alleles) ***(1–4)***. These alleles are characterized by the lack of a charged amino acid at position 57 of the MHC class II ß chain ***(5,6)***. Conversely, inheritance of MHC class II alleles containing a charged amino acid at position 57 of the ß chain is associated with protection from disease. In some populations, as many as 28% of individuals who inherit particular HLA-DQ alleles such as DQB1*0302 and 34% who inherit DQB1*0201 develop diabetes ***(7)***. The MHC class II region in NOD mice, a model of spontaneous type 1 diabetes, encodes a single MHC class II molecule, I-Ag7 (the mouse homologue of HLA-DQB1), which contains polymorphisms that are strikingly similar to those in human alleles associated with type 1 diabetes susceptibility ***(8)***.

At-risk alleles are structurally distinct from diabetes-resistant alleles, and both humans and NOD mice lack a charged amino acid at position 57 of the ß chain, which prevents the formation of a salt bridge with arginine 76 of the α chain ***(9)***. The introduction of an MHC class II I-Aß chain molecule that contains a charged amino acid at position 57 through genetic engineering of hematopoietic bone marrow stem cells is sufficient to prevent insulitis and destruction of ß cells in the pancreas, and completely prevents the occurrence of spontaneous diabetes in NOD recipients ***(10)***. This suggests that genetic engineering of autologous hematopoietic stem cells could be used to prevent diabetes occurrence in high-risk individuals who inherit diabetes-permissive MHC class II alleles.

Retroviral vectors derived from murine leukemia virus (MuLV) have been widely used for gene transfer into dividing cells, and have the advantage of allowing for the integration of DNA encoding the therapeutic proteins of interest into the host genome [reviewed in ***(11)***]. This allows for the long-term expression of the target protein, which may be critical in gene therapy applications that aim to modify the immune repertoire ***(12)***. To deliver protective MHC class II genes in vivo, we generated retroviruses carrying the gene encoding the MHC class II alleles associated with protection from disease (*see* **Fig. 1**) ***(10)***. To track gene expression of the retrovirally encoded gene, we created a fusion protein in which the MHC class II β chain was fused to green fluorescent protein (GFP). We focused on the MHC class II β chain, because the α chain is relatively non-polymorphic and is not associated with susceptibility to diabetes ***(13)***. The DNA encoding this fusion protein was cloned into the pMMP vector, an MFG derivative, containing the DNA encoding viral long terminal repeats and a packaging signal. The structural proteins required for viral assembly, gag and pol, are provided by the vector, pMD.MLV gag pol. The envelope (Env) G protein from the vesicular stomatitis virus (VSV-G), encoded by pMD.G,

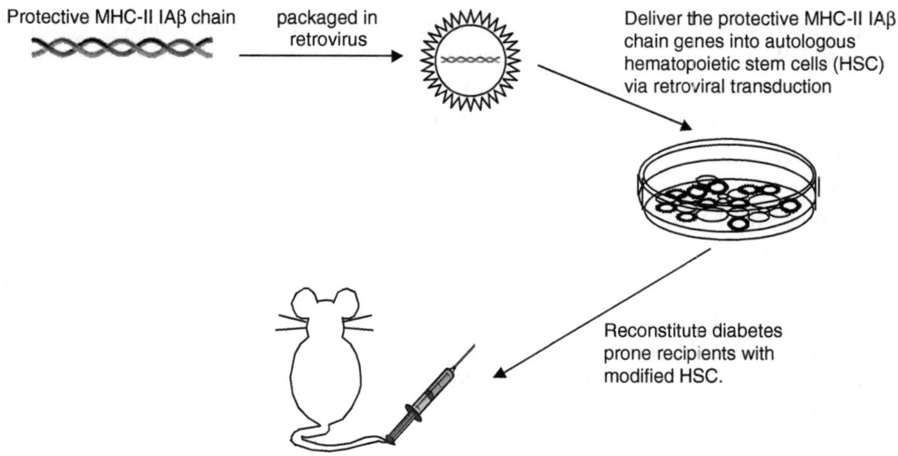

Fig. 1. Schematic illustrating strategy to prevent type 1 diabetes through genetic engineering of bone marrow stem cells. MHC class II chain associated with protection from diabetes is cloned into a retroviral vector and packaged into virions. Viral particles are used to deliver gene to bone marrow progenitors. Bone marrow cells are then re-infused into conditioned recipients with susceptibility to type 1 diabetes.

is used to construct a "pseudotyped" MuLV virus expressing the VSV-G Env protein on its surface. The pseudotyped virus has a broader host range than viruses expressing the amphotrophic Env protein. When these three plasmids are co-transfected into 293T cells, viral titers of 3×10^6 to 10^7 particles/ml can be achieved. Virus preparations can then be used to efficiently transduce hematopoietic stem cells present in bone marrow. Mice reconstituted with bone marrow transduced with virus encoding MHC class II β chains display long-term expression of the target protein and are protected from the onset of spontaneous diabetes.

2. Materials

2.1. Virus Production

1. 293T cells: 293T is a highly transfectable derivative of the 293 cell line into which the temperature-sensitive gene for SV40 T-antigen was inserted. 293T cells were cloned and the clones tested to obtain a line capable of producing high titers of infectious retrovirus. Available from ATCC (Manassas, VA, USA).
2. 293 medium: Dulbecco's modified Eagle's medium (DMEM, Gibco BRL, Gaithersburg, MD, USA) with 10% heat-inactivated fetal bovine serum (FBS, Sigma Aldrich, St Louis, MO, USA), 4 mM L-glutamine, 1 mM sodium pyruvate (Invitrogen, Grand Island, NY, USA), 0.1 mM MEM non-essential amino acids

(Invitrogen), 50 U/ml penicillin and 50 μg/ml streptomycin (Mediatech, Herndon, VA, USA).

3. 2× HEPES-buffered saline: 281 mM NaCl, 100 mM HEPES (Fisher Scientific, Fair Lawn, NJ, USA), 1.5 mM Na_2HPO_4, pH 7.12, sterile filtered through a 0.22-μm filter cellulose acetate low protein-binding membrane (Corning, Corning, NY, USA).
4. 10 mM HEPES in 293 medium.
5. Transfection medium: Iscove's modified Dulbecco's medium (IMDM, Gibco BRL) with 10% FBS, 25 mM HEPES, 4 mM L-glutamine, 50 U/ml penicillin and 50 μg/ml streptomycin.
6. Tissue culture dishes, 6 or 10 cm (BD Biosciences, Bedford, MA, USA).
7. 0.5% Trypsin–EDTA 4Na (Gibco BRL).
8. 2M $CaCl_2$.

2.2. Titration of Active Recombinant Retrovirus

1. NIH 3T3 cells, ATCC.
2. 3T3 medium: DMEM (Gibco BRL) with 10% FBS, 4 mM L-glutamine, 1 mM sodium pyruvate, 0.1 mM MEM non-essential amino acids, 50 U/ml penicillin and 50 μg/ml streptomycin.
3. 6-Well tissue culture plates (BD Biosciences).
4. Polybrene (hexadimethrine bromide, Sigma Aldrich). Dilute in sterile water to 16 mg/ml, store at 4 °C.

2.3. Transduction of Murine Bone Marrow

1. 5-Fluorouracil provided as a sterile solution at 10 mg/ml (InvivoGen, San Diego, CA, USA).
2. RetroNectin: RetroNectin is a recombinant human fibronectin fragment, CH-296, produced in *Escherichia coli.* When coated on the surface of plates or flasks, it enhances retroviral mediated gene transfer into mammalian cells by co-localization of retrovirus particles and target cells (Takara Bio, Inc., Shiga, Japan).
3. Hanks balanced salt solution (HBSS, HyClone, Logan, UT, USA).
4. Non-tissue culture-treated 6-well plates (BD BioSciences).
5. Bovine serum albumin, fraction 5 (BSA, Sigma Aldrich).
6. 0.22-μm filter, cellulose acetate low protein-binding membrane.
7. HBSS/HEPES: HBSS with 2.5% v/v HEPES.
8. ACK lysis buffer (Cambrex BioSciences, Walkersville, MD, USA).
9. Bone marrow culture medium: DMEM containing 15% FBS, 24 mM HEPES, 1 mM sodium pyruvate, 0.1 mM MEM non-essential amino acids, 4 mM L-glutamine and 0.1% gentamicin.
10. 100-μm filter mesh (Lab Pak, Sefar America, Kansas City, MO, USA). Mesh may be purchased sterile or sterilized by autoclaving.

11. 2× transduction medium: Bone marrow medium supplemented with thrombopoietin (TPO, R&D Systems, Minneapolis, MN, USA), Flt-3 ligand (FLT-3L, R&D Systems), stem cell factor (SCF, R&D Systems), and IL-6 (R&D Systems). Stock solutions should be prepared and stored according to the manufacturer's instructions.

Components	2×	Final concentration
TPO (human)	100 ng/ml	50 ng/ml
mFlt-3L	100 ng/ml	50 ng/ml
m SCF	200 ng/ml	100 ng/ml
IL-6	200 ng/ml	100 ng/ml

2.4. Monitoring Recipients of Transduced Bone Marrow

1. Monoclonal antibodies specific for CD4 (RM4-5), CD8 (53-6.7), CD3 (145-2C11), Ly-6G (Gr-1, RB6-8C5), B220 (RA3-6B2), CD11b (Mac-1, M1/70), NK cells (DX5), CD11c (HL3), I-Ad (39–10–8), and I-Aßk (10-3.6) are available from BD Biosciences.
2. Glucometer (BD Logic, BD Medical, Franklin Lakes, NJ, USA) and glucose test strips.

3. Methods

3.1. Production of Recombinant Retroviruses by Calcium Phosphate Transfection

1. Divide the culture of 293T cells 20–24 h before transfection. Plate 2×10^6 cells/4 ml 293 medium in 6-cm plates, or 4.5×10^6 cells/10 ml 293 medium in 10-cm plates (*see* **Note 1**).
2. 17–21 h later, and 3 h before transfection, change the medium to transfection medium.
3. Prepare DNA calcium mix for transfection.

For	6-cm plates	10-cm plates
pMD.MLV gag.pol	6 μg	15 μg
pMMP	8 μg	20 μg
pMD.G	2 μg	5 μg
2M $CaCl_2$	31 μl	62 μl
ddH_2O	219 μl	438 μl

4. Add 250 μl 2× HEPES-buffered saline to the DNA/calcium mix slowly. Flick the tube several times to mix well and let stand at room temperature undisturbed for 20 min (*see* **Note 2**).
5. Slowly add the solution to the cells in transfection medium, in a dropwise fashion, mixing gently.
6. 12–16 h after transfection, replace the transfection medium with 10 mM HEPES/293 medium.
7. 40–48 h after transfection, harvest the supernatant. Pass the supernatant through a 0.45-μm cellulose acetate low protein-binding filter. The supernatant can be directly frozen and stored at –80 °C.

3.2. Titration of Active Retroviral Particles by Infection of NIH 3T3 Cells

1. Divide culture of NIH 3T3 cells the day before infection. Plate 10^5cells/3 ml 3T3 medium per well in 6-well plates, or 10^6cells/10 ml 3T3 medium per 10-cm plate.
2. The next day, supplement 10 mM HEPES in 293 medium with polybrene to a final concentration of 8 μg/ml.
3. Add 1 ml polybrene supplemented medium/well for a 6-well plate, or 4 ml polybrene supplemented medium/10-cm plate.
4. Add appropriate amounts of virus preparation to NIH 3T3 cells. Usually, 100, 30, 10, 3, and 1 μl will generate a titration curve (*see* **Note 3**).
5. Incubate the cells for 4–6 h and replace the supernatant with fresh 3T3 medium.
6. After 48 h, harvest cells and analyze for green fluorescent protein (GFP) expression by flow cytometry (*see* **Note 4**).

3.3. Transduction of Murine Bone Marrow

1. Seven days before bone marrow harvest, inject 4-week-old NOD/LtJ mice intravenously with 150 mg/kg 5-fluorouracil (*see* **Note 5**).
2. On the day of bone marrow harvest, prepare tissue culture plates coated with RetroNectin. First, dilute RetroNectin to 46 ug/ml (2.5 ml stock + 2.5 ml H_2O) dissolve very well, and bring volume to 54 ml with HBSS. Filter this solution through 0.22-μm sterile filter. Dispense 1 ml of sterile RetroNectin solution into each well of a 6-well non-tissue culture-treated plate and allow the covered plate to stand for 2 h at room temperature. Remove RetroNectin solution and add 2 ml of sterile 2% bovine serum albumin in HBSS into each well. Allow the plate to stand at room temperature for 1 h. Remove the BSA solution and wash once with 3 ml of HBSS/Hepes.
3. Harvest bone marrow from femur and tibia of 5-flurouracil-treated NOD/LtJ mice. Kill mice and remove leg bones by sterile dissection. Cut bones below the hip joint, above and below the knee joint and above the ankle. Flush bone marrow cells from femur with a 27-G needle and bone marrow culture medium (*see* **Note 6**). Lyse red cells with ACK lysis buffer for 3 min at room temperature. Wash

cells twice by with re-suspending in medium and centrifugation to remove all traces of ACK lysis buffer.

4. Plate bone marrow cells at 5–9 × 10^6 per well with 1 ml fresh retroviral supernatant and 1 ml 2× transduction medium (*see* **Note 7**). Incubate at 37 °C, 5% CO_2.
5. After 24 h, add 1 ml fresh viral supernatant.
6. 72 h after transduction, harvest bone marrow from each well, re-suspend with 1 ml fresh virus supernatant, and 1 ml 2× transduction medium.
7. 24 h after re-plating bone marrow, harvest bone marrow from plates, and count. Generally, bone marrow cells will expand 30–50% in culture.
8. Filter bone marrow cells through mesh and resuspend in HBSS. Cell concentration should not exceed 1.5×10^7 cells/ml (*see* **Note 8**).
9. Monitor transduction efficiency by detection of GFP expression by flow cytometry.

3.4. Reconstitution of Recipients

1. Glucose levels in young NOD/Lt mice should be monitored weekly using a standard glucometer. Briefly, a drop of blood should be extracted from the tail vein of recipient mice and applied to the glucometer test strip. Strips should then be read according to the manufacturer's instructions.
2. Non-diabetic 4-week-old NOD/Lt mice should receive 10.25 Gy irradiation 1 day before bone marrow transplant (*see* **Note 9**).
3. Inject at least 4×10^6 transduced bone marrow cells in the tail vein of each recipient. Cells will home to the bone marrow of conditioned recipients.

3.5. Monitoring Gene Expression Levels and Diabetes Occurrence

1. Glucose levels in recipients of transduced bone marrow should be monitored weekly using a standard glucometer. Mice are considered diabetic after two consecutive readings >200 mg/dl. Tests should be done at the same time daily to avoid variations because of feeding.
2. Gene expression levels can be monitored by expression of GFP, which will capture the total percentage of surviving transduced hematopoietic cells (*see* **Note 10**). When combined with lineage-specific antibody staining, one can determine the number of MHC class II expressing cells that are the progeny of transduced progenitors and thus the levels of cell surface expression of transduced MHC class II molecules ***(10)*** (*see* **Note 11**).

4. Notes

1. The number of 293T cells to be plated before transfection will vary depending on the rate of growth of the cell cultures and should be confirmed experimentally. Optimal virus production will occur when cells are 65–75% confluent before transfection.

2. The efficiency of transfection is greatly affected by the quality of the DNA preparation. All DNA should be supercoiled, desalted, and concentrated to a minimum of 1 mg/ml. Cesium chloride gradient preparation generally yields the best quality DNA for transduction; however, DNA preparations obtained using column extraction (e.g., Qiagen) can also be used.
3. Calculation of titer. Percentage of GFP-positive cells can be multiplied by the total number of cells infected on the day of transduction. This, divided by the volume of viral supernatant, will yield the number of infectious particles/volume.
4. Viral titers should be at least 2×10^6 viral particles/ml for efficient transduction of bone marrow in this protocol (multiplicity of infection, MOI = 1). If these titers cannot be achieved, viral supernatants can be concentrated by spinning viral supernatants in swinging bucket rotors at $50{,}000\times g$ force, 4 °C for 1.5 h. Viral pellets can then be re-suspended in DME medium at 4 °C overnight. These viral supernatants should then be tested for their ability to successfully transduce bone marrow cells as previously described as they often display greater toxicity for bone marrow than un-concentrated stocks.
5. The tail veins of mice are readily accessible for intravenous injection. Mice should be gently heated for 3–5 min to allow veins to dilate. Mice should then be restrained and the tail wiped with 70% ethanol. Volumes of up to 1 ml can then be injected into the tail vein. Pressure should then be applied to the site of injection to halt bleeding.
6. During bone marrow harvest, remove leg bones and flush bone marrow from groups of no more than five mice to minimize the time bone marrow cells are allowed to sit without medium. Numbers of bone marrow cells obtained per mouse vary widely amongst the strains and should be determined empirically.
7. A minimum of 10^6 bone marrow cells can rescue recipients from lethal irradiation; however, gene expression levels will be significantly lower than those achieved with 4×10^6 transduced bone marrow cells.
8. Bone marrow cell aggregates, or a high concentration of cells, can cause pulmonary embolism in recipients, so it is highly recommended that the cells be no greater than the indicated concentration, and filtered through sterile mesh before injection as described. In addition, cells should be injected into recipients slowly to minimize shock.
9. Recipients can also be conditioned with 0.5 mg of the anti-CD4 antibody GK1.5 ***(14)***, and 0.5 mg of the anti-CD8 antibody 116–13.1 ***(15)***; however, this results in a substantial delay in the onset of spontaneous diabetes.
10. Gene expression levels will be low for the first several weeks as bone marrow reconstitutes, peak at week 4 after reconstitution, and remain stable indefinitely after 8 weeks post-reconstitution.
11. It should be noted that anti-MHC class II antibody specific for IA^k 10-3.6 cannot be used to distinguish between endogenous and retrovirally transduced MHC class II.

References

1. McDevitt, H.O. (1998). The role of MHC class II molecules in susceptibility and resistance to autoimmunity. *Curr Opin Immunol* **10**, 677–681.
2. Morel, P.A., J.S. Dorman, J.A. Todd, H.O. McDevitt, and M. Trucco. (1988). Aspartic acid at position 57 of the HLA-DQ beta chain protects against type I diabetes: a family study. *Proc Natl Acad Sci USA* **85**, 8111–8115.
3. Nepom, G.T., and W.W. Kwok. (1998). Molecular basis for HLA-DQ associations with IDDM. *Diabetes* **47**, 1177–1184.
4. Wicker, L.S., J.A. Todd, and L.B. Peterson. (1995). Genetic control of autoimmune diabetes in the NOD mouse. *Annu Rev Immunol* **13**, 179–200.
5. Davies, J.L., Y. Kawaguchi, S.T. Bennett, J.B. Copeman, H.J. Cordell, L.E. Pritchard, P.W. Reed, S.C. Gough, S.C. Jenkins, S.M. Palmer, and et al. (1994). A genome-wide search for human type 1 diabetes susceptibility genes. *Nature* **371**, 130–136.
6. Mein, C.A., L. Esposito, M.G. Dunn, G.C. Johnson, A.E. Timms, J.V. Goy, A.N. Smith, L. Sebag-Montefiore, M.E. Merriman, A.J. Wilson, L.E. Pritchard, F. Cucca, A.H. Barnett, S.C. Bain, and J.A. Todd. (1998). A search for type 1 diabetes susceptibility genes in families from the United Kingdom. *Nat Genet* **19**, 297–300.
7. Ilonen, J., M. Sjoroos, M. Knip, R. Veijola, O. Simell, H.K. Akerblom, P. Paschou, E. Bozas, B. Havarani, A. Malamitsi-Puchner, J. Thymelli, A. Vazeou, and C.S. Bartsocas. (2002). Estimation of genetic risk for type 1 diabetes. *Am J Med Genet* **115**, 30–36.
8. Hattori, M., J.B. Buse, R.A. Jackson, L. Glimcher, M.E. Dorf, M. Minami, S. Makino, K. Moriwaki, H. Kuzuya, H. Imura, and et al. (1986). The NOD mouse: recessive diabetogenic gene in the major histocompatibility complex. *Science* **231**, 733–735.
9. Lee, K.H., K.W. Wucherpfennig, and D.C. Wiley. (2001). Structure of a human insulin peptide-HLA-DQ8 complex and susceptibility to type 1 diabetes. *Nat Immunol* **2**, 501–507.
10. Tian, C., J. Bagley, N. Cretin, N. Seth, K.W. Wucherpfennig, and J. Iacomini. (2004). Prevention of type 1 diabetes by gene therapy. *J Clin Invest* **114**, 969–978.
11. Pages, J.C., and T. Bru. (2004). Toolbox for retrovectorologists. *J Gene Med* **6**(Suppl 1), S67–82.
12. Bagley, J., and J. Iacomini. (2003). Gene therapy progress and prospects: gene therapy in organ transplantation. *Gene Ther* **10**, 605–611.
13. Tian, C., J. Bagley, and J. Iacomini. (2005). Preventing autoimmune diabetes through gene therapy. *Discovery Medicine* **5**, 186–190.
14. Dialynas, D.P., Z.S. Quan, K.A. Wall, A. Pierres, J. Quintans, M.R. Loken, M. Pierres, and F.W. Fitch. (1983). Characterization of the murine T cell surface molecule, designated L3T4, identified by monoclonal antibody GK1.5: similarity of L3T4 to the human Leu-3/T4 molecule. *J Immunol* **131**, 2445–2451.
15. Shen, F.-W. 1983. Monoclonal antibodies to mouse lymphocyte differentiation alloantigens. In *Monoclonal Antibodies and T-Cell Hybridomas* (U.H. G. J. Hammerling and J. F. Kearney, editors), Elsevier, Amsterdam, pp. 25–31.

18

Lentiviral Vector Delivery of siRNA and shRNA Encoding Genes into Cultured and Primary Hematopoietic Cells

Mingjie Li and John J. Rossi

Summary

Lentiviral vectors are able to transduce non-dividing cells and maintain sustained long-term expression of the transgenes. Many cell types including brain, liver, muscle, and hematopoietic stem cells have been successfully transduced with lentiviral vectors carrying a variety of genes. These properties make lentiviral vectors attractive vehicles for delivering small interfering RNA (*siRNA*) genes into mammalian cells. RNA polymerase III (Pol III) promoters are most commonly used for expressing siRNAs from lentiviral vectors. Pol III promoters are relatively small, have high activity, and use simple termination signals of short stretches of U. It is possible to include several Pol III expression cassettes in a single lentiviral vector backbone to express different siRNAs or to combine siRNAs with other transgenes. This chapter describes the delivery of Pol III promoted siRNAs by HIV-based lentiviral vectors and covers vector design, production, and verification of siRNA expression and function. This chapter should be useful for establishing a lentiviral vector-based delivery of siRNAs in experiments that require long-term gene knockdown or developing siRNA-based approaches for gene therapy applications.

Key Words: Lentiviral vector; RNA interference; siRNA; shRNA; hematopoietic cells; HIV.

1. Introduction

Small interfering RNAs (siRNAs) mediate RNA interference (RNAi) providing a powerful new tool for gene silencing in a sequence specific manner. RNAi is not only a useful tool for studying gene function and regulation, but

From: *Methods in Molecular Biology, vol. 433: Volume 1: Production and In Vivo Applications of Gene Transfer Vectors*
Edited by: J. M. Le Doux © Humana Press, Totowa, NJ

also has potential for therapeutic applications. In many cases, a long-term effect of RNAi is required, such as in HIV-1 infected individuals. Lentiviral vectors are able to transduce non-dividing cells with sustained long-term expression of genes contained in the vector backbone.

The majority of target cell types for gene therapy are non-dividing or slowly dividing. These include brain, liver, muscle, and hematopoietic stem cells. Each of these cell types has been efficiently transduced by lentiviral vectors ***(1–3)***. We and others ***(4–6)*** have reported that lentiviral vectors can efficiently deliver si/shRNA expression cassettes into various cells with sustained expression and potent function of the encoded siRNAs.

The current third generation replication-defective and self-inactivating (SIN) lentiviral vectors have minimized the potential risk of generating replication-competent helper virus. We used pHIV-7-GFP, a typical third generation replication-defective SIN vector, as the transfer vector for delivering shRNAs. As illustrated in **Fig. 1**, this vector contains a hybrid 5′ LTR, in which the U3 region is replaced with the cytomegalovirus (CMV) promoter and enhancer sequence. Use of this promoter to make the vector RNA makes the transcription of the vector sequence independent of HIV-1 Tat function which is normally required for HIV gene expression ***(7)***. The packaging signal (Ψ) is essential for encapsidation and the Rev-responsive element (RRE) is required for producing high titer vectors. The flap sequence or central polypurine tract (cPPT) and the central termination sequence (CTS) are important for nuclear import of the vector DNA, a feature required for transducing non-dividing cells ***(8)***. The enhanced green fluorescent protein (EGFP) reporter gene is driven by an internal CMV promoter. This is useful for vector titration, measurements of transduction efficiency, and selection of transduced cells. In the 3′ LTR, the *cis*-regulatory sequences were completely removed from the U3 region. This deletion is copied to 5′ LTR after reverse transcription, resulting in transcriptional inactivation of both LTRs. To produce the packaged vectors, the vector DNA along with three other plasmids (*see* **Fig. 1**) is co-transfected into 293T cells. The Gag/Pol proteins are supplied by the pCgp plasmid, pCMV-rev encodes Rev that binds to the RRE for efficient RNA export from the nucleus and pCMV-G encodes the vesicular stomatitis virus glycoprotein (VSV-G) that replaces HIV-1 Env. VSV-G expands the tropism of the vectors and allows concentration through ultracentrifugation. The siRNA sequences along with a Pol III promoter are inserted directly upstream of the CMV-EGFP sequence.

The human U6 small nuclear RNA and human H1 promoters are among the common Pol III promoters used for expressing siRNAs. These promoters have relatively small sizes, and transcription is conveniently terminated within a stretch of four or more uridines. In our experience, at least three Pol III expression cassettes can be delivered by a single lentiviral vector backbone. The sense and antisense sequences of siRNAs can be expressed from separate

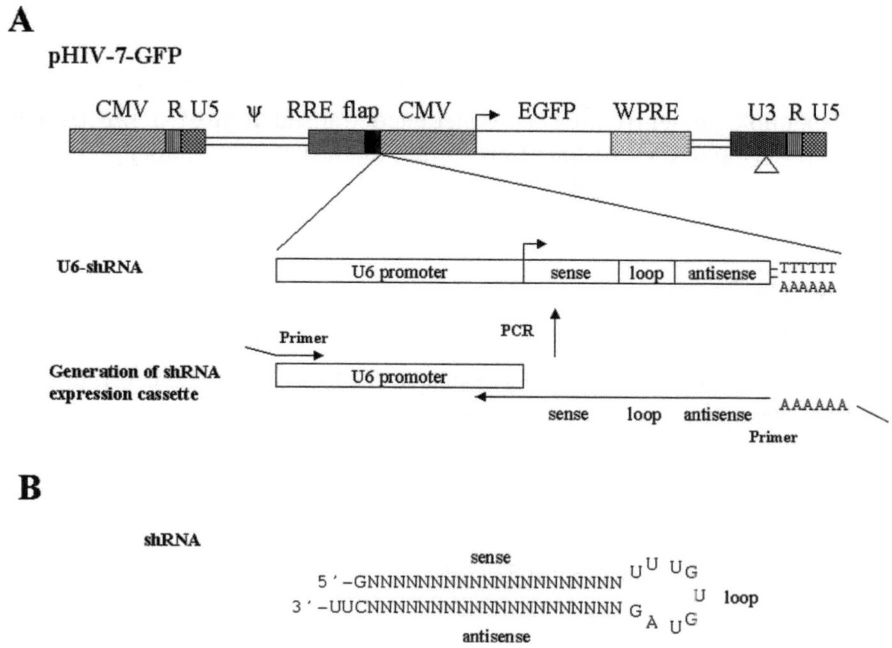

Fig. 1. Construction of lentiviral vector expressing shRNA. (**A**) Top. The transfer vector, pHIV-7-GFP contains a hybrid 5′ LTR in which the U3 region is replaced with the CMV promoter, the packaging signal (Ψ), the RRE sequence, the flap sequence, the *EGFP* gene driven by CMV promoter, the woodchuck hepatitis virus posttranscriptional regulatory element (WPRE) and the 3′ LTR in which the *cis*-regulatory sequences are completely removed from the U3 region. The genes of interest along with a Pol III promoter can be inserted directly upstream of the CMV promoter of EGFP in pHIV-7-GFP vector. Bottom. The U6-shRNA expression cassette is constructed by PCR with primers shown. The PCR product contains restriction sites at both ends for cloning into the lentiviral vector. The shRNA expression cassette, including the U6 Pol III promoter, the sense and antisense sequence of the shRNA separated by a 9-base loop, and a terminator composed of 6 thymidines is inserted directly upstream of the CMV promoter of EGFP in pHIV-7-GFP vector. Arrows indicate the orientation of transcription for a given gene. (**B**). The putative structure of the shRNA expressed in target cells.

promoters or from a single promoter directing a short hairpin (shRNA) structure. We previously demonstrated that anti-rev siRNAs expressed from separate promoters showed marked target downregulation as well as anti-HIV-1 activity *(9)*. We subsequently found that the shRNAs exhibited even more potent RNAi activity than the separately expressed siRNAs designed to inhibit the same target sequences *(4)*. In addition, shRNA expression cassettes are easier to construct than the siRNA counterparts. Therefore, we have focused on shRNA constructs in the lentiviral vector delivery system incorporating the human U6 Pol III promoter to direct shRNA expression. In this chapter, we describe

(1) the construction of lentiviral vectors containing *shRNA* genes, (2) production of the vectors, (3) transduction of the vectors in target cells, (4) detection of the expression of the shRNA, and (5) verification of shRNA functions. As an example of lentiviral vector-mediated transduction of a *shRNA* gene, we also describe RNAi-mediated protection against HIV-1 challenge in stably transduced hematopoietic cells.

2. Materials

2.1. Construction of Lentiviral Vectors Carrying shRNA Sequences

1. Plasmid DNA: lentiviral vector backbone pHIV-7-GFP, pTZ-U6+1 (available from the authors) or other U6 promoter-containing plasmids.
2. Polymerase chain reaction (PCR) primers. See **Subheading 3.1.2** and **Fig. 1** for primer design.
3. Taq DNA polymerase, Restriction enzymes, T4 DNA ligase.
4. QIAEX II gel extraction kit (QIAGEN, Valencia, CA, USA).
5. QIAGEN Plasmid Maxi Kit (QIAGEN).

2.2. Production of Lentiviral Vectors

2.2.1. Packaging of the Vectors

1. 293T cell (human embryonic kidney cells containing SV40 large T antigen).
2. Dulbecco's modified Eagles medium (DMEM) with high glucose (4500 mg/l) supplemented with 10% fetal bovine serum (FBS), 100 units/ml penicillin, 100 μg/ml streptomycin.
3. Plasmid DNA: the transfer vector containing shRNA sequences, pCgp, pCMV-rev, and pCMV-G (available from the authors).
4. TE 79/10: 1 mM Tris–HCl, 0.1 mM EDTA, pH 7.9.
5. 2 M $CaCl_2$.
6. 2x HBS: 0.05 M HEPES, 0.28 M NaCl, 1.5 mM Na_2HPO_4, pH 7.12.
7. 0.6 M sodium butyrate.

2.2.2. Concentration of the Vectors

1. 30-ml syringe.
2. 0.2-μl syringe filter.
3. Ultracentrifuge and SW28 rotor (Beckman, Fullerton, CA, USA).
4. 1×3.5 inch polyallomer centrifuge tubes (Beckman).

2.2.3. Titration of the Vectors

1. Lentiviral vector stock.
2. HT1080 (a human fibrosarcoma cell line).

3. DMEM supplemented with 10% FBS, 100 units/ml penicillin, 100 μg/ml streptomycin.
4. 4 mg/ml Polybrene stock.
5. 3.7% formaldehyde in PBS.

2.3. Transduction of Lentiviral Vectors to Target Cells

1. Concentrated lentiviral vector supernatant.
2. Target cells: K562 cells, a chronic myelogenous leukemia cell line (ATCC, #CCL-243). $CD34^+$ hematopoietic stem cells, purchased from AllCells (Berkley, CA, USA) or purified from umbilical cord blood or bone marrow by anti-CD34 antibody-coupled magnetic beads (Miltenyi Biotech, Aubum, CA, USA).
3. Culture Media. For K652 cells: RPMI 1640 medium supplemented with 10% FBS, 100 units/ml penicillin, 100 μg/ml streptomycin. For $CD34^+$ hematopoietic stem cells: Iscove's modified Dulbecco's medium (IMDM) supplemented with 20% BIT9500 (Stem Cell Technology, Vancouver, BC, Canada), 40 μg/ml human low density lipoproteins (Sigma, St. Louis, MO, USA), 10^{-4} M 2-mercaptoethanol, 100 ng/ml stem cell factor, 100 ng/ml flt3-ligand, 10 ng/ml TPO (PeproTech, Rocky Hill, NJ, USA), 20 ng/ml interleukin-3 (IL-3), 20 ng/ml IL-6 (R & D Systems, Minneapolis, MN, USA).
4. Polybrene.
5. RetroNectin (Takara Mirus Bio Inc., Madison, WI, USA).

2.4. Detection of the Expression of shRNAs

1. STAT-60 reagent (Tel-Test, Inc., Friendswood, TX, USA).
2. Polyacrylamide gel containing 7 M urea.
3. RNA loading buffer: 95% deionized formamide, 0.025% bromophenol blue, 0.025% xylene cyanol, 0.5 mM EDTA, 0.025% SDS.
4. 0.5 M TBE buffer: 45 mM Tris-borate, 1 mM EDTA.
5. Hybond-N nylon membrane (Amersham, Arlington Heights, IL, USA).
6. Electrotransfer apparatus.
7. 20× SSPE stock: 3M NaCl, 0.2 M NaH_2PO_4, 0.02M EDTA, pH 7.4.
8. Pre-hybridization buffer: 6× SSPE, 5× Denharts, 0.5% SDS, and carrier DNA.
9. $\gamma^{32}P$-labeled oligonucleotide probe.

3. Methods

3.1. Construction of Lentiviral Vectors Carrying shRNA Sequences

3.1.1. The Lentiviral Vector Backbone and the Packaging System

The vector backbone we used as a transfer vector of shRNAs is pHIV-7-GFP ***(10)***, which is a typical third generation, replication-defective, and self-inactivating (SIN) vector. This vector contains a hybrid 5' LTR, in which the U3 region is replaced with the CMV promoter and enhancer sequence. This strong promoter makes the transcription of the vector sequence Tat-independent. The psi element (Ψ) is essential for packaging the vector sequence. The RRE is required for producing high-titer vectors. The flap sequence is important for nuclear import, a feature required for transducing non-dividing cells *(8)*. The EGFP reporter gene driven by an internal CMV promoter. The woodchuck hepatitis virus posttranscription regulation element (WPRE) *(11)* following the EGFP sequence improves the expression of the reporter gene by promoting RNA nuclear export and/or polyadenylation. In the 3′ LTR, the *cis*-regulatory sequences were completely removed from the U3 region resulting in transcriptional inactivation of both LTRs. To produce the viral vector, a cotransfection with the transfer vector and three other plasmids into 293T cells is carried out. The pCgp encodes the Gag/Pol protein products. The pCMV-rev provides Rev protein, which binds to the RRE for efficient RNA export from the nucleus. The pCMV-G encodes the VSV-G that replaces HIV-1 env for increasing tropism and allowing ultracentrifugation for concentration of the vector supernatant.

3.1.2. Construction of shRNA Expression Cassette

The following steps describe construction of a shRNA cassette with human U6 promoter (*see* **Note 1**). The human U6 small nuclear RNA promoter is among the common Pol III promoters used for expressing shRNAs. This promoter is relatively small in size and the transcription is conveniently terminated within a stretch of four or more uridines. In our experience, at least three Pol III expression cassettes can be delivered by a single lentiviral vector backbone *(4)*.

1. PCR amplification of plasmid containing the human U6 promoter sequence is accomplished following standard procedures. Use a plasmid containing the U6 promoter sequence (e.g., pTZ U6+1) as the template. The upstream primer is complementary to the 5′ end of the U6 promoter. This can serve as a universal primer for all *shRNA* gene constructs. The downstream primer should be complementary to 3′ end of the U6 promoter and include the complement to the sense, loop, antisense, and transcriptional terminator (*see* **Fig. 1**) (*see* **Note 2**) *(12)*. Include a restriction site at the 5′ end of both primers for ligating the PCR products to the vector (*see* **Note 3**).
2. Gel-purify the PCR product with a QIAEX II gel extraction kit following the manufacturers instructions.

3.1.3. Cloning the shRNA Gene into the Transfer Vector

1. Digest the PCR product and the transfer vector with appropriate restriction enzymes and ligate the insert into the multiple cloning site located between the flap and CMV-EGFP in the lentiviral vector backbone.
2. Verify the cloning product by restriction analysis and DNA sequencing.
3. Amplify and purify the correct cloning product and the three packaging-required plasmids with Plasmid Maxi Kit (QIAGEN).

3.2. Production of Lentiviral Vectors

3.2.1. Packaging of the Vectors

1. The packaging cell line 293T is maintained in DMEM, supplemented with 10% FBS, 100 units/ml penicillin, 100 μg/ml streptomycin in 37 °C incubator with 5% CO_2. Plate the 293T cells in 100-mm tissue culture dishes 24 h before transduction. We usually prepare at least five dishes per vector to obtain the amount of vector needed to repeat the experiments. The cells are about 30–40% confluent when they are seeded and about 80% confluent when they are transfected.
2. Change the culture medium with 10 ml of fresh medium 5 h before transfection.
3. Prepare 1 ml of calcium phosphate-DNA suspension for each 100-mm plate of cells as follows:

    a. Set up two sterile tubes for transfection of one plate. Label the tubes 1 and 2.
    b. Add 0.5 ml 2× HBS to tube 1.
    c. Add TE 79/10 to tube 2. The volume of TE 79/10 = 440 μl – the volume of DNA.
    d. Add 15 μg the transfer vector containing the shRNA expression cassette, 15 μg pCgp, 5 μg pCMV-rev, and 5 μg pCMV-G (*see* **Note 4**) to tube 2 and mix.
    e. Add 60 μl 2 M $CaCl_2$ solution to tube 2, gently mix.
    f. Transfer the contents from tube 2 to tube 1, dropwise with gently mixing.
    g. Allow the suspension to sit at room temperature for at least 30 min.

4. Mix the precipitation well by pipetting or vortexing.
5. Add 1 ml of suspension to a 100-mm plate containing cells. The suspension must be added slowly and dropwise while gently swirling the media in the plate. Return the plates to the incubator and leave the precipitation for 5–6 h.
6. Replace the media with 6 ml complete media. Add 60 μl 0.6 M sodium butyrate (*see* **Note 5**). Return to culture.
7. After 24 h of culture, collect the supernatant and freeze it at –80 °C. Add 6 ml complete media to each plate. Add 60 μl 0.6 M sodium butyrate. Return to culture.
8. After 12 h of culture, collect supernatant. Freeze it at –80 °C or go to next step.

3.2.2. Concentration of the Vectors

1. Centrifuge the supernatant freshly collected or thawed from freezing at 2000 rpm (900 g) for 10 min to remove any cell debris in the supernatant. Filter the supernatant with 0.2-μm syringe filter (*see* **Note 6**). Transfer the supernatant to polyallomer tubes (*see* **Note 7**). Concentrate the supernatant by ultracentrifugation at 24,500 rpm (108,000 g) at 4 °C for 1.5 h with a Beckman SW28 swing rotor or equivalent.
2. Remove the supernatant, add an appropriate amount of culture media (e.g., 150 μl for 30 ml of original supernatant if a 200-fold concentration is desired) and resuspend the pellet by repeated pipetting (*see* **Note 8**).
3. The concentrated vector should be divided into 10–50 μl aliquots and stored at –80 °C until use. To prevent degradation of the vector, avoid freeze-thaw cycles.

3.2.3. Titration of the Vectors

1. Seed 1×10^5/well of HT1080 in 6-well plate in DMEM supplemented with 10% FBS and culture overnight.
2. Add serial diluted vector stock and 4 μl/ml Polybrene (*see* **Note 9**) to the cultured cells. Continue culture for 48 h.
3. Trypsinize the cells. Following centrifuge, remove the supernatant and resuspend the pellet with 300 μl of 3.7% formaldehyde in PBS.
4. Determine the percentage of EGFP-positive cells by FACS analysis (*see* **Note 10**). The titer will be represented as transduction unit (TU) per milliliter concentrated vector (TU/ml).

$$\text{Titer} = \frac{\text{cell number} \times \text{percentage of EGFP}^{+}\ \text{cells} \times \text{dillution}}{\text{vector volume (ml)} \times 100}$$

In this formula, the cell number stands for the cell count when the vector was added.

3.3. Transduction of Lentiviral Vectors to Target Cells

The following are transduction protocols for some commonly used target cells in shRNA research and therapeutic applications.

3.3.1. Transducing Monolayer Cultured Cells

For monolayer cultured cells, seed cells to culture plates 24 h before transduction. At the time of transduction, the cell density should be 30–40% confluent. Add a vector at an appropriate multiplicity of infection (MOI) (*see* **Note 11**) and Polybrene at a final concentration of 4 μg/ml and return the cells into incubator. After overnight culture, replace the culture medium. For many monolayer cultured cell lines, an MOI of 5 can achieve over 90% transduction efficiency. For a new cell line to be used, a series of different MOIs should be tested to find the minimum, effective MOI.

3.3.2. Transducing Suspension-Cultured Cell Lines

To transduce suspension-cultured cell lines, seed 2×10^5/well cells into a 24-well plate. Add appropriate amount of vector and 4 μg/ml Polybrene. After overnight incubation, replace the media. For K562 cell line (a leukemia cell line from a CML patient), the transduction efficiency is close to 100% at a MOI of 10. For some suspension-cultured cells such as CEM (a human T cell line), centrifugation can remarkably enhance transduction efficiency. Place 2×10^5 cells in 1 ml culture medium in 15-ml centrifuge tube. Add 4 μg/ml Polybrene. Centrifuge at 2,000 rpm (900 g) and 20 °C for 30 min. Resuspend the cell pellet with a pipette and transfer the cells to the culture plate. After overnight culture, replace the media. Transduction efficiencies can be determined by FACS analyses 48 h after transduction.

3.3.3. Transducing Hematopoietic Stem Cells

For transduction of hematopoietic stem cells, $CD34^+$ cells are enriched from umbilical cord blood or bone marrow by anti-CD34 antibody coupled magnetic beads. Forty-eight hours before transduction, the $CD34^+$ cells are cultured in the medium containing cytokines. Coat 24-well non-tissue culture-treated plates with RetroNectin following the manufacturer's instruction. The lentiviral vector stock is adjusted to 40 MOI in 200 μl culture medium and loaded to the wells of the coated plate. After incubation at 37 °C for 4 h, the vector supernatant is removed and the well washed with PBS. The pre-stimulated $CD34^+$ cells are added to the well at 5×10^4/ml in the growth medium. After overnight culture, centrifuge the cells, change the media, and return the cells back to culture.

3.4. Detecting the Expression of shRNA Delivered by Lentiviral Vector

Northern blotting is among the most reliable methods to verify whether the cloned shRNA sequence can be expressed and for determining the level of expression in the target cells (*see* **Note 12**).

1. Extract total RNA from transduced cells with STAT-60 reagent (Tel-Test, Inc.) according to the manufacturer's protocol.
2. Prepare polyacrylamide gel containing 7 M urea. A 15% polyacrylamide gel is commonly used for detecting shRNA. An 8% polyacrylamide gel can resolve the shRNA band equally well and is more efficient to transfer. In addition, an 8% polyacrylamide gel allows shRNAs and other larger RNAs up to several hundred nucleotide length be detected in a single blotting.
3. Fifteen micrograms of total RNA solution (we were able to detect shRNA using 3.5 μg of total RNA extracted from sorted $EGFP^+$ population of $CD34^+$ cells) is

mixed with an equal volume of RNA loading buffer. Heat the samples at 95 °C for 4 min and then place on ice.

4. Load the samples in the gel and electrophoresis them in 0.5× TBE buffer until the bromophenol blue dye migrates to about two-thirds of the gel plate.
5. Transfer the RNA to Hybond-N nylon membrane by electroblotting in 0.5× TBE buffer.
6. Fix the RNA to the membrane with a UV cross-linker (Stratagene Inc., La Jolla, CA, USA).
7. Pre-hybridize in buffer containing 6× SSPE, 5× Denharts, 0.5% SDS, and carrier DNA for 2 h.
8. Add the γ^{32}P-labeled oligonucleotide probe complementary to the antisense sequence of the shRNA and hybridize overnight at 37 °C.
9. The membrane is washed with 6× SSPE and 0.1% SDS, at 37 °C for 10 min then washed with 2× SSPE and 0.1% SDS twice at 37 °C for 10 min each and exposed to an X-ray film in –20 °C freezer. The exposure time varies from a few hours to a few days depending on cell type, amount of RNA, vector design, probe labeling efficiency, etc.

3.5. Determining the Efficacy of RNA Cleavage by shRNAs in Lentiviral Vector Transduced Cells

For detecting the cleavage of endogenous mRNAs by shRNAs, reverse transcriptase (RT)–PCR, real-time PCR, and Northern blotting are among the choices. To determine the knockdown of protein levels, Western blotting and immunofluorescence assays are the most common approaches. For functional assay, in the case of anti-HIV shRNAs, assays for HIV-1 p24 antigen and RT are used. For shRNA targeting fusion mRNA in malignant cells, cell proliferation and apoptosis assays are good choices.

3.6. Experimental Example

In the following example, human embryonic cord blood derived CD34+ hematopoietic progenitor cells were transduced with HIV-7 harboring an shRNA targeting HIV-1 rev or the HIV-7 vector backbone as a control. Eleven days after transduction, the cells were sorted by FACS, and EGFP$^+$ cells were collected. After recovery from sorting, 5×10^5 EGFP$^+$ cells were exposed to the monocytotropic JR-FL strain of HIV-1 at an MOI of 0.01 overnight. The infected cells were washed four times with HBSS and the cultures were in media for CD34$^+$ cells as described above. The culture supernatants were collected on a weekly basis. The p24 antigen analyses were performed using a Coulter HIV-1 p24 Antigen Assay (Beckman Coulter, Brea, CA) according to the manufacturer's instructions. From the data of **Fig. 2**, it can be seen that the cells expressing only EGFP were highly susceptible to HIV-1 replication,

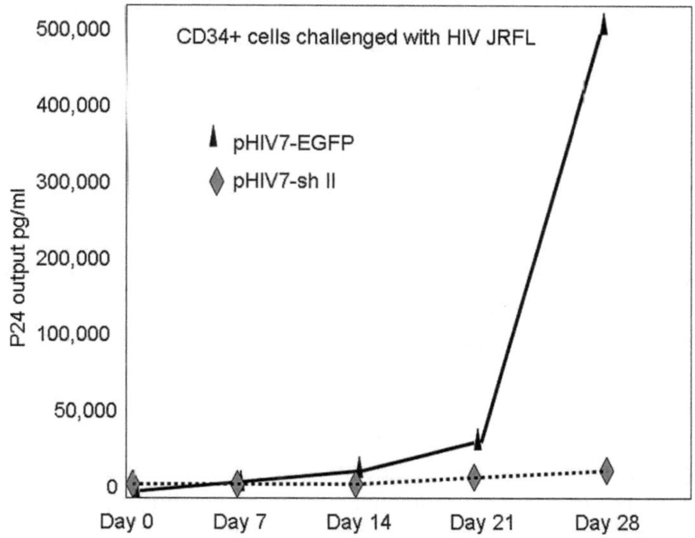

Fig. 2. HIV-1 challenge assay of anti-HIV rev shRNA transduced cells. Embryonic cord blood derived $CD34^{+}$ hematopoietic progenitor cells were transduced with either the parental HIV-7 vector or HIV-7-U6-shRNArev. Transduced cells were FACS sorted for EGFP expression and following a short period of recovery challenged with HIV-1 JRFL at a multiplicity of infection of 0.01. The supernatants were withdrawn on the indicated days and assayed for HIV-1 secreted p24 antigen as described in the text.

whereas the shRNA transduced cells were completely protected from HIV-1 replication as indicated by the baseline levels of p24 antigen production. The protection lasted for over one month before the experiment was terminated. This type of protection is commonly observed in anti-HIV shRNA expressing primary cells. There are many potential target sites in HIV for RNAi, and we highly recommend multiplexing shRNAs targeting two or more different HIV sequences to minimize viral escape mutants.

In summary, lentiviral vectors can be effective vehicles for the delivery of *shRNA* genes into various cultured and primary cells. Our example demonstrates that human hematopoietic progenitor cells are good targets for lentiviral vector-mediated transduction, thus providing the foundation for future hematopoietic stem cell gene therapy against viral diseases and perhaps some leukemias.

4. Notes

1. In addition to the human U6 promoter, other Pol III promoters, such as the human H1 promoter ***(13)***, and tRNA promoter ***(14)*** can also be used for expressing siRNAs or shRNAs.

2. The first position in the transcript of U6 promoter should be guanine (G) to ensure efficient transcription. If the first base in the target sequence is not G, an extra G can be added to the sense strand transcript without affecting the function of the shRNA.
3. Alternatively, the restriction sites in the primers can be omitted by ligating the PCR product to pCR2.1 using the TA cloning kit (Invitrogen, Carlsbad, CA, USA).
4. These amounts of DNA were arrived at by empirical optimization of transfections, and some variation of these numbers may be required for individual labs to optimize their transfections.
5. Sodium butyrate can enhance promoter activity and in turn enhance the vector titer.
6. It may also be filtered with 0.2 μm cellulose acetate bottle-top filter for preparing large quantity and reducing the loss of the vector.
7. The polyallomer tubes can be autoclaved. Using autoclaved centrifuge tubes and 0.2 μl filters can minimize contamination during handling the vector supernatant. This is especially important if the transduced cells are used for long-term culture or engrafting animals.
8. Sometimes, the pellet is difficult to see. When removing the supernatant, be careful not to touch the center of the centrifuge tube where the concentrated vector particles are located.
9. Polybrene can enhance transduction efficiency in many cell types by neutralizing negative charges on both the cell membrane and the vector envelope.
10. If the transfer vector used does not contain a reporter gene, the vector titer may be determined as the number of integrated vector DNA/ml of vector by real-time PCR with primers complementary to the psi element of the vector ***(15)***.
11. MOI determines the amount of vector per cell used for a transduction. An MOI of 1 means 1 TU of vector per cell. For example, if the cell number used for transduction is 1×10^5, an MOI of 5 means 5×10^5 TU of vector is needed.
12. ShRNA transcripts are too short to be detected by most PCR-based methods.

Acknowledgments

We thank Dr. Jiing-Kaun Yee for providing pHIV-7-GFP and the packaging plasmids. This work was supported by NIH grants AI29329 and AI42552 and HL074704 to J.J.R.

References

1. Naldini, L., Blomer, U., Gage, F. H., Trono, D., and Verma, I. M. (1996) Efficient transfer, integration, and sustained long-term expression of the transgene in adult rat brains injected with a lentiviral vector. *Proc Natl Acad Sci USA* **93**, 11382–8.

2. Kafri, T., Blomer, U., Peterson, D. A., Gage, F. H., and Verma, I. M. (1997) Sustained expression of genes delivered directly into liver and muscle by lentiviral vectors. *Nat Genet* **17**, 314–7.
3. Uchida, N., Sutton, R. E., Friera, A. M., He, D., Reitsma, M. J., Chang, W. C., Veres, G., Scollay, R., and Weissman, I. L. (1998) HIV, but not murine leukemia virus, vectors mediate high efficiency gene transfer into freshly isolated G0/G1 human hematopoietic stem cells. *Proc Natl Acad Sci USA* **95**, 11939–44.
4. Li, M.-J., Bauer, G., Michienzi, A., Yee, J.-K., Lee, N.-S., Kim, J., Li, S., Castanotto, D., Zaia, J., and Rossi, J. J. (2003) Inhibition of HIV-1 infection by lentiviral vectors expressing Pol III-promoted anti-HIV RNAs. *Mol Ther* **8**, 196–206.
5. Qin, X. F., An, D. S., Chen, I. S., and Baltimore, D. (2003) Inhibiting HIV-1 infection in human T cells by lentiviral-mediated delivery of small interfering RNA against CCR5. *Proc Natl Acad Sci USA* **100**, 183–8.
6. Scherr, M., Battmer, K., Ganser, A., and Eder, M. (2003) Modulation of gene expression by lentiviral-mediated delivery of small interfering RNA. *Cell Cycle* **2**, 251–7.
7. Arya, S. K., Guo, C., Josephs, S. F., and Wong-Staal, F. (1985) Trans-activator gene of human T-lymphotropic virus type III (HTLV-III). *Science* **229**, 69–73.
8. Sirven, A., Pflumio, F., Zennou, V., Titeux, M., Vainchenker, W., Coulombel, L., Dubart-Kupperschmitt, A., and Charneau, P. (2000) The human immunodeficiency virus type-1 central DNA flap is a crucial determinant for lentiviral vector nuclear import and gene transduction of human hematopoietic stem cells. *Blood* **96**, 4103–10.
9. Lee, N. S., Dohjima, T., Bauer, G., Li, H., Li, M. J., Ehsani, A., Salvaterra, P., and Rossi, J. (2002) Expression of small interfering RNAs targeted against HIV-1 rev transcripts in human cells. *Nat Biotechnol* **20**, 500–5.
10. Yam, P. Y., Li, S., Wu, J., Hu, J., Zaia, J. A., and Yee, J. K. (2002) Design of HIV vectors for efficient gene delivery into human hematopoietic cells. *Mol Ther* **5**, 479–84.
11. Zufferey, R., Donello, J. E., Trono, D., and Hope, T. J. (1999) Woodchuck hepatitis virus posttranscriptional regulatory element enhances expression of transgenes delivered by retroviral vectors. *J Virol* **73**, 2886–92.
12. Castanotto, D., Li, H., and Rossi, J. J. (2002) Functional siRNA expression from transfected PCR products. *Rna* **8**, 1454–60.
13. Brummelkamp, T. R., Bernards, R., and Agami, R. (2002) A system for stable expression of short interfering RNAs in mammalian cells. *Science* **296**, 550–3.
14. Kawasaki, H., and Taira, K. (2003) Short hairpin type of dsRNAs that are controlled by tRNA(Val) promoter significantly induce RNAi-mediated gene silencing in the cytoplasm of human cells. *Nucleic Acids Res* **31**, 700–7.
15. Sastry, L., Johnson, T., Hobson, M. J., Smucker, B., and Cornetta, K. (2002) Titering lentiviral vectors: comparison of DNA, RNA and marker expression methods. *Gene Ther* **9**, 1155–62.

19

Nanoparticle-Mediated Gene Delivery to the Lung

Rajagopal Ramesh

Summary

Drug delivery, especially gene delivery to the lung, has been a challenge. Numerous gene delivery systems to the lung have been developed and tested in preclinical studies. However, only a few of them have been successfully tested in the clinic and shown promise. The reasons for failure to translate preclinical findings into clinical setting include inefficient gene delivery, toxicity, stability, and other factors related to scaling and manufacturing of the gene delivery vehicle. Therefore, there is a need for developing and testing of new gene delivery systems that can overcome some of the existing limitations. Preclinical studies from our laboratory using a cationic lipid (1,2-Dioleoyl-3-Trimethylammonium-Propane (DOTAP):cholesterol)-based nanoparticle have shown efficient and effective gene delivery to the lung especially to tumor-bearing lungs of mice. Based on the efficacy and toxicity studies observed in preclinical studies, we have recently initiated a Phase I clinical trial for systemic treatment of non-small cell lung cancer (NSCLC). In this clinical trial, a tumor suppressor gene encapsulated in the lipid-based nanoparticle will be delivered intravenously to determine the maximum-tolerated dose (MTD). The results from this clinical trial will provide a basis for conducting subsequent Phase II trial that will focus on determining toxicity and therapeutic efficacy. In this chapter, the details for the synthesis and testing of the lipid-based nanoparticle for systemic gene delivery to the lung with emphasis on lung cancer is provided.

Key Words: Lung cancer; nanoparticle; liposomes; systemic; gene delivery; tumor suppressor genes.

1. Introduction

Cancer is a leading cause of death in the USA. It was estimated that 1,372,910 humans would be diagnosed with various forms of cancer in 2005

From: *Methods in Molecular Biology, vol. 433: Volume 1: Production and In Vivo Applications of Gene Transfer Vectors*
Edited by: J. M. Le Doux © Humana Press, Totowa, NJ

and approximately 570,280 (50%) of the patients would be dying of the disease *(1)*. Additionally, the probability of developing invasive cancer at any bodily site increases with age for both men and women, with a probability of developing cancer being one in two for men and one in three for women from birth to death *(1)*. Although the incidence and mortality rates are high when all types of cancer are included, they vary when individual tumor types based on body sites are examined. For example, the incidence of lung cancer in men has declined during the past decade whereas it has increased in women for the same period. Associated with the reduced incidence is the decline in death rate, though not very significant. This is partly due to the dissemination of cancer awareness and the establishment new guidelines for early detection and development of new therapies. Despite these advances made to reduce the suffering from cancer and improve patient survival, the mortality rate remains high with 5-year survival being less than 15% for most epithelial cancers *(2)*.

One of the reasons for the failure to treat cancer such as lung cancer is the inability to treat metastases. Effective treatment of metastatic cancers require the following: ability of drug to be delivered systemically, reach the cancer site effectively, accumulate selectively inside the cancer cells, and kill the cancer cells with minimal to no toxic effects on adjacent normal cells. Many of the drugs currently available in the market do not meet these requirements and hence cannot effectively kill cancer cells present at a distant site in the body. Therefore, development of novel drug delivery agents that will have the desirable properties and aid in the effective delivery of drugs to the cancer site is warranted.

One agent that has been developed and being tested as drug delivery vehicle for therapy is nanoparticles *(3–5)*. Nanoparticles are small particles that are 5–500 nanometers (nm) in size. The nanoparticles being developed and tested currently vary in their size, structure, and composition. A majority of the nanoparticles currently being developed focus on diagnosis and detection of a disease *(4–8)*. However, nanoparticles are also being tested to deliver drugs such as chemotherapy and therapeutic genes to treat diseases such as cancer *(9–11)*. A major emphasis in the field of nanotechnology is to develop nanoparticles that can be administered systemically to deliver drugs to the target site. The disease of interest to several investigators including our laboratory is in the treatment of metastatic cancers such as lung cancer.

Preclinical studies from our laboratory using cationic lipid-based, DOTAP:cholesterol (DOTAP:Chol) nanoparticles have shown these nanoparticles carrying therapeutic genes (e.g., *p53*, *Fus1*, *mda-7/IL-24*) when injected systemically into mice effectively deliver the therapeutic gene to the lung tumors resulting in a therapeutic effect *(12–16)*. Although these untargeted nanoparticles have been shown to be effectively taken up by the tumor cells in vivo (*17,18*) the opportunity to attach targeting moieties to the surface of these nanoparticles and thereby increase specificity exists. The development and testing of targeted DOTAP:Chol nanoparticles to lung cancer is currently under investigation in our laboratory.

On the basis of the therapeutic efficacy and toxicity studies observed in our preclinical studies using DOTAP:Chol-based nanoparticles, we have recently initiated a Phase I clinical trial for systemic treatment of non-small cell lung cancer (NSCLC). In this clinical trial, a tumor suppressor gene encapsulated in the lipid-based nanoparticle will be delivered intravenously to determine the maximum-tolerated dose (MTD). The results from this clinical trial will provide a basis for conducting subsequent Phase II trial that will focus on determining toxicity and therapeutic efficacy.

In this chapter, a step-by-step procedure for the synthesis and testing of DOTAP:Chol-based nanoparticles for systemic gene delivery to the lung with emphasis on lung cancer is provided. Experimental procedures and assays for testing the effectiveness of gene delivery by these nanoparticles both in vitro and in vivo are provided. It is anticipated that by following the described steps the reader will easily be able to synthesize the DOTAP:Chol-based nanoparticles and test in the laboratory.

It is to be noted that the methods provided in this chapter are primarily for development and testing of the nanoparticles in preclinical studies. The procedure described can also be followed and applied for clinical development and testing. However, several additional steps such as scaling and manufacturing procedures (e.g., quality control and quality assurance) under good manufacturing practice (GMP) are involved in the development of these nanoparticles for clinical testing and need to be approved by the Federal and Drug Agency (FDA). Because of the large number of steps involved in production of these nanoparticles for clinical testing, it is not possible for the author to cover all of the information in this chapter. The procedures described below are therefore recommended as a guideline for clinical development of these nanoparticles for systemic delivery of therapeutic genes for treatment of lung cancer.

2. Materials

2.1. Nanoparticle Synthesis

1. 5% Dextrose in water (D5W) for injection (Baxter Healthcare Company, Deerfield, IL).
2. 0-grade argon gas, Air Liquid or equivalent.
3. 1× solution of RBS pF detergent concentrate (Pierce, Rockford, IL).
4. 10-cc disposable sterile luer-lok syringes (Becton Dickinson, Franklin Lakes, NJ).

5. BlueMax® Falcon 15- and 50-ml polystyrene conical tubes (Becton Dickinson).
6. Rotary water bath.
7. Dimethyldioctadecylammonium bromide (1,2-Dioleoyl-3-Trimethylammonium-Propane (DOTAP)) (Avanti Polar Lipids, Alabaster, AL).
8. Ethanol (EtOH).
9. High Pressure Liquid Chromatography (HPLC)-grade chloroform (Mallinckrodt, Hazelwood, MO).
10. Lyophilizer.
11. Parafilm.
12. Synthetic cholesterol (Sigma Chemicals, St. Louis, MO).
13. Sterile glass tubes.
14. Sterile glass vials and bottles with caps.
15. Weigh boats.
16. Weigh Balance (1–100 mg range; OHAUS, Pine Brook, NJ).
17. Whatman Anotop sterile disposable syringe filters (0.1 and 0.2 μm) (Fisher Scientific, Pittsburgh, PA).
18. Whatman polysulfone sterile disposable syringe filters (0.45 and 1.0 μm pore size) (Fisher Scientific).

2.2. Optical Density Determination

1. 1.5- and 0.5-ml Microcentrifuge tubes (Pyrogen and metal free, DNase/RNase free).
2. Beckman DU640 Spectrophotometer with Turbidity Cell Holder (Beckman Coulter, Fullerton, California).
3. Eppendorf pipettes (P-20, P-200, and P-1000).
4. Freshly prepared DOTAP:Chol nanoparticles.
5. Lens paper.
6. Masked Semi-microcell with UV-silica windows (Beckman/Coulter).
7. Sterile disposable pipette tips (1–200 μl and 1 ml).
8. Sterile Water for Irrigation (USP) (American Pharmaceutical Partners, Inc., Schaumburg, IL).

2.3. Particle Size Analysis

1. Freshly prepared DOTAP:Chol nanoparticles.
2. Disposable plastic cuvettes with caps (Beckman/Coulter).
3. Sterile disposable pipette tips (1–200 μl and 1 ml).
4. Sterile water (American Pharmaceutical Partners, Inc.).

2.4. Charge Potential Analysis

1. 10-cc disposable sterile luer-lok syringes (Becton Dickinson).
2. EtOH.
3. Double distilled water (dH_2O).

4. Coulter Delsa 440SX (Beckman Coulter).
5. Freshly prepared DOTAP:Chol nanoparticles.

2.5. Preparation of DNA-Containing Nanoparticles

1. 1.5- and 0.5-ml Microcentrifuge tubes (Pyrogen and metal free, DNase/RNase free).
2. 70% EtOH.
3. D5W for Injection (Baxter Healthcare Company).
4. Beckman DU640 Spectrophotometer with Turbidity Cell Holder (Beckman).
5. DNA at 2–10 mg/ml.
6. DOTAP:Chol nanoparticles.
7. Eppendorf pipettes (P-1000, P-200, and P-20).
8. Glass vials with crimp tops (Teflon must face the solution) (Fisher Scientific).
9. Hand crimper (Kimble, Vineland, NJ).
10. Masked Semi-microcell (cuvette) with UV-silica windows (Beckman-Coulter).
11. Pipet Aid (Drummond Scientific, Broomall, PA).
12. Sterile disposable pipette tips (1–200 μl and 1 ml).
13. Sterile Water for Irrigation (USP) (American Pharmaceutical Partners, Inc.).

2.6. *In Vitro* Transduction

2.6.1. Cell Culture

1. Bright-field inverted microscope (Nikon, Melville, NY).
2. BCA Protein Assay Reagent Kit (Pierce).
3. Cell lysis buffer: 0.125 M Tris–HCl (pH 6.8), 2% (w/v) sodium dodecyl sulfate (SDS), 10% (w/v) glycerol, 6 M urea, 5% (w/v) 2-β-mercaptoethanol (2-ME), 0.1 ml of 5% (w/v) bromophenol blue. Store at –20°C.
4. Cold calcium-free phosphate-buffered saline (PBS).
5. DNA containing DOTAP:Chol nanoparticles.
6. DOTAP:Chol nanoparticles.
7. Falcon® six-well tissue culture plates (Becton Dickinson).
8. Fetal Bovine Serum (FBS; Gibco/BRL, Invitrogen Corporation, Grand Island, NY).
9. H1299 and A549 lung cancer cells (American Type Culture Collection, Rockville, MD).
10. Plasmid DNA expressing a marker or therapeutic gene [e.g., *Lac-Z*, Green fluorescent protein (*GFP*), *p53*, *mda-7*, *Fus1*, etc.].
11. Tissue culture medium: RPMI-1640, Hams-F-12 (Cellpro, Mediatech Inc., Herndon, VA).
12. Trypsin containing EDTA (1×; Gibco/BRL).
13. Eppendorf pipettes (P-1000, P-200, and P-20).
14. Sterile disposable pipette tips (1–200 μl and 1 ml).
15. Sterile disposable serological pipettes (10 ml, 5 ml, and 1 ml; Corning Inc., Corning, NY).

2.6.2. Immunohistochemical Analysis for Transgene Expression

1. 0.5% glutaraldehyde (made from 25% glutaraldehyde stock solution by diluting 1 ml of stock solution with 49 ml of cold PBS) stored in +4°C.
2. X-Gal staining solution: 26 µl 1 M $MgCl_2$, 60 µl 5 M NaCl, 0.88 ml 1 M HEPES (pH 7.4), 2.0 ml 30 mM potassium ferricyanide, 2.0 ml of 30 mM potassium ferrocyanide, 50 µl of 2% X-Gal solution [200 mg X-Gal powder dissolved in 10 ml N_1N-Dimethyl-Formamide (Sigma Chemicals)]. Make the final volume to 10 ml with water. Always prepare a fresh solution for use.

2.6.3. Western Blotting Analysis for Transgene Expression

2.6.3.1. Sodium Dodecyl Sulfate–Polyacrylamide Gel Electrophoresis

1. Ammonium persulfate (APS): Prepare 10% (w/v) APS solution and aliquot in 0.5 ml volume and store at –20°C.
2. Bis-acrylamide: Prepare 30% bis-acrylamide solution in water and store at +4°C.
3. Prestained molecular weight markers: The markers can be purchased from several commercial vendors (e.g., BioRad, Hercules, CA; Santa Cruz Biotechnology, Santa Cruz, CA).
4. Power/Pac 200 (BioRad).
5. Running buffer: Prepare a 10× stock by mixing 0.025 M Tris, 0.192 M glycine, 0.1% SDS solution. Store the solution at room temperature.
6. Sample loading buffer: Mix 5 ml of cell lysis buffer with 0.5 ml of 2-ME and adjust the final volume to 10 ml with water.
7. Separating buffer: 3.0 M Tris–HCl (pH 8.8). Store the solution at room temperature.
8. Stacking buffer: 0.5 M Tris–HCl (pH 6.8). Store the solution at room temperature.
9. SDS: 10% (w/v). Store the SDS solution at room temperature.
10. N, N, N, N´-Tetramethyl-ethylenediamine (TEMED).
11. Water-saturated isobutanol: Mix dH_2O with isobutanol in 1:1 ratio (v/v) and allow the two phases to separate. Use the upper phase of the water saturated isobutanol.
12. Western Blot apparatus with transfer system (BioRad).

2.6.3.2. Protein Transfer and Detection

1. Bio-Max ML film (Kodak, Rochester, NY).
2. Enhanced chemiluminescent (ECL) reagent (Amersham Biosciences, Piscataway, NJ).
3. Horse-radish peroxidase (HRP)-conjugated secondary antibody (Jackson Immunoresearch, West Grove, PA).
4. Nitrocellulose membrane (Millipore, Bedford, MA).
5. PBS-T blocking buffer: Add 25 g of fat-free milk powder to 450 ml of PBS and mix thoroughly using magnetic stirrer. Add 0.5 ml of Tween-20, continue to mix, and adjust the final volume to 500 ml. Store the solution at +4°C.

6. Primary Antibody (Commercial source varies depending on the target protein being investigated; e.g., Santa Cruz Biotechnology; Cell Signaling, Worcester, MA).
7. Stripping buffer: Mix equal volumes of H_2O_2 and PBS in 1:1 ratio. Always prepare the solution fresh and store at room temperature until use.
8. Tris-buffered saline (TBS-T) blocking buffer: Add 25 g of fat free milk powder to 450 ml of TBS and mix thoroughly using magnetic stirrer. Add 0.5 ml of Tween-20, continue to mix, and adjust the final volume to 500 ml. Store the solution at +4°C.
9. Transfer buffer: 25 mM Tris, 0.192 mM glycine, 20% (v/v) methanol. Store the buffer solution at room temperature.
10. Whatmann chromatography paper (3M; Whatmann, Maidstone, UK).

2.7. In Vivo Nanoparticle Delivery

1. 1-cc disposable syringe (Becton Dickinson).
2. 150-mm CellStar® tissue culture plates (Greiner Bio-One GmBH, Frickenhausen, Germany).
3. 15% India Ink: To 85 ml dH_2O, add four drops of ammonium hydroxide (NH_4OH) and 15 ml of Higgins India Ink. Mix and store the solution at room temperature. Solution can be used over several months.
4. Bright-field inverted microscope (Nikon).
5. Blunt and pointed end forceps (Fisher Scientific).
6. Calipers.
7. Formalin.
8. Cold calcium-free PBS.
9. Dissection board.
10. Dissection scissors.
11. DOTAP:Chol nanoparticles.
12. Face mask.
13. Fekete's solution: Mix 700 ml EtOH, 300 ml dH_2O, 100 ml formaldehyde, and 50 ml acetic acid in a glass beaker. The solution is light sensitive and is stored in lightproof dark glass bottle at room temperature. Solution can be used over several months.
14. Female nude mice or Severe Combined Immunodeficient (SCID)/Beige mice (4–6 weeks old; Charles River Laboratory, Wilmington, MA).
15. FBS (Gibco/BRL).
16. Freshly prepared DNA containing DOTAP:Chol nanoparticles.
17. H1299 and A549 lung cancer cells (American Type Culture Collection).
18. Infrared heat lamp (Fisher Scientific).
19. Latex hypoallergen gloves.
20. Mouse restrainer.
21. Needles ($25G^{5/8}$ and $27G^{1/2}$; Becton Dickinson).
22. Plasmid DNA expressing a marker or therapeutic gene (e.g., *Lac-Z*, *GFP*, *p53*, *mda-7*, *Fus1*, etc.).

23. Rib-Back Carbon Steel Scalpel blade (Becton Dickinson Acute Care).
24. Stereomicroscope (Leica Microscopes, Barrington, NJ).
25. Tissue culture medium: RPMI-1640, Hams-F-12 (Cellpro, Mediatech Inc.).
26. Trypsin containing EDTA (1×; Gibco/BRL).
27. Tyvek suit.

3. Methods

3.1. Synthesis of Nanoparticles

1. Prior to start of the experiment, take out all of the necessary reagents required for synthesis and stored at different temperatures to come to room temperature.
2. Weigh 140 ± 1 mg of DOTAP using a clean spatula.
3. Transfer the DOTAP into a 1-liter round-bottom glass flask, being careful to not touch the sides of the flask with the DOTAP.
4. Weigh 69.3 ± 1 mg of cholesterol and place into the round-bottom flask containing DOTAP.
5. Pipet 10 ml of HPLC-grade chloroform to the bottom of the flask and quickly cover the mouth of the flask with Parafilm.
6. Gently swirl the flask in a circular motion to completely dissolve the DOTAP and cholesterol powder present at the bottom of the flask.
7. Connect the flask to a rotary evaporator.
8. Set the rotary water bath to 30°C.
9. Place ice under the condensation flask to facilitate chloroform condensation.
10. Turn the rotary evaporator under vacuum (approximate pressure is 30 Hg) and allow the flask to rotate for 30 min.
11. After 30 min, stop the vacuum and remove the flask from the rotary evaporator.
12. All of the chloroform should have evaporated from the flask and a thin uniform white layer should be visible at the bottom and along the sides of the flask.
13. Attach the flask to a lyophilizer and turn on the instrument.
14. After 15 min, turn the instrument off and remove the flask from the lyophilizer.
15. Any traces of chloroform present should have evaporated leaving behind a dry white thin film at the bottom and along the sides of the flask.
16. Add 10 ml of D5W to the bottom of the flask.
17. Cover the mouth of the flask with Parafilm and gently swirl the flask to hydrate the lipid film present at the bottom of the flask.
18. Make a few pin holes in the Parafilm.
19. Place the flask in a water bath set at 50°C, and rotate the flask at different angles for 45 min or until all of the lipid film is dissolved.
20. Appearance of a white viscous solution indicates lipid dissolution.
21. Place the flask in a water bath set at 37°C and gently swirl the flask as in **step 19** for an additional 10 min.
22. Remove the flask from the water bath.
23. Remove the Parafilm with holes and cover the mouth of the flask with new Parafilm to prevent evaporation.

24. Leave the covered flask to sit at room temperature (25–30°C) overnight.
25. The next day, turn on the sonicating water bath and set the temperature to 50°C.
26. Puncture holes in the Parafilm covering the mouth of the flask.
27. Hold the flask with clamps attached to a stand and submerge the flask such that the bottom of the flask in immersed in the water bath.
28. Set the sonicator at low frequency (approximately 35 kHz) and sonicate the lipid solution in flask for 5 min. Rotate the flask during sonication. Ripple waves should be visible on the surface of the lipid solution indicating effective sonication.
29. After 5 min, remove the flask from the sonicator.
30. Place the flask inside a biological safety tissue culture hood and remove the Parafilm from the flask.
31. Using a sterile pipette, pipet the nanoparticle solution (10 ml volume) into a sterile 50 ml BioMax Falcon tube (*see* **Note 1**).
32. Warm the nanoparticle solution by placing the tube into a circulating water bath set at 50°C for 10 min.
33. While the nanoparticle solution is warming, take four 10-cc disposable sterile luer-lok syringes and attach to each syringe a 1.0-, 0.45-, 0.2-, and 0.1-μm sterile Whatman filter under the biological safety tissue culture hood.
34. After 10 min of warming of the nanoparticle solution, filter the solution in a sequential order passing through the syringes attached with decreasing filter size—1.0, 0.45, 0.2, and 0.1 μm (*see* **Note 2**).
35. Collect the filtrate passing through 0.1-μm filter into a 15-cc BioMax Falcon tube.
36. When filtration is complete, aliquot the nanoparticle solution using a 1-ml sterile pipette into sterile glass vials.
37. Flush the surface of each glass vial with filtered argon gas, then cap and seal it.
38. Appropriately label the vials and store the vials in a refrigerator at 2–6°C (*see* **Note 3**).

3.2. Optical Density Determination

1. Turn on the instrument and allow it to stabilize for 1 h.
2. While the instrument is stabilizing, take two 0.5-ml microcentrifuge tubes and place them inside the biological safety culture hood.
3. Label one of the microcentrifuge tubes as "Blank" and the other as "Sample."
4. Pipet 100 μl of D5W into the tube labeled "Blank."
5. Pipet 95 μl of D5W into the tube labeled "Sample." Add 5 μl of freshly prepared nanoparticle into this tube and gently mix by pipetting up and down five times.
6. Set the instrument at 400 nm wavelength.
7. Fill a quartz cuvette used for spectrophotometric reading with the sample labeled "Blank."
8. Place the cuvette inside the spectrophotometric cell holder, close the lid, and read the blank. The values normally obtained are zero. If some numbers appear, then this number is considered as background noise.

9. Remove the cuvette from the holder and discard the D5W.
10. Rinse the well of the cuvette with dH_2O two times and wipe any traces of water droplets present.
11. Pipet 100 μl of sample from the tube labeled "Sample" into the cuvette.
12. Wipe the outside surface to remove traces of water or sample droplets and fingerprints that may be present.
13. Place the cuvette into the cell holder, close the lid, and read the sample.
14. The optical density (OD) reading for the sample should be in range of 0.30–0.45.

3.3. Particle Size Analysis

1. Turn the instrument on and allow it to stabilize for 1 h.
2. Take plastic cuvettes and label them as "Blank" and as "Sample." If there are more than one sample to be measured, use the appropriate number of cuvettes required and label accordingly.
3. Into each cuvette, pipet 4 ml of dH_2O.
4. To the cuvette labeled "Blank," pipet 10 μl of D5W.
5. Close the cuvette with the white caps supplied with the cuvette and mix by gently inverting the cuvette three to four times. Set aside the cuvette until ready for reading.
6. Pipet 10 μl of freshly prepared nanoparticle into the cuvettes labeled "Sample."
7. Close the cuvette with cap and mix the contents by inverting the cuvette three to four times.
8. The control and sample is now ready for analysis.
9. Set the appropriate parameters for the instrument to analyze the sample.
10. Place the cuvette labeled "Blank" into the cell holder, close the lid, and read the blank.
11. Remove the cuvette labeled "Blank" from the cell holder, and replace with the cuvette labeled "Sample."
12. Read the "Sample" by clicking start.
13. The results of the analysis can be saved on the computer or printed if the instrument is connected to a printer.
14. The size for empty nanoparticles should be in the range of 50–100 nm and for DNA-nanoparticles in the range of 100–400 nm (*see* **Note 4**).
15. Clean the area around the instrument and turn off the instrument.

3.4. Net Charge Analysis

1. Turn the instrument (Delsa 440SX) on and allow it to warm up for 1 h.
2. Test the instrument settings with the appropriate controls provided by the manufacturer prior to measuring the charge of the nanoparticles.
3. Set the required parameters in the instrument for measuring the net charge of the nanoparticles.

4. Take the freshly prepared nanoparticle solution and carefully fill the cell assembly avoiding any bubbles (*see* **Note 5**).
5. Wipe the outside of the sample assembly for any dust or liquid that may be present.
6. Place the sample cell assembly into the sample chamber compartment.
7. Analyze the sample for the zeta potential.
8. The zeta potential for the nanoparticles will be in the range of 50–80 mV and for DNA-nanoparticles in the range of 35–60 mV.
9. Save the results on the computer or print if the instrument is attached to a printer.
10. Clean the area around the instrument and turn off the instrument.

3.5. Preparation of DNA-Containing Nanoparticles

1. Take out the reagents (D5W, nanoparticles, and DNA) that are required for preparing the DNA-nanoparticles from its storage area.
2. Allow the reagents to sit inside the biological safety culture hood for 1 h to reach room temperature.
3. While the reagents are thawing, take two 1.5-ml microcentrifuge tubes and label one of the tubes as "nanoparticle" and the other tube as "DNA."
4. Wipe the outside of the reagent vials with rubbing alcohol.
5. In the tube labeled "nanoparticle," pipet 85 μl of D5W to the bottom of the tube.
6. Into the same tube, pipet 65 μl of nanoparticle. Gently mix by pipetting up and down five times to give a final volume of 150 μl.
7. Take the tube labeled "DNA," and pipet appropriate volume of D5W followed by addition of the appropriate volume of DNA solution. The amount of D5W solution to be added depends on the stock DNA concentration. The final volume in this tube after addition of DNA and D5W volumes should be 150 μl.
8. Pipet the contents (150 μl) from the tube labeled "nanoparticle" into the tube labeled "DNA."
9. Gently mix by pipetting up and down five times using a 1-ml Eppendorf pipette.
10. A turbid solution should result after mixing and should be clear of any precipitates or particulates. If precipitates appear, discard the sample and prepare a new mix.
11. From the mixture, take 5 μl and add to 95 μl of D5W contained in a 0.5-ml microcentrifuge tube for OD measurement; take 10 μl and add to 4 ml of water contained in a plastic cuvette for particle size analysis; take 50 μl for charge analysis.
12. Subject the diluted sample from **step 11** to spectrophotometric analysis, particle size analysis, and charge analysis.
13. Spectrophotometric analysis of the sample should provide a sample reading between 0.65 and 0.95 at OD 400 nm.
14. Particle size analysis should provide a size range between 100 and 400 nm (*see* **Note 6**).

15. Charge analysis of the DNA-nanoparticle should be in the range of 30–50 mV (*see* **Note 7**).
16. If the reading for any of the analysis is outside the expected range, discard the sample and repeat the mixing procedure from **steps 5–15** (*see* **Note 8**).
17. When a large volume of DNA-nanoparticles is required, repeat **steps 5–15** for preparing multiple samples. Pool all of the DNA-nanoparticle samples thus prepared in several tubes into a 5-ml sterile glass vial and subject the pooled sample for analysis as described in **steps 13–15**.
18. The resulting pooled DNA-nanoparticle solution is ready for use in in vitro and in vivo studies.

3.6. In Vitro Transfection

1. From a tissue culture incubator, take out a tissue culture plate (100 mm^2) actively growing a monolayer of cells that are approximately 70–80% confluent.
2. Trypsinize cells by adding 0.5 ml of 1% tryspin containing 0.1% EDTA and incubating at room temperature for 1–2 min.
3. Neutralize the trypsin activity by adding 0.5 ml of serum-free culture medium and gently tap the culture plate to loosen all of the attached cells.
4. Collect the detached cells using a 1-ml pipette and transfer into a 2-ml culture grade sterile polystyrene tube.
5. Gently pipet 10 times to make single cell suspension.
6. Take 100 μl of the cell suspension and add to tube containing 100 μl of Trypan blue. Gently mix by pipetting three times.
7. Transfer 10 μl of the cell suspension in 90 μl of Trypan blue into the chambers of a micrometer that is routinely used for cell counting.
8. Place the micrometer under an inverted bright-field microscope and adjust the focus to see the center of the micrometer that contains the "box." The box contains four equal squares.
9. Count the number of cells present in each of the four squares, and add the total number of cells present in the four squares.
10. Divide the total number of cells by 4 to get the average number of cells.
11. Calculate the total number of cells per milliliter using the formula: Average number of cells × dilution in Trypan blue × 10^4 = number of cells/ml. For example, if the number of cells counted were 40, then the average number of cells is 40/4 = 10. Then the total number of cells present is: 10 × dilution factor = 10 (10 μl of cell suspension diluted in 90 μl of Trypan blue) × 10^4 = 100 × 10^4 = 1 × 10^6 cells/ml.
12. Seed cells (5 × 10^5/well) in 2 ml of culture medium in a six-well tissue culture plate (*see* **Note 9**).
13. Incubate the culture plate in a tissue culture incubator set at 37°C temperature and with 5% CO_2 for 18–24 h.
14. The next day remove the plate, aspirate the medium with a 5-ml sterile pipette, and replace with 1 ml of sterile PBS. Gently swirl the plate to remove traces of culture medium.

15. Remove PBS and add 1.0 ml of serum-free culture medium to the wells and leave the plate inside the biological safety hood for 10 min.
16. Take out a freshly prepared vial of DNA-containing nanoparticles (described in **Subheading 3.5.**), and using a 20-µl pipette, take different volumes (1, 2.5, 5, and 10 µl) of the nanoparticles and add to each well. Gently swirl the plate for the nanoparticles to mix the monolayer of attached cells.
17. Return the culture plate to the incubator and incubate the plate for 3 h.
18. After incubation is complete, take out the plate from the incubator and place inside the biological safety hood. Add 2 ml of 10% serum containing culture medium and return the plate to the incubator.
19. Incubate the plate for 24 h at 37°C with 5% CO_2.
20. The next day, remove the plate from the incubator and analyze for transduction efficiency or transgene expression as described below.

3.6.1. Transduction Efficiency

1. Aspirate the culture medium from the culture wells using a 5-ml pipette and add 2 ml of sterile cold PBS. Gently swirl the plate to wash off any remaining culture medium.
2. Remove the PBS and replace with 2 ml of ice-cold 0.5% glutaraldehyde.
3. Incubate for 5 min at room temperature.
4. Prepare fresh X-Gal staining solution and add 1–2 ml of the solution to each well.
5. Incubate the culture plate in the dark for 6–24 h.
6. After incubation, remove the culture plate and aspirate the X-gal solution using a 5-ml pipette.
7. Add 1 ml of cold PBS to each well and gently swirl to remove traces of X-Gal solution. Repeat this step three times.
8. Add 1.0 ml of cold PBS and visualize blue-stained cells under a bright-field inverted microscope (*see* **Fig. 1**). The number of blue-stained cells present in each well can be qualitatively and semi-quantitatively determined.

3.6.2. Transgene Expression

1. To the wells, add 0.5 ml of trypsin and incubate for 1 min at room temperature. Add 0.5 ml of serum-free medium; gently tap the plate along the sides to dislodge any loosely attached cells.
2. Collect the cells into a 1.5-ml Eppendorf tube.
3. Place the Eppendorf tube in a tabletop microcentrifuge and spin for 10 min at 10,000 g, room temperature.
4. Cells will appear as a pellet at the bottom of the tube with a clear supernatant at the top.
5. Carefully aspirate the upper supernatant without disturbing the pellet.
6. To the cell pellet, add 0.5 ml of cell lysis buffer and gently pipette up and down 10 times.

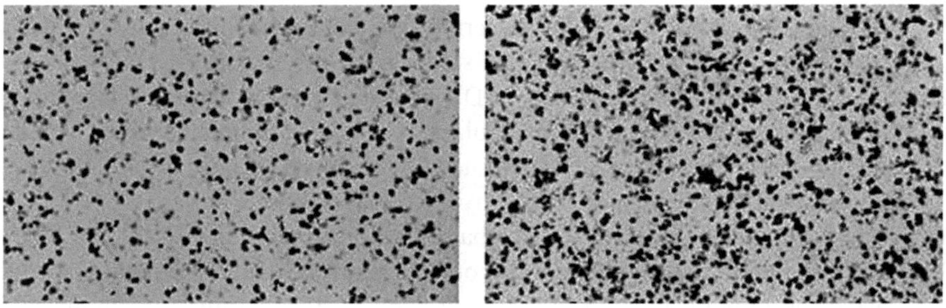

Fig. 1. Determination of transduction efficiency. Lung tumor cells transfected with DNA-nanoparticles carrying the *Lac-Z* gene were stained for Lac-Z expression. Cells expressing the Lac-Z protein are stained blue. Approximately **(A)** 40% of cells and **(B)** 70% of cells are transfected as determined by the number of blue stained cells (×100 magnification).

7. Place the tube in a microcentrifuge and spin for 10 min at 10,000 g, room temperature.
8. Collect the upper layer of supernatant that contains the total cellular protein using a 1-ml pipette and transfer into a new 1.5-ml Eppendorf tube. Be careful not to collect the cell debris present at the bottom of the tube.
9. The collected supernatant can be stored at –80°C until use. Prior to storing, an aliquot of the supernatant is used to determine the protein concentration.
10. Protein concentration is determined using the bicinchoninic acid (BCA) method (Pierce) and expressed as microgram (μg) of protein per microliter (μl).
11. When ready for use, appropriate volume of the supernatant that contains 50 μg of protein is mixed with a gel-loading dye and subjected to gel electrophoresis and western blotting analyses as described below.

3.6.2.1. Sodium Dodecyl Sulfate–Poly Acrylamide Gel Electrophoresis

1. Wipe the glass plates with EtOH and mount them into the stand.
2. Add dH_2O to check for leakage.
3. Drain the dH_2O and dry using a Whatman 3M blotting paper.
4. Prepare the resolving gel according to the molecular weight of the protein to be investigated. For example, to detect low molecular weight proteins, prepare a 12.5% gel; for high molecular weight protein, prepare 7.5% gel.

	7.5%	10%	12.5%
Acrylamide-bis (30%)	3.75 ml	5 ml	6.25 ml
3 M Tris–Hcl (pH 8.8)	3.75 ml	3.75 ml	3.75 ml
dH_2O	7.25 ml	6.0 ml	4.75 ml
TEMED	12.0 μl	12.0 μl	12.0 μl
10% SDS	150 μl	150 μl	150 μl
10% APS	150 μl	150 μl	150 μl

5. Mix the gel by pipetting and then add 7.4 ml of the gel between the glass plates.
6. Add 300 µl water-saturated isobutanol to remove the bubbles.
7. Let stand for 45 min to 1 h till the gel polymerizes.
8. Prepare the stacking gel:

Acrylamide-bis (30%)	1.3 ml
0.5 M Tris–HCl (pH 6.8)	2.5 ml
dH_2O	6.1 ml
TEMED	15.0 µl
10% SDS	100 µl
10% APS	150 µl

9. After 1 h, drain the isobutanol and wash with dH_2O to remove the remaining isobutanol.
10. Carefully remove excess dH_2O using a blotting paper, so as not to touch the gel.
11. Mix the stacking gel and add on top of the separating gel.
12. Put the comb to make the wells (1.5-mm 10-well comb or 15-well comb can be used depending on the number of wells required).
13. Let stand for 45 min to 1 h till the gel polymerizes.
14. After 1 h, remove the combs carefully and gently wash with dH_2O to remove excess gel.
15. Suction out the dH_2O carefully, remembering not to touch the gel.
16. Remove the glass *plates* from the gel stand and put it into the electrophoresis stand.
17. Pour 1× running buffer between the *glass plates.*
18. Mix the loading buffer and protein *sample* according to the prepared chart to get equal loading concentration of the proteins.
19. Using a sampler, comb load 20 µl of the prepared proteins to each well.
20. Add 10 µl of protein ladder to one of the wells.
21. Run the gel at 100 volts, monitoring the blue sample buffer line.
22. Stop the gel once the blue line reaches at the bottom of the gel.
23. Drain of the running buffer and remove the slides from the stand.
24. Carefully open the two slides and remove the gel for transfer.
25. Transfer the proteins from the polyacrylamide gel to nitrocellulose membrane.

3.6.2.2. Western Blotting

1. Add 1× transfer buffer in a dish and place the transfer case in the buffer.
2. Place a sponge on the case followed by blotting paper, nitrocellulose paper, polyacrylamide gel, 3M Whatman blotting paper, sponge (so called "sandwiching" between two electrodes).
3. Put the transfer case into the transfer stand with the black side of the case towards the red side of the transfer kit/stand (depending on the direction you are running the current).
4. Put the kit in a plastic container and pour 1× transfer buffer.

5. Put in a magnetic stirrer and an ice pack to keep the gel cool.
6. Transfer the gel at 100 Volts for 1–2 h (depending on the molecular weight of the proteins we want to check—longer transfer may be required for high molecular weight proteins) replacing the ice pack whenever needed.
7. Following transfer, remove the stand and cut the nitrocellulose membrane at the gel marks with a scalpel.
8. Place this membrane in a blocking dish and block with PBS-T or TBS-T blocking buffer depending on the antibody to be used for 1 h to block non-specific binding sites.
9. After 1 h, drain the blocking buffer.
10. Add appropriate primary antibody diluted 1:500 to 1:1000 in PBS-T or TBS-T blocking buffer.
11. Leave the membrane on a shaker set at *slow speed* overnight at +4°C.
12. Next morning remove the PBS-T containing the primary antibody.
13. Add PBS-T blocking buffer to the membrane and wash the membrane for 15 min by placing it on a *rotary shaker.*
14. Repeat **step 13** described in **Subheading 3.6.2.2.** two more times with washing time 5 min for each wash.
15. Drain the washing buffer.
16. Add appropriate secondary antibody at a dilution of 1:5000 in PBS-T blocking buffer.
17. Place on a rotary shaker set at *slow speed* and allow shaking for 1 h.
18. Wash the membrane as indicated in **steps 13** and **14** described in **Subheading 3.6.2.2.**
19. Wash the membrane with PBS twice. Each wash includes shaking for 5 min on a rotary shaker.
20. Drain the PBS.
21. Add 10 ml of Amersham's chemiluminescence ECL Reagent (Detection Reagent 1 and Detection Reagent 2 mixed in a ratio of 1:1) and shake for 1 min on the rotary shaker.
22. Drain the ECL reagent and blot dry the membrane using a Whatman blotting paper.
23. Place the membrane in the X-ray film cassette and expose the membrane using Biomax ML film or any other appropriate film in dark room.
24. Develop the film in the dark room after exposing for few minutes or as appropriate (**Fig. 2**). Exposure times vary depending on the signal obtained. If the signal is too weak, expose the film overnight.

3.6.2.3. Stripping and Reprobing

1. After obtaining the appropriate result, take out the membrane from the X-ray film cassette.
2. Place the membrane in a dish previously used for washing the membrane.
3. Add 10 ml of stripping buffer to the membrane contained in the dish.
4. Place the dish on a rotary shaker and shake for 30 min.

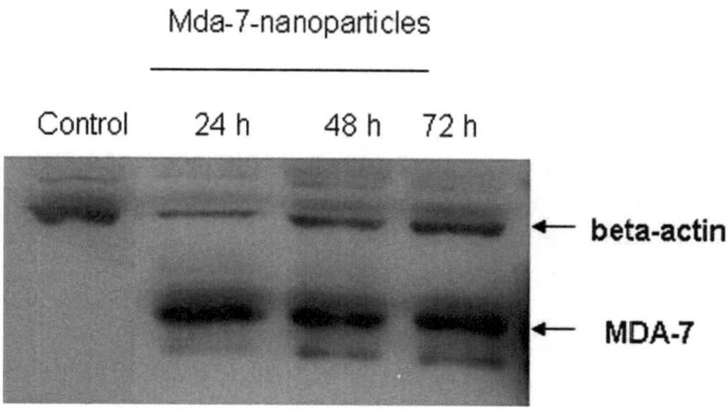

Fig. 2. Determination of transgene expression. Lung tumor cells transfected with DNA-nanoparticles carrying the *mda-7/IL-24* tumor suppressor/cytokine gene were harvested at 72 h after transfection, lysed, and subjected to western blotting. Cells expressing the exogenous MDA-7/IL-24 protein were detected using anti-mda-7 antibody. Beta-actin was used as internal loading control.

5. Drain the stripping buffer.
6. Add PBS-T buffer solution containing antibody to beta-actin (1:1000).
7. Repeat **steps 11–24** described in **Subheading 3.6.2.2**.
8. Detection of a signal on the X-ray film corresponding to the expected protein size for beta-actin will allow the investigator to determine equal loading of protein samples and serve as internal control. Detection of the equal levels of beta-actin will reflect true changes in the protein of interest investigated in **Subheading 3.6.2.2**.

3.7. In Vivo Nanoparticle Delivery

3.7.1. Subcutaneous Tumor Model in Mice

1. Seed H1299 or A549 lung tumor cells (1×10^6) suspended in appropriate tissue culture medium in 150-mm^2 tissue culture plates (*see* **Note 10**).
2. Incubate the plates in an incubator set at 37°C with 5% CO_2.
3. Monitor the plates every day for cell growth and confluency.
4. When the cells are at 70–80% confluency, remove the plates from the incubator and take it to the biological safety hood.
5. Harvest cells by trypsinization (*see* **Subheading 3.6.**), transfer to 50-ml Falcon tube, and suspend in 5 ml of serum-free medium for cell counting.
6. Perform cell counting by Trypan-blue assay method (*see* **Subheading 3.6.**).
7. Place the 15-ml Falcon tube containing the cells in a tabletop clinical centrifuge and spin at 500 g for 10 min at 4°C.
8. Take out the 50-ml Falcon tube and place it inside the culture hood.
9. Aspirate the upper layer of culture medium, and retain the cell pellet visible at the bottom of the tube.

10. Add 10 ml of sterile cold PBS to the cell pellet and gently pipet 10 times using a 5-ml pipette to disassociate the cell pellet.
11. Centrifuge as indicated in **step 7**.
12. Repeat **steps 8–10** two times.
13. Resuspend the cell pellet in an appropriate volume of cold PBS to give a final cell concentration of 5×10^6 cells per 200 µl of PBS.
14. Place the Falcon tube on ice and take it to the animal room where the mice are housed and are to be injected with the tumor cells.

3.7.1.1. Preparation of Mice

1. Place the request for appropriate number of female nude mice (4–6 weeks old) to be used for the experiment 1 week prior to start of the experiment (*see* **Note 11**).
2. One day prior to injection of tumor cells, check the number of mice required for the study and their health.
3. The number of mice per cage should not be more than five.
4. Take the mice to a source of gamma irradiation available in the institute.
5. Subject individual mice to total body radiation at 350 rads (*see* **Notes 12**).
6. After irradiation, return the mice to the cages and return the cages to the assigned animal room in the veterinary department.
7. Allow the mice to recover from the radiation-induced stress for 24 h.
8. The mice are now ready to be injected with the tumor cells.

3.7.1.2. Injection of Tumor Cells to Mice

1. Before entering the animal room, wear shoe covers, tyvek body suits, caps to cover the head, mask, and gloves.
2. In the animal room, clean the biological safety hood with alcohol (70%) thoroughly.
3. Spread a disposable diaper on the floor of the safety hood and arrange all of the materials needed (e.g., 25-gauge needle, 1-ml tuberculin syringe, vials containing the cells, etc.).
4. Take out the vial containing tumor cells and mix the suspension by gently flushing with a pipet or with a syringe and needle.
5. Take 1 ml of the cell suspension into a tuberculin syringe and keep it aside.
6. Take one cage containing five mice and place inside the biological safety hood.
7. Open the cage and take out one mouse and sedate the mice by anesthesia (isofluorane) until the respiration slows down.
8. Take out the mice and gently place on the floor of the hood and identify the lower right flank.
9. Wipe the flank with cotton gauze dipped in rubbing alcohol.
10. Slightly stretch the flank with your left hand, take the syringe containing the tumor cells (**step 5**) with your right hand, and slowly insert the needle underneath the skin of the flank (*see* **Note 13**).
11. Slowly inject 200 µl (5×10^6) of the cell suspension.

12. Retain the needle inside the skin for few seconds after injection to minimize reverse flow and leakage.
13. Return the mice to the cage and monitor for recovery. Mice will recover in few minutes.
14. Repeat **steps 7–13** for injecting cells into the remaining mice.
15. Monitor the mice every day for their health and tumor growth.
16. Measure the tumor size with a caliper.
17. Palpable tumors (5–6 mm^2) should be observed in 10–14 days and are ready for treatment (*see* **Fig. 3**).

3.7.1.3. Treatment of Subcutaneous Tumor

1. When the tumor is palpable, tag the mice with ear tags to allow easy identification of the individual mice.
2. Determine the tumor size in all of the mice injected with tumor cells using a caliper by measuring the longest (a) and the shortest (b) length of the tumor.
3. Calculate the tumor volume using the formula:

$$\frac{a \times b^2}{2}(\mathrm{mm}^3)$$

4. The tumor volume of each mouse should be between 50 and 100 mm^3 (*see* **Note 14**).
5. The tumor-bearing mice are now ready to receive the treatment.

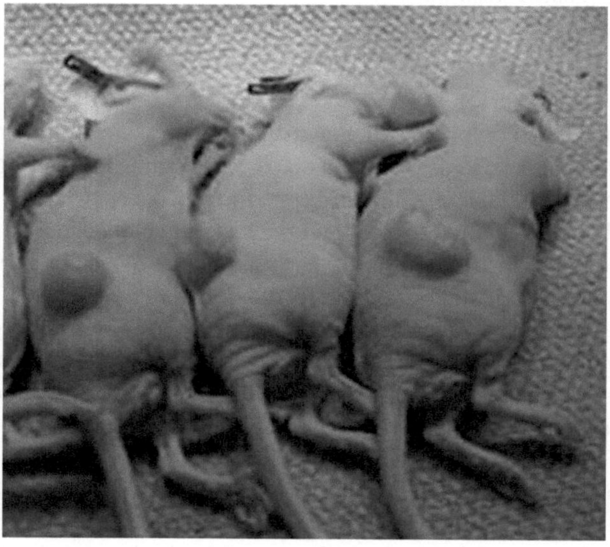

Fig. 3. Establishment of subcutaneous lung tumor xenograft. Human H1299 lung tumor xenografts were established in female nude mice by injecting 5×10^5 cells on the lower right flank. Subcutaneous tumors were established and ready for treatment in 10–14 days after tumor cell injection.

6. Randomize the mice to the different treatment groups.
7. Label the cage cards with the tumor type injected, treatment received, and the ear tag numbers.
8. Freshly prepared DNA-containing nanoparticles as described in **Subheading 3.5** is aspirated into 1-ml tuberculin syringe and kept ready for injection into the tumor.
7. Anesthetize the mice using isofluorane as described in **step 7 (Subheading 3.7.1.2.)**.
8. Gently hold the edges of the tumor between the thumb and forefinger of your left hand.
9. Take the syringe containing DNA-nanoparticles in your right hand and slowly insert the needle on one side of the tumor and pass to the middle of the tumor.
10. Slowly inject the DNA-nanoparticles (100 μl) into the tumor.
11. Retain the needle inside the tumor after completing the injection for 1 min to prevent reverse flow of the material and leakage.
12. Gently withdraw the needle from the tumor and wipe the injection site with rubbing alcohol.
13. Note the ear tag number in the record book and return the mice to its cage.
14. Repeat **steps 7–13** for injecting DNA-nanoparticles to other tumor-bearing mice.
15. For repeated treatment per experimental design, repeat **steps 8–13** (*see* **Note 15**).
16. Discard used needles and syringes in Sharps biohazard container; discard used plastic vials and tubes in biohazard bags that are placed in the animal room.
17. Remove all of clothing (gloves, shoe covers, tyvek body suit, etc.) and discard in the designated trash bins.
18. Wash hands with soap.

3.7.1.4. Analysis of Subcutaneous Tumors for Therapeutic Effect

1. During and after completion of the treatment, measure tumor growth in each mouse using calipers as described in **steps 2** and **3** of **Subheading 3.7.1.3**.
2. Measurement of tumor growth can be performed once, twice, or thrice a week and is a choice of the investigator and the experimental design.
3. Record the tumor measurements for each individual mouse in a record book.
4. Calculate the average tumor volume for each treatment group per time point.
5. Enter the values of the average tumor volume for each treatment group per time point in Microsoft Excel spreadsheet or other statistical software spreadsheets.
6. Calculate the difference in the tumor volume among the treatment groups for significance using the appropriate statistical method.
7. Statistical analysis can also be performed with the help of a statistician if available in the department or institute.
8. Euthanize the animals by CO_2 inhalation or by any other procedure as recommended and approved by the investigators of Institutional Animal Care and Use Committee (IACUC) and by the American Association of Animal Welfare.
9. Place the euthanized mice in biohazard bags and discard in appropriate containers designated by the department of veterinary medicine in the investigators' institute.

3.7.2. Experimental Lung Metastasis Model in Mice

1. Follow **steps 1–12** as described in **Subheading 3.7.1**.
2. Resuspend the cell pellet in an appropriate volume of cold PBS to give a final cell concentration of 1 × 10^6 cells per 100 μl of PBS.
3. Place the Falcon tube on ice and take it to the animal room where the mice are housed and are to be injected with the tumor cells.

3.7.2.1. Preparation of Mice

1. Follow **steps 1–8** described in **Subheading 3.7.1.1** if the animals to be used are nude mice.
2. If studies are designed to use SCID/Beige mice, then these mice need not be irradiated and are ready to receive tumor cells.

3.7.2.2. Injection of Tumor Cells to Mice

1. Follow **steps 1–6** as described in **Subheading 3.7.1.2**.
2. Open the cage and take out one mouse and retain the mice by its tail in a restrainer.
3. Hold the tail under an infrared heat lamp source for 2–3 min to enable the lateral vein visible.
4. Wipe the tail with cotton gauze dipped in rubbing alcohol.
5. Slightly stretch the tail with your left hand, take the syringe containing the tumor cells with your right hand, and slowly insert the needle into the vein (*see* **Note 16**).
6. Slowly inject 100 μl (1 × 10^6) of the cell suspension (*see* **Note 17**).
7. Retain the needle inside the tail vein for few seconds after injection to minimize reverse flow and leakage.
8. Hold the injection site pressed with cotton gauze dipped in rubbing alcohol for 1–2 min to stop bleeding.
9. Return the mice to the cage and monitor for recovery. Mice will recover in few minutes.
10. Repeat **steps 2–9** for injecting tumor cells into the tail vein of remaining mice.
11. Monitor the mice every day for their health.
12. Any mice showing signs of sickness (watery eyes, immobile, skin wounds, loss of body weight more than 20%, etc.) should be euthanized.
13. Micrometastatic tumors in the lungs should be established in 6–10 days and detectable histologically after hematoxylin and eosin (H&E) staining (*see* **Fig. 4**).
14. The animals are ready for treatment of micrometastatic tumors at this time.
15. Should the experimental design require treatment of macroscopic lung tumors, allow 16–18 days before initiating treatment.
16. In 5–7 weeks from the time of tumor cell inoculation, tumors would have spread outside the lung and established various organs (*see* **Fig. 5**).

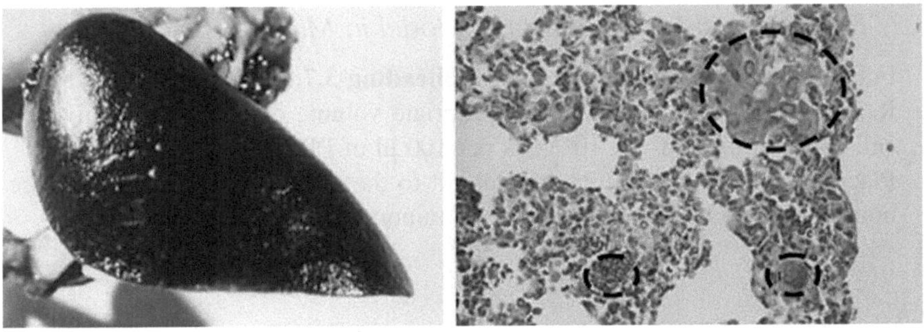

Fig. 4. Micrometastatic tumors in the lung. Human lung tumor (A549) cells were injected intravenously into female nude mice via the tail vein. Animals were sacrificed 6–10 days after tumor cell injection and analyzed for micrometastatic tumors by histopathology. Extrapulmonary tumor nodules are not visible macroscopically (left panel) while small micrometastatic tumors are visible histologically (right panel; ×200 magnification).

3.7.2.3. Treatment of Experimental Lung Metastasis

1. The time to initiate the treatment depends on the experimental design.
2. It is recommended that treatment be first tested against micrometastatic tumors for a therapeutic effect. The procedure described below is for treatment of micrometastatic tumors.
3. Six days after tumor inoculation randomize the mice into groups. The number of animals per group and number of groups will depend on the experimental design and the investigators' statistical requirements.
4. Ear tag the mice in each group and record the numbers in the notebook.
5. Label the cage cards for each group of mice with the tumor type injected, treatment to be received, and the ear tag numbers.
6. Freshly prepared DNA-containing nanoparticles as described in **Subheading 3.5** are aspirated into 1-ml syringe with a 27-gauge needle and kept ready for injection.
7. Bring a cage of mice inside the biological safety hood.
8. Subject the whole cage to heat by exposing to infrared heating lamp for 3–5 min.
9. Open the cage and take out one mouse and retain the mice by its tail in a restrainer.
10. Hold the mouse by its tail and wipe the tail with cotton gauze dipped in rubbing alcohol.
11. The lateral vein visible should be visible. If not visible, expose the tail to the infrared heat lamp source for additional 2–3 min.
12. When the vein is visible slightly stretch the tail with your left hand, take the syringe containing the DNA-nanoparticles with your right hand, and slowly insert the needle into the vein.

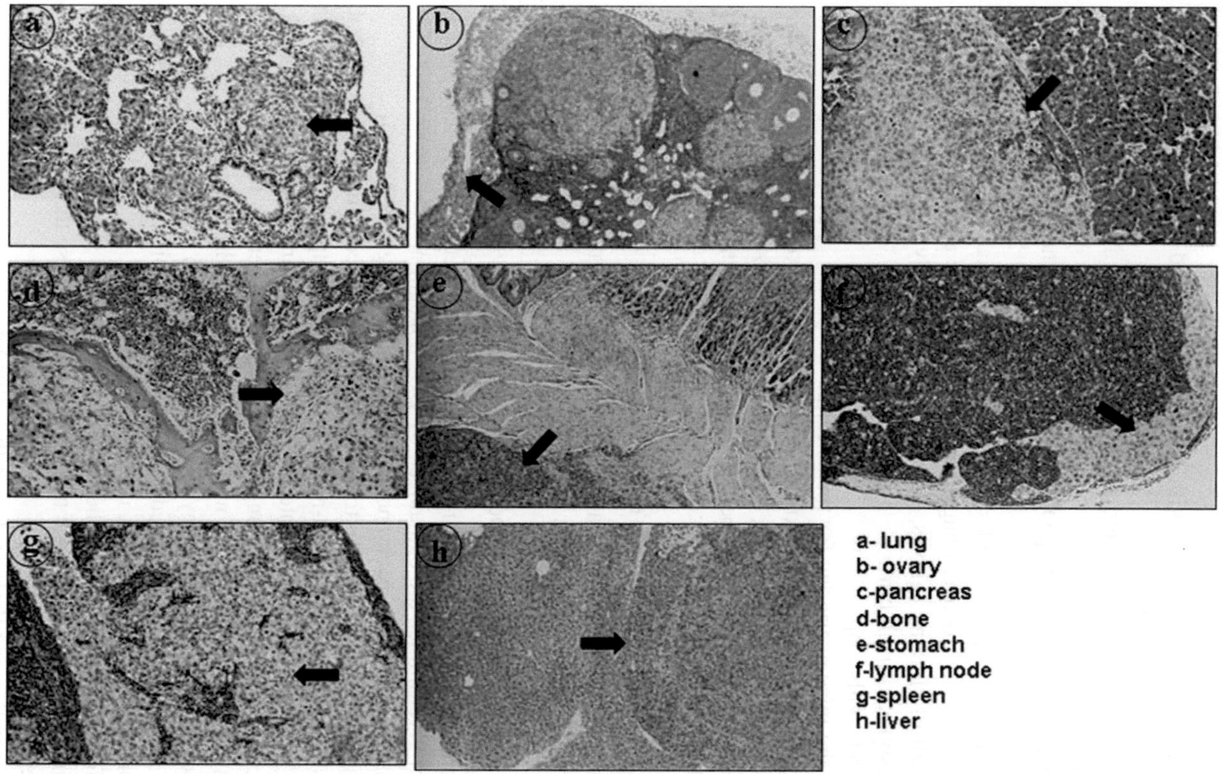

Fig. 5. Lung tumor disseminated to other organs. Human lung tumor (H1299) cells were injected intravenously into female SCID/Beige mice via the tail vein. Animals were sacrificed 6–7 weeks after tumor cell injection and analyzed for metastatic tumors in various organs by histopathology. Presence of metastatic tumors outside the lung as indicated by the arrows was detected in ovary, pancreas, bone, stomach, lymph node, spleen, and liver (×400 magnification).

13. Slowly inject 100–200 μl of the DNA-nanoparticles over 1-min period (*see* **Note 18**).
14. Retain the needle inside the tail vein for few seconds after injection to minimize reverse flow and leakage.
15. Hold the injection site pressed with cotton gauze dipped in rubbing alcohol for 1–2 min to stop the bleeding.
16. Return the mice to the cage and monitor for recovery. Mice will recover in a few minutes.
17. Repeat **steps 9–16** for injecting DNA-nanoparticles into the tail vein of remaining mice.
18. Discard any used needles and syringes in Sharps biohazard container, plastic vials, and tubes in biohazard bags that are placed in the animal room.
19. Remove all of clothing (gloves, shoe covers, tyvek body suit, etc.) and discard in the designated trash bins.
20. Wash hands with soap.
21. Monitor the mice every day for their health.
22. Any mice showing signs of sickness (watery eyes, immobile, skin wounds, loss of body weight more than 20%, etc.) should be euthanized.
23. For repeated treatment as per experimental design, repeat **steps 6–16** (*see* **Note 19**).

3.7.2.4. Analysis of Lungs for Therapeutic Effect

The therapeutic effect of DNA-nanoparticles on experimental lung metastasis can be determined by two methods as described below:

3.7.2.4.1. Counting the Number of Extrapulmonary Tumor Nodules.

1. At the end of the treatment (after the required number of DNA-nanoparticle treatment given), mice are euthanized by CO_2 inhalation.
2. The mouse is laid on its back and pinned to a dissecting board.
3. The area from the lower jaw up to the diaphragm is wiped with rubbing alcohol.
4. An incision is made in the mid-neck region with a scalpel blade, and the skin with the underlying fat is separated from the trachea using a forcep and scissor.
5. The trachea along with the lungs is exposed by opening the rib cage and extending the incision from the mid-neck region up to the diaphragm.
6. Any blood present is removed by absorbing with cotton gauze.
7. Using a 23-gauge needle attached to a 5-ml syringe, aspirate India ink and keep ready for injecting the India ink into the mouse trachea.
8. With a forcep, separate the trachea from the underlying skin and hold it up.
9. Slowly insert the needle into the trachea and slowly infuse the India ink solution from **step 7**.
10. The lungs will expand/inflate during the infusion.
11. Once when all of the lobes of the lung have expanded, remove the needle from the trachea and depress the injection site with a forcep for 2 min to prevent leakage.

12. With a scissor, cut the upper part of the trachea (above the injection site) and hold the trachea with a forcep.
13. Slowly pull out the trachea along with the lungs and cut any attached tissues to the lungs.
14. Remove the lungs and transfer into a labeled container (e.g., 50-ml Falcon tube) that contains Fekete's solution until cleared of excess India ink.
15. Lung metastases should appear as white nodules (*see* **Fig. 6**).
16. Dissect any additional organ that is required per experimental design and transfer into appropriate containers—drop in liquid nitrogen if frozen tissues (e.g., molecular analysis) are needed and in formalin for formalin-fixed tissues (histologic examination, immunohistochemistry, etc.).
17. Repeat **steps 2–15** for each animal and drop the organs in appropriately labeled containers for each animal in a group.
18. Care must be taken not to pool the lungs or other organs from all animals in a group into the same container as this will not reflect the difference in the individual mice.
19. Take the lungs and other organs collected back to the laboratory.
20. Take the containers with lungs to a room housing a stereomicroscope.
21. Wear a respiratory mask, gloves, and glasses to protect eyes.

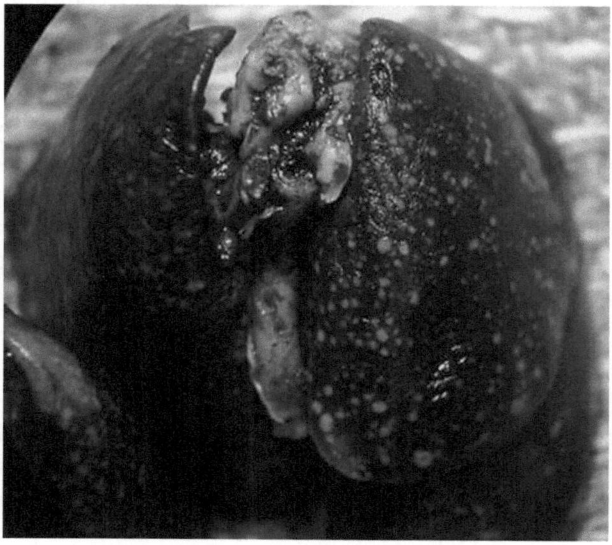

Fig. 6. Extrapulmonary lung tumor nodules. Human lung tumor (A549) cells were injected intravenously into female nude mice via the tail vein. Animals were sacrificed at the end of the experiment and analyzed for extrapulmonary lung tumors by injecting India ink. Presence of extrapulmonary tumors appears as white nodules on the surface of the lung.

22. Take out a lung of an individual mouse from one of the containers and transfer into a petri plate. Add 1 ml of PBS to avoid drying of the tissue.
23. With a forcep and scalpel blade, dissect individual lobes of the lung.
24. Take one of the lobes and place it into a petri plate and place the plate under the stereomicroscope.
25. Count the number of extrapulmonary tumor nodules that is visible (white in color) on both sides of the lobes.
26. Note the number of tumor nodules counted in a record notebook.
27. Transfer the lobe back to its original container.
28. Repeat **steps 22–26** for the remaining lobes of the lung.
29. Repeat **steps 22–27** for counting the tumor nodules in the lungs of each mice from each group.
30. When counting is completed, average the total number of tumor nodules per animal per treatment group.
31. Subject the data thus obtained for each treatment group to appropriate statistical test and analyze for statistical significance.

3.7.2.4.2. Animal Survival.

1. If animal survival is the measure for evaluating the therapeutic effect of DNA-nanoparticle treatment, the survival of each mouse per treatment group is monitored regularly (e.g., daily, three times a week, etc.).
2. The survival of each mouse identified by its ear tag is recorded from the date the first treatment was started.
3. Animals that are sick during or after completion of the treatment are euthanized to avoid discomfort and the data are recorded in record notebook.
4. The study is carried out until all of the animals have died or study terminated at a fixed time point (1, 2, 3 period or 6 month) as determined by the experimental design.
5. The data obtained at the end of the study are entered into a statistical program and appropriate statistical analysis (e.g., Log-rank test, Kaplan-Meir plot) is applied for significance.

4. Notes

1. Upon completion of **step 31**, it is necessary to clean the glass flask as quickly as possible. It is recommended that the user clean the glass flask on the same day of nanoparticle synthesis so as to remove any trace remnants that may be present. This will enable the user to reuse the flask and avoid potential problems such as contamination during subsequent synthesis.

   a. Rinse the flask thoroughly with hot water three times.
   b. Drain the hot water and add EtOH and wash for few minutes.
   c. Drain the used EtOH into a container labeled for safe disposal.
   d. Rinse the flask again with hot water two times.

e. Rinse the flask with dH_2O.
f. Completely fill the flask with 1× RBS detergent, cover the mouth of the flask with Parafilm, and leave the flask to sit at room temperature overnight.
g. The next day, drain the 1× RBS.
h. Rinse the flask with hot water three times.
i. Rinse the flask with dH_2O three times.
j. Allow the flask to air dry.
k. Rinse one final time with EtOH.
l. Allow the flask to air dry.
m. Cover the mouth of the flask with Parafilm and keep the flask in a dust-free place for future use.

2. Manual filtration is hard and time consuming. An alternate approach is to use a "Mini-Extruder" from Avanti Polar Lipids (Alabaster, AL). Using the extruder is fast and reliable and the extrusion is uniform.
3. The DOTAP:Chol nanoparticles when stored at +4°C are stable for 5–6 months. However, storage beyond this time will lead to slow oxidation leading to erroneous results when mixed with DNA.
4. If the size of the empty nanoparticles is not in the appropriate range, check the date on which the nanoparticles were synthesized. If the nanoparticles are old, discard and synthesize a fresh batch. If the size range for the DNA-nanoparticles is not within the appropriate limits, then check the concentration of the DNA, stability of the DNA (circular closed, circular relaxed, nicked or degraded, etc.), and the size of the DNA. These features of the DNA will influence the DNA-nanoparticle size.
5. Ensure that no bubbles are in the hemispheres or channel. Presence of bubbles interferes with instrument reading. Bubbles can be removed by carefully withdrawing the sample back into the syringe and then refilling the sample cell assembly again.
6. The size of nanoparticle cannot be controlled by the method described as the size is partly dependent on the number of DNA molecules encapsulated. Because the DNA and nanoparticle interaction is based on the charge, the recommended size is the average mean size of the preparation. Nanoparticles that are larger than 400 nm in size are not ideal for intravenous injection due to potential cause of embolism in mice leading to animal death.
7. Interaction of the nanoparticles (positively charged) with cells (negative charged) in vivo is based on net charge. We recommend that the positive charge on the nanoparticles be around 30–50 mV. In our experience, the recommended positive charge on the nanoparticle has found to be optimal and effective.
8. We recommend not using DNA-nanoparticles that are not in the recommended range for either particle size or charge. Deviations from the recommended parameters can be due to degradation of the DNA, oxidation of the nanoparticles, improper storage of DNA and nanoparticles, change in mixing conditions, etc. Should this occur, try using another batch or lot of plasmid DNA and/or

nanoparticles. Alternate is to synthesis nanoparticles fresh and purify a new batch of plasmid DNA.

9. The number of cells to be seeded for in vitro transfection depends on the cell type and their growth rate. Prior titration using different cell number is recommended. Very low cell number however will lead to toxicity as well as low transfection. Optimization is strongly recommended.
10. Prior to seeding the cells, make sure the cells are healthy and around 70% confluent. Check for presence of mycoplasma and bacterial contamination as this will interfere with tumor formation when implanted in mice. Recommended passage number for H1299 and A549 cells is about 20–50. However, the reader is recommended to use these parameters as a guideline when testing other cell lines.
11. Mice can be purchased from different vendors. However, the age of the mice is an important factor in establishing the tumors. Animals that are very young are sensitive to treatments while old animals are resistant and have developed some level of immunity to reject xenograft tumors. Additionally, female mice are relatively easy to handle. Male mice are aggressive and often result in fighting and inflicting wounds on mice within a cage. This will result in isolating mice into separate cages and also elimination of mice from a study.
12. Nude mice although indicated as immunodeficient (T- and B-cell deficient) have natural killer (NK) cells. Additionally, some B-cell activity exists. As a result, rejection of tumor xenografts is commonly observed. Therefore to increase the engraftment of xenograft tumors, whole-body sub-lethal radiation (350 rads) results in 90–100% tumor cells uptake and tumor formation.
13. During subcutaneous implantation of tumor cells into mice, care need to be taken not to inject the cells into the underlying muscle. Whether cells are being injected subcutaneously or intramuscularly can be easily realized based on the pressure in the syringe. Free flow of cells from the syringe indicates subcutaneous injection. However, injection of cells with pressure indicates intramuscular injection. Should cells be injected intramuscularly, the animals need to be excluded from the study immediately as tumor formation in these mice are not easily detectable, tumors cannot be easily treated, and tumors cannot be easily measured.
14. It is often recommended that an extra number of animals be included as the tumor size in each animal can vary with some animals not forming the tumors while in others being less than the required size. Inclusion of extra mice will avoid falling short of the required number of mice at the start of the experiment. Ten percent wastage of animals is often observed.
15. When repeated treatments are planned, care needs to be taken in determining the injection sites as well as determining if the previously injected site has completely healed. Presence of unhealed sites will lead to leakage of the injected material. Care also needs to be taken not to pierce the needle through the tumor. Additionally, tumors become hard as they grow at which time it is difficult to inject DNA-nanoparticles as realized by the intratumoral pressure. In such

events, it is recommended that the material be injected from two sides of the tumor. Penetration of the needle into the tumor vasculature when tumors are highly vascular will result in hemorrhage and death.

16. Do not insert the needle into the tail vein unless clearly visible. Always start from the end of the tail when ready to inject. Successful intravenous injection is indicated often by the lack of syringe pressure and bleeding upon withdrawal of the needle from the tail vein.
17. Make sure that the cells are not clumped and are in single cell suspension. Clumped cells can be made into single cells by gently pipetting few times. Injection of clumped cells or injecting rapidly will cause embolism and animal death.
18. Care needs to be taken when injecting DNA-nanoparticles intravenously. Rapid injection will lead to embolism and animal death. Injecting over a period of 1 min is optimized and safe.
19. When repeated injections are required, always start the treatment by injecting the first dose close to the tail tip with subsequent treatments progressing upwards. This will allow quick healing of the injected site as well minimize leakage of the injected material.

Acknowledgments

The author thanks past and present members of the laboratory for their contribution in the development and testing of nanoparticles for gene delivery to the lung. This work was supported in part by the W.M. Keck Gene Therapy grant, by the Texas Higher Education Coordinating Board ARP/ATP grant 003657-0078-2001, by a Career Development award from the University of Texas SPORE in Lung Cancer (P50 CA70907), and by a sponsored research agreement with Introgen Therapeutics, Inc.

References

1. Edwards, B.K., Brown, M.L., Wingo, P.A., Howe, H.L., Ward, E., Ries, L.A., Schrag, D., Jamison, P.M., Jemal, A., Wu, X.C., Friedman, C., Harlan, L., Warren, J., Anderson, R.N., and Pickle, L.W. (2005) Annual report to the nation on the status of cancer, 1975–2002, featuring population-based trends in cancer treatment. *J Natl Cancer Inst* **97**, 1407–1427.
2. Greenlee, R., Murray, T., Bolden, S., and Wingo, P.A. (2000) Cancer statistics. *CA Cancer J Clin* **50**, 7–33.
3. Cuenca, A.G., Jiang, H., Hochwald, S.N., Delano, M., Cance, W.G., and Grobmyer, S.R. (2006) Emerging implications of nanotechnology on cancer diagnostics and therapeutics. *Cancer* Jun 22, **107**, 459–466.
4. Kubik, T., Bogunia-Kubik, K., and Sugaisaka M. (2005) Nanotechnology on duty in medical applications. *Curr Pharm Biotechnol* **6**, 17–33.

5. Yezhelyev, M.V., Gao, X., Xing, Y., Al-Hajj, A., Nie, S., and O'Regan, R.M. (2006) Emerging use of nanoparticles in diagnosis and treatment of breast cancer. *Lancet Oncol* **7**, 657–667.
6. Kaul, Z., Yaguchi, T., Kaul, S.C., and Wadhwa, R. (2006) Quantum dot-based protein imaging and functional significance of two mitochondrial chaperones in cellular senescence and carcinogenesis. *Ann N Y Acad Sci* **1067**, 469–473.
7. Kobayashi, H., Kawamoto, S., Brechbiel, M.W., Bernardo, M., Sato, N., Waldmann, T.A., Tagaya, Y., and Choyke, P.L. (2005) Detection of lymph node involvement in hematologic malignancies using micromagnetic resonance lymphangiography with a gadolinum-labeled dendrimer nanoparticle. *Neoplasia* **7**, 984–991.
8. Smith, A.M., Dave, S., Nie, S., True, L., and Gao, X. (2006) Multicolor quantum dots for molecular diagnostics of cancer. *Expert Rev Mol Diagn* **6**, 231–244.
9. Devalapally, H., Shenoy, D., Little, S., Langer, R., and Amiji, M. (2006) Poly(ethylene oxide)-modified poly(beta-amino ester) nanoparticles as a pH-sensitive system for tumor-targeted delivery of hydrophobic drugs: part 3. Therapeutic efficacy and safety studies in ovarian cancer xenograft model. *Cancer Chemother Pharmacol* Jul 22, 2007, **59**, 477–404.
10. Reddy, L.H., Vivek, K., Bakshi, N., and Murthy, R.S. (2006) Tamoxifen citrate loaded solid lipid nanoparticles (SLN): preparation, characterization, in vitro drug release, and pharmacokinetic evaluation. *Pharm Dev Technol* **11**, 167–177.
11. Farokhzad, O.C., Karp, J.M., and Langer, R. (2006) Nanoparticle-aptamer bioconjugates for cancer targeting. *Expert Opin Drug Deliv* **3**, 311–324.
12. Ramesh, R., Saeki, T., Templeton, N.S., Ji, L., Stephens, L.C., Ito, I., Wilson, D.R., Wu, Z., Branch, C.D., Minna, J.D., and Roth, J.A. (2001) Successful treatment of primary and disseminated human lung cancers by systemic delivery of tumor suppressor genes using an improved liposome vector. *Mol Ther* **3**, 1–14.
13. Ito, I., Ji, L., Tanaka, F., Saito, Y., Gopalan, B., Branch, C.D., Xu, K., Atkinson, E.N., Bekele, B.N., Stephens, L.C., Minna, J.D., Roth, J.A., and Ramesh, R. (2004) Liposomal vector mediated delivery of the 3p FUS1 gene demonstrates potent antitumor activity against human lung cancer *in vivo*. *Cancer Gene Ther* **11**, 733–739.
14. Gopalan, B., Ito, I., Branch, C.D., Stephens, C., Roth, J.A., and Ramesh, R. (2004) Nanoparticle based systemic gene therapy for lung cancer: molecular mechanisms and strategies to suppress nanoparticle-mediated inflammatory response. *Technol Cancer Res Treat* **3**, 647–657.
15. Ramesh, R., Ito, I., Saito, Y., Wu, Z., Mhashikar, A.M., Wilson, D.R., Branch, C.D., Roth, J.A., and Chada, S. (2004) Local and systemic inhibition of lung tumor growth after nanoparticle-mediated mda-7/IL-24 gene delivery. *DNA Cell Biol* **23**, 850–857.
16. Templeton, N.S., Lasic, D.D., Frederick, P.M., Strey, H.H., Roberts, D.D., and Pavlakis, G.N.(1997) Improved DNA:liposome complexes for increased systemic delivery and gene expression. *Nat Biotechnol* **15**, 647–652.

17. Ito, I., Began, G., Mohiuddin, I., Saeki, T., Saito, Y., Branch, C.D., Vaporciyan, A., Stephens, L.C., Yen, N., Roth, J.A., and Ramesh, R. (2003) Increased uptake of liposomal-DNA complex by lung metastases following intravenous administration. *Mol Ther* **7**, 409–418.
18. Ito, I., Saeki, T., Mohuiddin, I., Saito, Y., Branch, C.D., Vaporciyan, A., Roth, J.A., and Ramesh R. (2004) Persistent transgene expression following intravenous administration of a liposomal complex: role of IL-10 mediated immune suppression. *Mol Ther* **9**, 318–327.

20

Retroviral-Mediated Gene Therapy for the Differentiation of Primary Cells into a Mineralizing Osteoblastic Phenotype

Jennifer E. Phillips and Andrés J. García

Summary

Bone tissue engineering has emerged as a promising strategy for the repair of critical-sized skeletal fractures. However, the clinical application of this approach has been limited by the availability of a robust mineralizing cell source. Non-osteogenic cells, such as skin fibroblasts, are an attractive cell-source alternative because they are easy to harvest from autologous donor skin biopsies and display a high capacity for in vitro expansion. We have recently demonstrated that retroviral gene delivery of the osteoblastic transcription factor Runx2/Cbfa1 promotes osteogenic differentiation in primary dermal fibroblasts cultured in monolayer. Notably, sustained expression of Runx2 was not sufficient to promote functional osteogenesis in these cells, and co-treatment with the steroid hormone dexamethasone was required to induce deposition of biologically-equivalent matrix mineralization. On the basis of these results, we then investigated the osteogenic capacity of these genetically engineered fibroblasts when seeded on polymeric scaffolds in vitro and in vivo. These experiments demonstrated that Runx2-expressing fibroblasts seeded on collagen scaffolds produce significant levels of matrix mineralization after 28 days in vivo implantation in a subcutaneous, heterotopic site. Overall, these results offer evidence that transcription factor-based gene therapy may be a powerful strategy for the conversion of a non-osteogenic cellular phenotype into a mineralizing cell source for bone repair applications. This concept may also be applied to control functional differentiation in a broad range of cell types and tissue engineering applications. The chapter below outlines detailed methods for the isolation and ex vivo genetic modification of primary dermal fibroblasts using retroviral-mediated delivery of the Runx2 transgene in both monolayer culture and three-dimensional scaffolds.

From: *Methods in Molecular Biology, vol. 433: Volume 1: Production and In Vivo Applications of Gene Transfer Vectors*
Edited by: J. M. Le Doux © Humana Press, Totowa, NJ

Key Words: Gene therapy; tissue engineering; bone regeneration; Runx2/Cbfa1; transcription factor; fibroblasts; osteoblasts; cellular therapy; regenerative medicine; skin; bone.

1. Introduction

1.1. Current State of the Art in Biological and Synthetic Bone Grafts

Six million skeletal fractures occur each year in the USA and greater than 500,000 of these are non-healing defects which require some form of orthopaedic grafting substitute *(1,2)*. As a result, approximately US$2.5 billion is spent annually to treat victims of age-related bone deterioration, traumatic injury, tumor resection, or osteolytic disease *(3)*. Traditional skeletal repair techniques rely on the use of autogenic or allogenic bone grafts. Autografts have shown great clinical success, but are limited by inadequate tissue supply, donor site morbidity, and inconsistent bone graft quality *(4)*. Allografts offer advantages for off-the-shelf tissue availability but carry the risk of disease transmission and display reduced biological activity and mechanical properties due to tissue processing *(5,6)*. Synthetic materials have also been explored for bone grafting applications, but generally show poor tissue integration and may incite a host inflammatory response *(7)*. More recently, osteoinductive recombinant proteins, such as bone morphogenetic protein-2 (BMP-2) or BMP-7, have been approved by the US Food and Drug Administration for the treatment of human non-healing fractures and spinal fusion *(8–11)*. Yet, even these growth-factor-based approaches continue to be hampered by suboptimal delivery vehicles and safety issues associated with the extremely high doses needed to achieve efficacy *(11,12)*. Therefore, it is clear that the limitations of existing grafting templates leave a pressing need for alternative bone graft substitutes.

1.2. Genetic and Tissue Engineering Approaches to Bone Grafting

Bone tissue engineering has recently been investigated to overcome complications associated with conventional skeletal repair therapies *(13–16)*. The general paradigm for these strategies involves the integration of bioactive factors and/or osteoblastic cells into a three-dimensional (3-D) scaffolding matrix. Notably, a large number of the cell-based approaches utilize bone marrow-derived stromal cells, which contain subpopulations of osteoprogenitors and mesenchymal stem cells (MSCs) with significant capacity for mineralization in vitro and in vivo *(17–22)*. Although these constructs have successfully healed bone defects in a wide range of preclinical animal models, it is still unknown whether MSC transplantation alone will be sufficient to induce robust healing of critical-sized bone defects in humans. Moreover, gene therapy strategies based on the overexpression of cDNA encoding for soluble factors, such as BMP-2 and/or BMP-4, have further enhanced the bone healing capacity of MSCs ***(23–29)***. These results suggest that ex vivo genetic manipulation of MSCs may provide a more robust strategy for bone formation than direct implantation of these cells alone. Nevertheless, despite these advances, the clinical use of these precursors is still hindered by their labor intensive harvesting procedure ***(30)***, low frequency in healthy marrow ***(31,32)***, and dedifferentiation associated with donor age and disease-state ***(20,33)***. Overall, the inadequate availability of a sustained mineralizing cell source has been a significant barrier in the clinical application of tissue engineering strategies toward the development of large, mechanically robust grafting templates.

1.3. Runx2 in Osteoblastic Differentiation

Differentiation of osteoblasts, the cells regulating bone mineral deposition, is tightly regulated by a complex spatiotemporal cascade of growth factors, hormones, transcription factors, and cell–matrix interactions. Osteoinductive factors, such as BMPs, bind to specific transmembrane receptors to initiate signaling cascades that converge to activate a program of downstream transcriptional regulators (e.g., Runx2/Cbfa1, Dlx-3, Dlx-5, Msx-2, AP-1, Osterix) ***(34)***. Among these, Runx2/Cbfa1 is an essential transcriptional regulator of osteoblast differentiation, matrix mineralization, and normal bone maintenance (*see* **Fig. 1**) ***(35)***. Runx2 functions as a transcriptional scaffolding protein that recruits a number of accessory proteins to the OSE2 *cis*-acting element found within the promoter region of osteogenic target genes to regulate their expression ***(36)***. Runx2 transactivation of osteoblast-specific genes, including alkaline phosphatase (ALP), osteocalcin (OCN), and bone sialoprotein (BSP), is a key step in the commitment of stem cells toward the osteoblastic lineage.

1.4. Dexamethasone in Osteoblastic Differentiation

Glucocorticoids (GCs) are cholesterol-derived hormones that have complex stimulatory and inhibitory effects on skeletal metabolism and bone formation ***(37)***. Long-term administration of GCs at pharmacological doses in vivo often has catabolic effects on the skeleton, resulting in osteoporosis and bone loss ***(38)***. By contrast, synthetic GCs, such as dexamethasone, promote osteogenic differentiation and maturation of primary human osteoblasts and bone marrow stromal cells when utilized at physiologic concentrations (10 nM) in vitro ***(39,40)***. The classical mechanism of GC action is primarily at the level of transcription, where the cholesterol-derived hormone freely diffuses into the cell and binds to its cytosolic GC receptor (GCR) (*see* **Fig. 1**). This complex can either enhance or attenuate gene expression by binding to a GC response

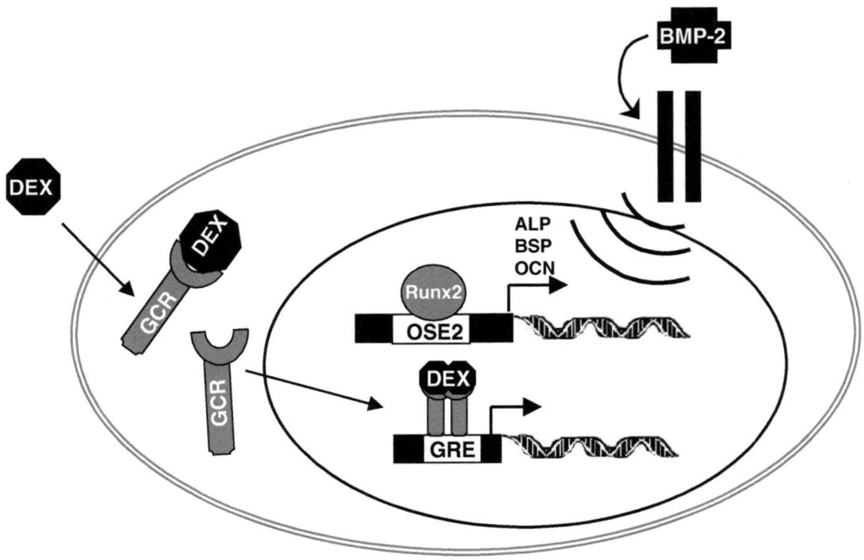

Fig. 1. Cartoon depiction of the Runx2/Cbfa1 transcription factor and the steroid hormone dexamethasone during osteogenic differentiation.

element (GRE) within the promoter region of target genes or by non-covalently associating with additional co-regulatory proteins *(41)*. In addition to this conventional mechanism, steroids also have rapid effects that are mediated by non-classical membrane-associated pathways *(41,42)*. However, beyond this general mechanistic knowledge, the cellular and molecular pathway(s) by which GCs induce osteogenic differentiation remains poorly understood, largely due to conflicting results associated with the various species and differentiation states of the model systems used to study this hormone in vitro *(41,43–48)*.

1.5. Conversion of Non-Osteogenic Cells into a Mineralizing Osteoblastic Phenotype

It is evident that numerous signaling molecules involved in osteogenic differentiation have potential as therapeutic targets for skeletal regeneration. In order to address cell-sourcing issues associated with bone tissue engineering, gene therapy strategies focused on these regulatory factors have been effectively applied to promote osteogenic differentiation in non-osteoblastic myogenic and fibroblastic cell types *(49–52)*. Primary skin fibroblasts are a particularly attractive cell source alternative because they are easily harvested with minimal donor site morbidity, available in large quantities, and susceptible to genetic manipulation by gene transfer vectors. Notably, dermal fibroblasts transduced with adenoviral vectors encoding for either BMP-2 or BMP-7 form significant amounts of bone in both ectopic implantation sites and critical-sized calvarial defects ***(53,54)***. Yet, despite promising results, the clinical feasibility of these growth-factor-based approaches may be hampered by complex release kinetics and paracrine signaling to neighboring non-osseous tissues.

In order to avoid detrimental effects associated with uncontrolled growth factor secretion, recent research initiatives have explored gene delivery of downstream transcriptional activators. It has been well documented that sustained overexpression of Runx2 upregulates osteogenic gene expression in a wide range of cell types, including MC3T3-E1 osteoblasts ***(55)***, primary bone marrow stromal cells ***(56,57)***, and non-osteoblastic skeletal myoblasts ***(58)***. Moreover, we and others have reported that forced Runx2 expression enhances mineralization in a cell-type-dependent manner, and it is insufficient to induce mineralized nodule deposition in fibroblasts and fibroblastic cell lines ***(52,53,55)***. These results suggest that Runx2-mediated mineralization requires additional cofactors, which may not be endogenously expressed in certain non-osteoblastic cells. In support of this idea, we have recently demonstrated that a combination of constitutive retroviral Runx2 expression and supplementation with dexamethasone synergistically induces osteogenic differentiation in primary dermal fibroblast monolayer cultures, including osteoblastic gene expression, ALP activity, and matrix mineralization ***(59)***. Furthermore, we utilized these Runx2-genetically engineered fibroblasts to create mineralized templates for bone repair in vitro and in vivo ***(60,61)***. Overall, these results establish Runx2-genetic engineering as a strategy for the conversion of a non-osteogenic cellular phenotype into a mineralizing cell source for bone tissue engineering applications. This work is significant because it concentrates on a hybrid gene therapy and tissue engineering approach to develop mineralized templates for skeletal regeneration. The concept of controlling expression of tissue-specific transcription factors to enhance cellular function may be applicable to overcome cell-sourcing limitations in other tissue engineering applications. Although our focus is on dermal fibroblasts, these techniques are applicable to other cell types.

In this chapter, we describe methods for the isolation and ex vivo genetic modification of primary dermal fibroblasts using retroviral-mediated delivery of the Runx2 transgene in both monolayer culture (*see* **Fig. 2**) and 3-D scaffolds (*see* **Fig. 3**). Briefly, primary fibroblasts were isolated from the skin of Wistar rats by enzymatic digestion of dermal tissue. Passage 4 cells were seeded in monolayer on 1 mg/ml collagen coated dishes and transduced with Runx2 retrovirus or left as unmodified controls. For 3-D experiments, Runx2-expressing or unmodified cells were seeded 2 days post infection onto fibrous collagen

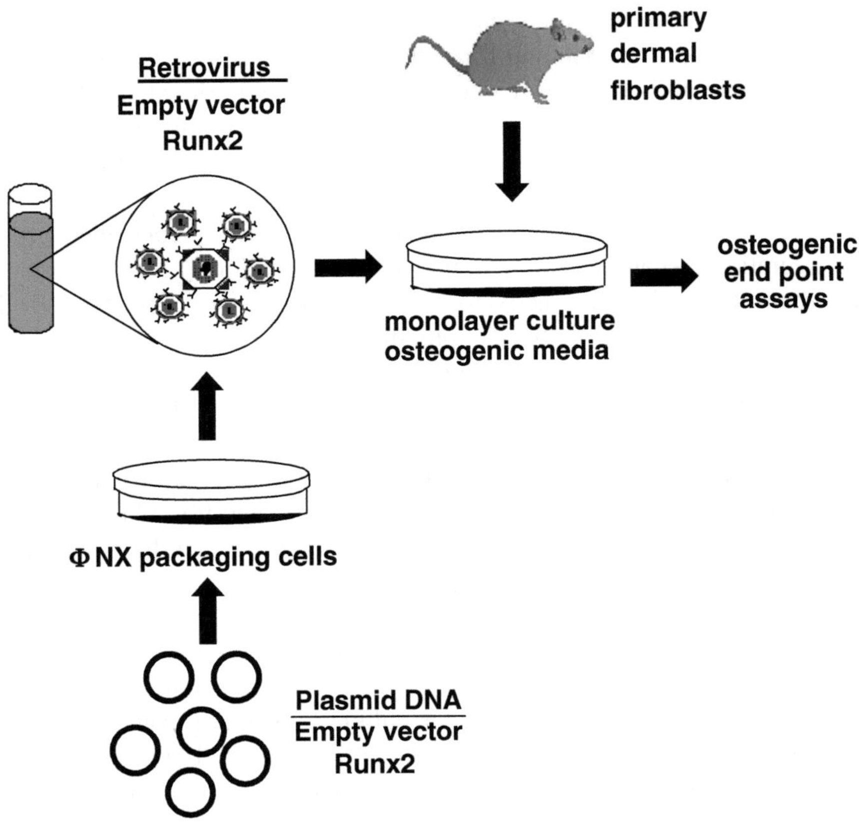

Fig. 2. Cartoon depiction of methodology for retroviral transduction of fibroblasts in monolayer. Runx2 retroviruses are packaged using ΦNX helper-free producer cells. Subconfluent cultures of primary dermal fibroblasts are transduced by incubation in retroviral supernatant. Following transduction, cell are maintained in osteogenic differentiation media supplemented with 50 μg/ml ascorbic acid, 3 mM β-glycerophosphate, and 10 nM dexamethasone until endpoint assays are performed.

scaffolds coated with 20 μg/ml fibronectin. In vitro constructs were cultured for 6 weeks in osteogenic differentiation media supplemented with 3 mM sodium β-glycerophosphate, 50 μg/ml ascorbic acid, and 10 nM dexamethasone. In vivo constructs were implanted 1 day post seeding into subcutaneous pockets made by blunt dissection in the backs of syngeneic Wistar rats in accordance to an Institutional Animal Care and Use Committee (IACUC)-approved protocol. Osteogenic differentiation of these genetically engineered cells in monolayer and on polymeric scaffolds was assessed with a number of endpoint assays as detailed elsewhere ***(58,59,62)***.

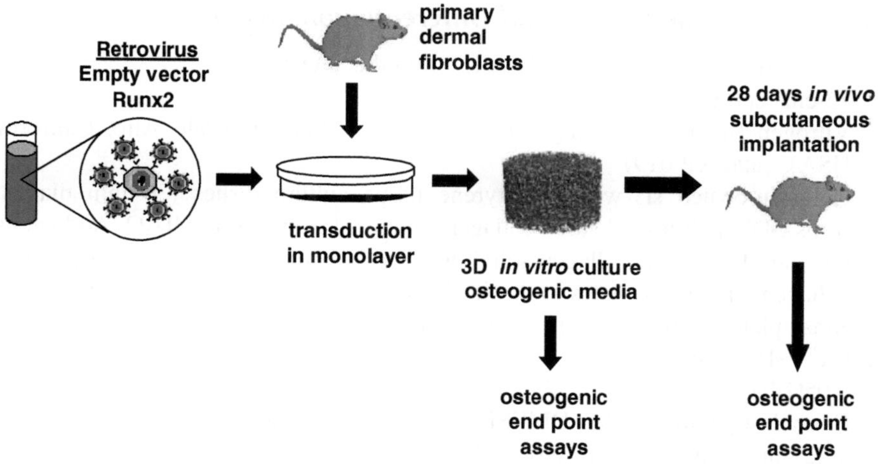

Fig. 3. Cartoon depiction of methodology for retroviral transduction of fibroblasts seeded within 3-D scaffolds. Subconfluent cultures of primary dermal fibroblasts are infected in monolayer culture by incubation in retroviral supernatant and seeded onto polymeric scaffolds. Following seeding, cells are either **(1)** cultured in vitro in osteogenic differentiation media supplemented with 50 μg/ml ascorbic acid, 3 mM β-glycerophosphate, and 10 nM dexamethasone or **(2)** implanted into an in vivo ectopic, subcutaneous site until endpoint assay.

2. Materials

2.1. Primary Dermal Fibroblast Isolation

1. Five 8–16 week old male Wistar rats.
2. Sterile phosphate-buffered saline without calcium or magnesium (Ca^{2+}/Mg^{2+}-free PBS).
3. 70% ethanol.
4. Fibroblast growth media: Dulbecco's Modified Eagle's Medium (DMEM), 10% fetal bovine serum (FBS), 100 U/ml penicillin G sodium, and 100 μg/ml streptomycin sulfate (1% pen-strep).
5. Collagenase-based tissue digestion solution: 0.3% type II collagenase (Worthington Biochemical Corporation (Lakewood, New Jersey, USA), cat#LS004176) + DMEM + 1% pen-strep.
6. Trypsin-based tissue digestion solution: 0.25% Trypsin + 1% pen-strep.
7. Sterile surgical tools (blades, blade handles, hemostats, scissors).
8. Sterilized metal cell strainer with 105-μm mesh.
9. 0.22-μm Polyethersulfone filter.
10. 75-cm^2 cell culture T flask.
11. 50-ml centrifuge tubes.
12. Shaker plate.

2.2. General Tissue Culture and Differentiation Media

1. Polystyrene tissue culture dishes (150 mm diameter).
2. 75-cm^2 cell culture T flask.
3. Vitrogen-100 type I collagen (Angiotech BioMaterials (Palo Alto, California, USA), cat#FXP-019).
4. Collagen-coated six-well polystyrene tissue culture dishes (35 mm diameter wells): Dilute 3 mg/ml stock collagen solution 1:3 in sterile ddH$_2$O. Add enough volume of 1 mg/ml collagen solution to cover bottom of each well, aspirate off collagen, and allow plates to dry in tissue culture hood for approximately 3–4 h. Rinse plates with PBS immediately before cell seeding.
5. Ca^{2+}/Mg^{2+}-free PBS.
6. 0.05% Trypsin.
7. Fibroblast growth media: DMEM, 10% FBS, 1% pen-strep.
8. 0.22-μm cellulose acetate syringe filter.
9. Osteogenic differentiation media: DMEM, 10% FBS, 1% pen-strep supplemented with 50 μg/ml L-ascorbic acid, 3 mM sodium β-glycerophosphate, and 10 nM dexamethasone (*see* **Note 1**).

   a. Ascorbic acid: Make a 100× stock concentration of 5 mg/ml by dissolving 0.5 g ascorbic acid into 100 ml sterile ddH$_2$O, sterile filter with a 0.22-μm filter, aliquot into 1 ml portions, and store at –80°C. Add 10 μl per ml growth media for a final concentration of 50 μg/ml (*see* **Note 2**).
   b. Sodium β-glycerophosphate: Make a 100× stock concentration of 300 mM by completely dissolving sodium β-glycerophosphate 6.53 g into 100 ml sterile ddH$_2$O, sterile filter with a 0.22-μm filter, aliquot into 1 ml portions, and store at –80°C. Add 10 μl per ml growth media for a final concentration of 3 mM (*see* **Note 3**).
   c. Dexamethasone: Make a 1000× stock concentration of 10 μM by first completely dissolving 0.007 g dexamethasone into 1 ml sterile ddH$_2$O for a 100,000× solution (1000 μM), then add 300 μl of the 100,000×solution to 30 ml ddH$_2$O for a 1000× (10 μM) solution, sterile filter with a 0.22-μm filter, aliquot into 1 ml portions, and store at –80°C. Add 1 μl per ml growth media for a final concentration of 10 nM (*see* **Note 4**).

2.3. Runx2 Retroviral Plasmid Vector

The murine cDNA for the type II MASNSLF Runx2 isoform was subcloned into the pTJ66 retroviral vector (*see* **Fig. 4**), which utilizes the promoter activity of a 5′ long terminal repeat to express a single, bicistronic mRNA encoding the therapeutic transgene followed by an internal ribosomal entry site and a Zeocin resistance-enhanced green fluorescent fusion protein. All plasmid DNA was purified from transformed *Escherichia coli* using Megaprep kits from Qiagen (Valencia, CA).

Fig. 4. Schematic diagram of the Runx2 retroviral expression vector.

2.4. Runx2 Retrovirus Production

1. PBS.
2. 15-ml conical tubes.
3. Vortexer.
4. Fibroblast growth media: DMEM, 10% FBS, 1% pen-strep.
5. ΦNX amphotropic cells (gift from Gary Nolan, Stanford University, Stanford, CA).
6. Prepare reagents ahead of time (Note: all reagent volumes for 150-mm plates):

   a. 25 mM chloroquine: Dissolve 0.0155 g chloroquine diphosphate salt in 1.2 ml of PBS. Sterile filter using a 0.22-μm cellulose acetate filter, aliquot into 150-μl aliquots, and store at –20°C.
   b. 2 M calcium chloride dihydrate ($CaCl_2 \cdot 2H_2O$): Dissolve 3.528 g calcium chloride dihydrate in 12 ml of ddH_2O. Sterile filter using a 0.22-μm cellulose acetate filter, aliquot into 1.5-ml aliquots, and store at –20°C.

2.5. Fibroblast Transduction

1. Retroviral stocks.
2. Beckman model GS-6R centrifuge with a swinging bucket rotor.
3. Polybrene (hexadimethrine bromide): Make a 100× stock concentration of 400 μg/ml by completely dissolving 20 mg hexadimethrine bromide into 50 ml sterile ddH_2O, sterile filter with a 0.22-μm filter, aliquot into 30-ml portions, and store at –20°C.
4. Becton-Dickinson FACS Vantage SE Cell Sorter.

2.6. Scaffold Seeding

1. Autoclaved Kimwipes®.
2. Sterile forceps and tweezers.
3. Scaffolds (cylindrical shape; 8 mm diameter × 2 mm thick).
4. Tissue culture materials as in **Subheading 2.2.**
5. Genetically engineered fibroblasts in T75 flasks (*see* **Subheading 3.2., step 8.b**).
6. Collagen scaffolds (*see* **Note 5**): 8 mm × 1.5 mm, average pore size 61.7 μm, 93.7% pore volume (Kensey Nash Corporation, Exton, PA).

2.7. In Vivo Subcutaneous Implantation

1. Surgical prep: sterile drapes, gowns, gloves, masks.
2. 70% isopropyl alcohol.
3. Sterile 4 × 4 guaze.
4. Betadine surgical scrub and prep solution.
5. Electric shaver.
6. CO_2 tank and chamber.
7. Sterile surgical equipment: large hemostats, tweezers, forceps, small sharp scissors, scalpel handles, scalpel blades, wound clips.

3. Methods

3.1. Primary Dermal Fibroblast Isolation and Expansion

1. Shave the backs of euthanized Wistar rats (*see* **Note 6**).
2. Sterilize harvest site by rinsing with 70% ethanol.
3. Remove full-thickness sections of skin from the harvest site with sterile blade/scissors and place into sterile 50-ml conical tubes filled with Ca^{2+}/Mg^{2+}-free PBS.
4. Incubate skin sections in trypsin-based tissue digestion solution overnight at 4°C in order to loosen the dermal–epidermal junction.
5. Mechanically separate the epidermis from the dermis with sterile hemostats/tweezers and cut dermis into small pieces.
6. Wash strips of dermis with Ca^{2+}/Mg^{2+}-free PBS and place approximately 1–2 g of diced tissue in each T75 flask.
7. Add 10 ml of collagenase-based digestion solution per 1 g of tissue to each T75 flask.
8. Secure T75 flasks firmly to a shaker plate and place entire assembly in incubator for 12–18 h at 300–500 Hz.
9. When tissue is fully digested, filter cell suspension with a sterile syringe attached to a sterile metal filter attached to a 105-μm mesh to remove excess debris.
10. Place filtered solution into 50-ml conicals and centrifuge at 1300 rpm (~270–300 × g) for 10 min.
11. Aspirate off supernatant and resuspend cells in fibroblast growth media.
12. Plate cells in T75 flasks at a density of 2000–5000 cells/cm^2 in a humidified 5% CO_2 atmosphere at 37°C until 75–80% confluent.
13. Expand cells up to seven passages by splitting 1:5 every 2–3 days or when the flask reaches 70–80% confluency (*see* **Note 7**).

3.2. General Tissue Culture and Cell Plating

1. When fibroblasts reach 75–80% confluence, aspirate off growth media.
2. Rinse cells 1× with 5 ml Ca^{2+}/Mg^{2+}-free PBS.

3. Add 0.05% trypsin warmed to 37°C (approximately 1 ml for 75 cm^2 surface area) and place in incubator for 3–5 min until cells round up.
4. Remove flasks/dishes from incubator and lightly tap on all sides until the majority of cells are dislodged from surface.
5. Inactivate trypsin with at least three to five times the volume of growth media.
6. Collect cell suspension into 50-ml centrifuge tubes.
7. Centrifuge cells at 1300 rpm (~270–300 × g) for 5–7 min.
8. Aspirate off supernatant from each tube while carefully avoiding disturbance of pellet.
9. Count cells and plate at a density of 5000 cells/cm^2 in fibroblast growth media.

   a. For monolayer cultures: Seed cells on collagen-coated six-well plates, approximately 5×10^4 cells in 2 ml of growth media per well, for a total of 5×10^5 cells per plate in 12 ml growth media.
   b. For 3-D experiments: Seed cells on collagen-coated T75 flasks, approximately 4×10^5 cells in 12 ml of growth media per flask.

3.3. Production of Retrovirus with the Transfection Method

1. Plate helper virus-free ΦNX amphotropic cells in growth media (DMEM, 10% FBS, and 1% pen-strep) at a density of 9×10^4 cells/cm^2, approximately 1.6×10^7 cells per 150 mm dish, 1 day prior to transfection.
2. Incubate cells for 24 h at 37°C in a humidified 5% CO_2 atmosphere.
3. Transfect each 150-mm dish with 0.5 μg/cm^2 of plasmid DNA using the calcium phosphate precipitation method.

   a. Re-feed cells with 18 ml growth media supplemented with 22.5 μl of 25 mM chloroquine stock for a final concentration of 25 μM (*see* **Note 8**). Upon completion of transfection (*see* **step 3**), the total volume in each 150-mm dish will be 22.5 ml.
   b. Prepare plasmid DNA solution: In a 14-ml conical tube, add 90.0 μg empty vector-TJ66 stock (6871 bp) or 112.8 μg Runx2-TJ66 stock (8544 bp) to sterile ddH_2O for a final volume of 1969 μl.
   c. Add 281 μl of 2 M calcium chloride dehydrate to plasmid DNA.
   d. Vortex this solution one tube at a time under the tissue culture hood and add 2.25 ml of 2× BES solution drop by drop.
   e. Add the plasmid transfection cocktail to ΦNX cell media and gently swirl dish.

4. Incubate cells at 37°C for 6–10 h and then re-feed with 35–40 ml of fresh growth media.
5. Twenty-four hours after the start of the transfection (and 14 h after removal of transfection media), re-feed cells with fresh growth media, and transfer plates to a humidified 5% CO_2 atmosphere at 32°C for enhanced stability of retroviral particles.

6. Collect 24–48 h supernatant, filter with a 0.45-μm cellulose acetate filter to remove cell debris, aliquot into freezing vials, and snap freeze in liquid nitrogen. Store at –80°C until use. Re-feed with 20 ml of growth media and incubate plates at 32°C.
7. Collect 48–60 h supernatant, filter with a 0.45-μm cellulose acetate filter to remove cell debris, aliquot into freezing vials, and snap freeze in liquid nitrogen. Store at –80 °C until use. Re-feed with 20 ml of growth media and incubate plates at 32°C.
8. Collect 60–72 h supernatant, filter with a 0.45-μm cellulose acetate filter to remove cell debris, aliquot into freezing vials, and snap freeze in liquid nitrogen. Store at –80 °C until use. Discard plates.

3.4. Retroviral Transduction of Fibroblasts

1. Twenty-four hours prior to retroviral transduction, plate cells on collagen-coated tissue culture dishes (according to **Subheading 3.2.**) (*see* **Note 9**).
2. Thaw retroviral supernatant in 37°C water bath.
3. Add 10 μl of 400 μg/ml hexadimethrine bromide (polybrene) stock solution to 0.2 ml/cm^2 of retroviral supernatant for a final polybrene concentration of 4 μg/ml.
4. For six-well plates:

   a. Add 2 ml of retroviral supernatant per well, wrap parafilm around dishes, and place in 32°C incubator for 15 min.
   b. Centrifuge dishes/plates at 1200 × *g* for 30 min in a Beckman model GS-6R centrifuge with a swinging bucket rotor.
   c. After first transduction, aspirate off retroviral supernatant, reapply growth media, and return cultures to 37°C.
   d. Transduce cells a second time 10–12 h later in order to increase infection efficiency. After this second transduction, replace retroviral supernatants with osteogenic differentiation media (DMEM, 10% FBS, 1% pen-strep supplemented with 50 μg/ml L-ascorbic acid, 3 mM sodium β-glycerophosphate, and 10 nM dexamethasone).

5. For T75 flasks:

   a. Add 16 ml of retroviral supernatant per flask and place in 32°C incubator for 15 min.
   b. Centrifuge flasks at 1200 × *g* for 30 min in a Beckman model GS-6R centrifuge with a swinging bucket rotor.
   c. After first transduction, aspirate off retroviral supernatant, reapply growth media, and return cultures to 37°C.
   d. Transduce cells a second time 10–12 h later in order to increase transduction efficiency. After this second transduction, replace retroviral supernatants with growth media.
   e. Allow transduced cells to expand for 72 h in a 37°C incubator.

6. Analyze cells at 72 h post transduction for eGFP expression by fluorescence microscopy and flow cytometry, using a Becton-Dickinson FACS Vantage SE Cell Sorter. Unmodified cells should be used as a control to monitor autofluorescence.

3.5. Scaffold Seeding

1. Scaffold preparation:
   a. Soak scaffolds in 70% ethanol for 30 min.
   b. Rinse scaffolds in ddH_2O for 10 min.
   c. Incubate scaffolds overnight in 10% FBS in PBS.
   d. Rinse scaffolds in PBS.
   e. Place coated scaffolds on sterile kimwipe or paper towel to wick off excess liquid.
2. Seed cells (*see* **Note 10**):
   a. Forty-eight hours after the second transduction, aspirate off growth media from T75 flasks.
   b. Rinse cells 1× with 5 ml Ca^{2+}/Mg^{2+}-free PBS.
   c. Add 0.05% trypsin warmed to 37°C (approximately 1 ml for 75 cm^2 surface area) and place in incubator for 3–5 min until cells round up.
   d. Remove flasks/dishes from incubator and lightly tap on all sides until the majority of cells are dislodged from surface.
   e. Inactivate trypsin with at least three to five times the volume of growth media.
   f. Collect cell suspension into 50-ml centrifuge tubes.
   g. Centrifuge cells at 1300 rpm (~270–300 × g) for 5–7 min.
   h. Aspirate off supernatant from each tube while carefully avoiding disturbance of pellet.
   i. Count cells and resuspend 1×10^6 cells/scaffold in 50 μl of growth media/scaffold.
   j. Place scaffolds in non-tissue culture-treated 24-well plates.
   k. Pipet 25 μl of cell suspension (5×10^5 cells) onto each side of scaffold.
3. Four hours later, immerse constructs in 2 ml of growth media.
4. After twenty-four hours, transfer scaffolds to tissue-culture-treated 12-well plates with sterile forceps and add 4 ml of differentiation media (DMEM, 10% FBS, 1% pen-strep, 50 μg/ml L-ascorbic acid, 3 mM sodium β-glycerophosphate, and 10 nM dexamethasone).
5. Replace media every 3 days until terminal assay.

3.6. In Vivo Subcutaneous Implantation

1. Surgeries are performed on 7-week-old syngeneic rats in accordance to an IACUC-approved protocol.

2. Under general anesthesia, shave the dorsal region of syngeneic rats and treat the shaved region with 70% isopropyl alcohol and Betadine®.
3. Make a 4-cm midline incision through the skin with sterilized instruments.
4. Separate the skin from the muscle via blunt dissection on both sides of the midline incision.
5. Insert one implant in each pocket and close the incision using wound clips.
6. Explant constructs after 4 weeks in vivo following euthanasia.

3.7. Osteogenic Endpoint Assays (Monolayer Cultures)

Endpoint assays for the evaluation of osteoblastic differentiation in monolayer have been described in detail by our group in previous publications ***(55,58)***. Briefly, osteogenic gene expression was investigated at 1, 3, and/or 7 days post transduction by quantitative RT–PCR using rat-specific primers ***(56)***. ALP activity was quantified at 7 days post transduction using 4-methyl-umbelliferyl-phosphate substrate and normalized to total protein (*see* **Fig. 5**). Matrix mineralization was assessed at 21 days post transduction by von Kossa histochemical staining for phosphate deposits (*see* **Fig. 6A**).

3.8. Osteogenic Endpoint Assays (3-D Constructs)

Endpoint assays for the evaluation of osteoblastic differentiation within 3-D scaffolds have also been described by our group in several manuscripts ***(60–62)***. Scaffold colonization and cellular viability were assessed at 1, 21, and 42

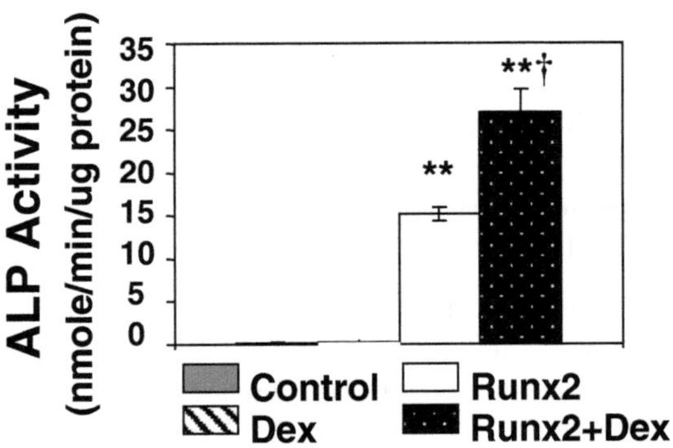

Fig. 5. Primary dermal fibroblasts were transduced with Runx2 or empty vector retrovirus and cultured in osteogenic differentiation media supplemented with or without 10 nM dexamethasone. Alkaline phosphatase (ALP) activity was examined by a biochemical assay at 7 days post transduction [mean ± SEM, $n = 12$; ANOVA: $p < 1E\text{-}8$; **different from empty vector and DEX controls, and †different from Runx2 ($p < 0.05$)]. Reprinted with permission from *Journal of Cell Science* ***(59)***.

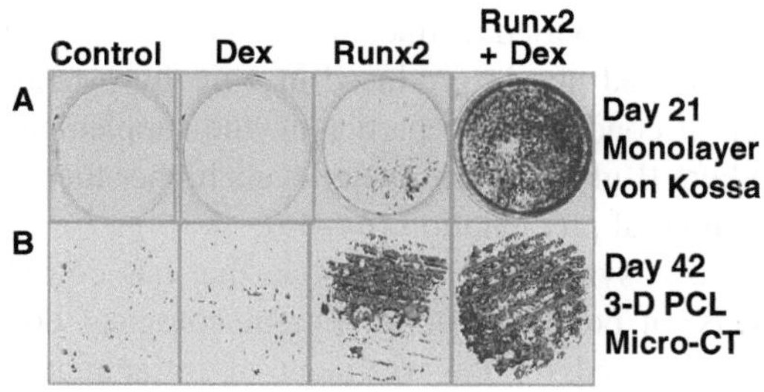

Fig. 6. (**A**) Mineralized matrix deposition by Runx2-expressing fibroblast monolayer cultures was assessed with von Kossa staining for phosphate-positive regions. (**B**) Representative micro-CT images of Runx2-expressing and unmodified fibroblasts seeded on fused deposition-modeled polycaprolactone (PCL) scaffolds and cultured in osteogenic media for 42 days. Reprinted with permission from *Journal of Cell Science* ***(59)***.

days post seeding by confocal microscopy and Live (green)/Dead (red) fluorescence staining. Scaffold seeding efficiency was evaluated by quantification of DNA content at 1 day post seeding. Mineral deposition was analyzed at 28 and 42 days post seeding by micro-CT imaging (*see* **Fig. 6B**). The distribution of cells and mineral deposits throughout tissue-engineered constructs was assessed by hematoxylin–eosin (H&E) and von Kossa staining of histological sections, respectively. Finally, the chemical composition of the mineral phase deposited on Runx2-engineered and control constructs was analyzed by Fourier Transform Infrared (FTIR) spectroscopy using a Nexus 470 FTIR spectrometer (*see* **Note 11**). This chemical signature was compared to the spectra for a cranial bone positive control in order to determine whether mineral deposition by Runx2-engineered cells displays the bands characteristic of a biologically-equivalent, poorly crystalline, carbonate-containing hydroxyapatite (e.g., an enhanced phosphate peak at 1100 cm^{-1}, a doublet split phosphate peak at 560 and 605 cm^{-1}, and a carbonate peak at 870 cm^{-1}) ***(63)***.

4. Notes

1. The potential for osteogenic differentiation by most primary cells/cell lines is highly dependent on the lot of serum used in the tissue culture media. We conduct all experiments with one specific serum lot that has been screened to maximize mineral deposition by genetically engineered fibroblasts and to minimize artifacts caused by lot-to-lot variability.
2. Ascorbic acid is light sensitive, so work quickly to make aliquots and avoid freeze–thaw cycles.

3. Previous studies have reported distinct differences in the chemical and biological makeup of mineralized matrix produced in vitro by osteoblastic cells cultured in a wide range of conditions. In particular, the supplementation of osteogenic media with sodium β-glycerolphosphate levels higher than 5 mM often leads to non-biological mineral precipitation ***(64–66)***.
4. Dexamethasone is a synthetic GC hormone that has complex stimulatory and inhibitory effects on bone formation. Pharmacological doses (≥1000 nM) of dexamethasone are used for the treatment of inflammatory and autoimmune diseases in vivo. However, long-term administration of this hormone in excess often has adverse side effects on the skeleton, inducing osteoporosis by impairing osteoblast activity. In contrast to these catabolic effects, physiologic levels (10 nM) of GC hormone have been shown to promote osteoblastic differentiation in vitro. Experimentally, we have observed that optimum cell viability and mineral deposition are induced at a concentration of 10 nM dexamethasone and concentrations ≥1000 nM lead to apoptosis in these cells.
5. We have evaluated the osteogenic potential of Runx2-expressing fibroblasts when cultured in vitro on three commercially available scaffolds with divergent properties: ***(1)*** fused deposition-modeled polycaprolactone (PCL) (8.1 mm diameter × 2.4 mm thick, 300–500 μm pore size, 66% porosity), ***(2)*** gas-foamed 75/25 polylactide-co-glycolide (PLGA) (8.1 mm diameter × 5 mm thick, 100–200 μm pore size, 85% porosity), and ***(3)*** fibrous collagen disks (8 mm × 1.5 mm, average pore size 61.7 μm, 93.7% pore volume). We have reported that the mineralization capacity of Runx2-engineered fibroblasts is scaffold dependent, with collagen foams exhibiting 10-fold higher mineral volume compared with PCL and PLGA matrices ***(61)***. For this reason, fibrous collagen disks were selected as the cell delivery vehicle for in vivo testing ***(60)***.
6. We have observed that the mineralization capacity of primary dermal fibroblasts varies in magnitude between skin harvests from different animals. In order to minimize this variability, We pooled dermal tissue from at least five animals for each harvesting procedure.
7. We have observed in experiments with three independent cell harvests that primary dermal fibroblasts can be expanded through at least eight passages without a marked detrimental effect on their mineralization capacity.
8. The addition of chloroquine to the transfection medium appears to increase retroviral titer by approximately twofold. However, it is extremely important that the length of chloroquine treatment does not exceed 12 h because longer periods of exposure may have a toxic effect on the cells. In our experience, 9–10 h of treatment gives the best results.
9. The highest transduction efficiency was observed when cells were 40–60% confluence at the time of exposure to the Runx2 retrovirus. At this cell density, eGFP expression was detected in ≥65% of primary dermal fibroblasts via flow cytometry at 72 h post transduction.
10. No selection protocols were performed after transduction in order to avoid further layers of complexity in the design/development of these tissue-engineered

constructs. Importantly, a significant amount of mineral was deposited by these cultures even though the Runx2-expressing cells were not purified from the non-transduced population.

11. von Kossa staining for phosphate-positive regions is typically used as the primary assay for quantifying mineralized matrix deposition in 2-D and 3-D histological sections. However, it has been well documented that in vitro culture conditions can lead to the artifactual detection of von Kossa-positive regions, which represent non-biological mineral ***(67)***. For this reason, characterization of the chemical composition of the mineral phase by FTIR spectroscopy is important to determine whether von-Kossa postitve regions display the chemical signature characteristic of carbonate-containing, poorly crystalline hydroxyapatite ***(63)***.

Acknowledgments

Collagen scaffolds were generously donated by Kensey Nash Corporation. This research was funded by the NIH (R01-EB003364), the Georgia Tech/Emory Engineering Research Center on the Engineering of Living Tissues (NSF EEC-9731643), the Emory-Georgia Tech Biomedical Technology Research Center, and a National Science Foundation Graduate Research Fellowship to J.E.P.

References

1. Lane, J. M., Tomin, E. and Bostrom, M. P. (1999) Biosynthetic bone grafting. *Clin. Orthop. Relat. Res.* S107–S117.
2. Bostrom, M. P., Saleh, K. J. and Einhorn, T. A. (1999) Osteoinductive growth factors in preclinical fracture and long bone defects models. *Orthop. Clin. North Am.* **30**, 647–658.
3. Bucholz, R. W. (2002) Nonallograft osteoconductive bone graft substitutes. *Clin. Orthop. Relat. Res.* 44–52.
4. Finkemeier, C. G. (2002) Bone-grafting and bone-graft substitutes. *J. Bone Joint Surg. Am.* **84-A**, 454–464.
5. Khan, S. N., Tomin, E. and Lane, J. M. (2000) Clinical applications of bone graft substitutes. *Orthop. Clin. North Am.* **31**, 389–398.
6. Perry, C. R. (1999) Bone repair techniques, bone graft, and bone graft substitutes. *Clin. Orthop. Relat. Res.* 71–86.
7. Vaccaro, A. R. (2002) The role of the osteoconductive scaffold in synthetic bone graft. *Orthopedics* **25**, s571–s578.
8. Friedlaender, G. E., Perry, C. R., Cole, J. D., Cook, S. D., Cierny, G., Muschler, G. F., Zych, G. A., Calhoun, J. H., LaForte, A. J. and Yin, S. (2001) Osteogenic protein-1 (bone morphogenetic protein-7) in the treatment of tibial nonunions. *J. Bone Joint Surg. Am.* **83-A Suppl 1**, S151–S158.

9. Geesink, R. G., Hoefnagels, N. H. and Bulstra, S. K. (1999) Osteogenic activity of OP-1 bone morphogenetic protein (BMP-7) in a human fibular defect. *J. Bone Joint Surg. Br.* **81**, 710–718.
10. Johnson, E. E., Urist, M. R. and Finerman, G. A. (1988) Bone morphogenetic protein augmentation grafting of resistant femoral nonunions. A preliminary report. *Clin. Orthop. Relat. Res.* 257–265.
11. Wozney, J. M. (2002) Overview of bone morphogenetic proteins. *Spine* **27**, S2–S8.
12. Uludag, H., Gao, T., Porter, T. J., Friess, W. and Wozney, J. M. (2001) Delivery systems for BMPs: factors contributing to protein retention at an application site. *J. Bone Joint. Surg. Am.* **83-A Suppl 1**, S128–S135.
13. Bruder, S. P. and Fox, B. S. (1999) Tissue engineering of bone. Cell based strategies. *Clin. Orthop. Relat. Res.* S68–S83.
14. Calvert, J. W., Weiss, L. E. and Sundine, M. J. (2003) New frontiers in bone tissue engineering. *Clin. Plast. Surg.* **30**, 641–648, x.
15. Cowan, C. M., Soo, C., Ting, K. and Wu, B. (2005) Evolving concepts in bone tissue engineering. *Curr. Top. Dev. Biol.* **66**, 239–285.
16. Mistry, A. S. and Mikos, A. G. (2005) Tissue engineering strategies for bone regeneration. *Adv. Biochem. Eng. Biotechnol.* **94**, 1–22.
17. Arinzeh, T. L., Peter, S. J., Archambault, M. P., van den, B. C., Gordon, S., Kraus, K., Smith, A. and Kadiyala, S. (2003) Allogeneic mesenchymal stem cells regenerate bone in a critical-sized canine segmental defect. *J. Bone Joint Surg. Am.* **85-A**, 1927–1935.
18. Bruder, S. P., Fink, D. J. and Caplan, A. I. (1994) Mesenchymal stem cells in bone development, bone repair, and skeletal regeneration therapy. *J. Cell Biochem.* **56**, 283–294.
19. Mauney, J. R., Volloch, V. and Kaplan, D. L. (2005) Role of adult mesenchymal stem cells in bone tissue engineering applications: current status and future prospects. *Tissue Eng.* **11**, 787–802.
20. Kahn, A., Gibbons, R., Perkins, S. and Gazit, D. (1995) Age-related bone loss. A hypothesis and initial assessment in mice. *Clin. Orthop. Relat. Res.* 69–75.
21. Ohgushi, H., Goldberg, V. M. and Caplan, A. I. (1989) Repair of bone defects with marrow cells and porous ceramic. Experiments in rats. *Acta Orthop. Scand.* **60**, 334–339.
22. Quarto, R., Mastrogiacomo, M., Cancedda, R., Kutepov, S. M., Mukhachev, V., Lavroukov, A., Kon, E. and Marcacci, M. (2001) Repair of large bone defects with the use of autologous bone marrow stromal cells. *N. Engl. J. Med.* **344**, 385–386.
23. Wang, J. C., Kanim, L. E., Yoo, S., Campbell, P. A., Berk, A. J. and Lieberman, J. R. (2003) Effect of regional gene therapy with bone morphogenetic protein-2-producing bone marrow cells on spinal fusion in rats. *J. Bone Joint Surg. Am.* **85-A**, 905–911.
24. Lieberman, J. R., Le, L. Q., Wu, L., Finerman, G. A., Berk, A., Witte, O. N. and Stevenson, S. (1998) Regional gene therapy with a BMP-2-producing murine stromal cell line induces heterotopic and orthotopic bone formation in rodents. *J. Orthop. Res.* **16**, 330–339.

25. Lieberman, J. R., Daluiski, A., Stevenson, S., Wu, L., McAllister, P., Lee, Y. P., Kabo, J. M., Finerman, G. A., Berk, A. J. and Witte, O. N. (1999) The effect of regional gene therapy with bone morphogenetic protein-2-producing bone-marrow cells on the repair of segmental femoral defects in rats. *J. Bone Joint Surg. Am.* **81**, 905–917.
26. Blum, J. S., Barry, M. A., Mikos, A. G. and Jansen, J. A. (2003) In vivo evaluation of gene therapy vectors in ex vivo-derived marrow stromal cells for bone regeneration in a rat critical-size calvarial defect model. *Hum. Gene Ther.* **14**, 1689–1701.
27. Park, J., Ries, J., Gelse, K., Kloss, F., von der, M. K., Wiltfang, J., Neukam, F. W. and Schneider, H. (2003) Bone regeneration in critical size defects by cell-mediated BMP-2 gene transfer: a comparison of adenoviral vectors and liposomes. *Gene Ther.* **10**, 1089–1098.
28. Chang, S. C., Chuang, H. L., Chen, Y. R., Chen, J. K., Chung, H. Y., Lu, Y. L., Lin, H. Y., Tai, C. L. and Lou, J. (2003) Ex vivo gene therapy in autologous bone marrow stromal stem cells for tissue-engineered maxillofacial bone regeneration. *Gene Ther.* **10**, 2013–2019.
29. Gysin, R., Wergedal, J. E., Sheng, M. H., Kasukawa, Y., Miyakoshi, N., Chen, S. T., Peng, H., Lau, K. H., Mohan, S. and Baylink, D. J. (2002) Ex vivo gene therapy with stromal cells transduced with a retroviral vector containing the BMP4 gene completely heals critical size calvarial defect in rats. *Gene Ther.* **9**, 991–999.
30. Bruder, S. P. and Fox, B. S. (1999) Tissue engineering of bone. Cell based strategies. *Clin. Orthop. Relat. Res.* S68–S83.
31. Haynesworth, S. E., Goshima, J., Goldberg, V. M. and Caplan, A. I. (1992) Characterization of cells with osteogenic potential from human marrow. *Bone* **13**, 81–88.
32. Bruder, S. P., Jaiswal, N. and Haynesworth, S. E. (1997) Growth kinetics, self-renewal, and the osteogenic potential of purified human mesenchymal stem cells during extensive subcultivation and following cryopreservation. *J. Cell Biochem.* **64**, 278–294.
33. Quarto, R., Thomas, D. and Liang, C. T. (1995) Bone progenitor cell deficits and the age-associated decline in bone repair capacity. *Calcif. Tissue Int.* **56**, 123–129.
34. Harada, S. and Rodan, G. A. (2003) Control of osteoblast function and regulation of bone mass. *Nature* **423**, 349–355.
35. Ryoo, H. M., Lee, M. H. and Kim, Y. J. (2006) Critical molecular switches involved in BMP-2-induced osteogenic differentiation of mesenchymal cells. *Gene* **366**, 51–57.
36. Ducy, P., Zhang, R., Geoffroy, V., Ridall, A. L. and Karsenty, G. (1997) Osf2/Cbfa1: a transcriptional activator of osteoblast differentiation. *Cell* **89**, 747–754.
37. Ishida, Y. and Heersche, J. N. M. (1998) Glucocorticoid-induced osteoporosis: both in vivo and in vitro concentrations of glucocorticoids higher than physiological levels attenuate osteoblast differentiation. *J. Bone Miner. Res.* **13**, 1822–1826.
38. Canalis, E. and Delany, A. M. (2002) Mechanisms of glucocorticoid action in bone. *Ann. N Y Acad. Sci.* **966**, 73–81.

39. Bellows, C. G., Aubin, J. E. and Heersche, J. N. M. (1987) Physiological concentrations of glucocorticoids stimulate formation of bone nodules from isolated rat calvaria cells – in vitro. *Endocrinology* **121**, 1985–1992.
40. Cheng, S. L., Yang, J. W., Rifas, L., Zhang, S. F. and Avioli, L. V. (1994) Differentiation of human bone-marrow osteogenic stromal cells in vitro – induction of the osteoblast phenotype by dexamethasone. *Endocrinology* **134**, 277–286.
41. Brann, D. W., Hendry, L. B. and Mahesh, V. B. (1995) Emerging diversities in the mechanism of action of steroid-hormones. *J. Steroid Biochem. Mol. Biol.* **52**, 113–133.
42. Boyan, B. D. and Schwartz, Z. (2004) Rapid vitamin D-dependent PKC signaling shares features with estrogen-dependent PKC signaling in cartilage and bone. *Steroids* **69**, 591–597.
43. Chen, T. L., Cone, C. M. and Feldman, D. (1983) Glucocorticoid modulation of cell-proliferation in cultured osteoblast-like bone-cells – differences between rat and mouse. *Endocrinology* **112**, 1739–1745.
44. Rickard, D. J., Sullivan, T. A., Shenker, B. J., Leboy, P. S. and Kazhdan, I. (1994) Induction of rapid osteoblast differentiation in rat bone-marrow stromal cell-cultures by dexamethasone and Bmp-2. *Dev. Biol.* **161**, 218–228.
45. Prince, M., Banerjee, C., Javed, A., Green, J., Lian, J. B., Stein, G. S., Bodine, P. V. and Komm, B. S. (2001) Expression and regulation of Runx2/Cbfa1 and osteoblast phenotypic markers during the growth and differentiation of human osteoblasts. *J. Cell Biochem.* **80**, 424–440.
46. Viereck, V., Siggelkow, H., Tauber, S., Raddatz, D., Schutze, N. and Hufner, M. (2002) Differential regulation of Cbfa1/Runx2 and osteocalcin gene expression by vitamin-D3, dexamethasone, and local growth factors in primary human osteoblasts. *J. Cell Biochem.* **86**, 348–356.
47. Pei, W., Yoshimine, Y. and Heersche, J. N. M. (2003) Identification of dexamethasone-dependent osteoprogenitors in cell populations derived from adult human female bone. *Calcif. Tissue Int.* **72**, 124–134.
48. Shui, C. X., Spelsberg, T. C., Riggs, B. L. and Khosla, S. (2003) Changes in Runx2/Cbfa1 expression and activity during osteoblastic differentiation of human bone marrow stromal cells. *J. Bone Miner. Res.* **18**, 213–221.
49. Franceschi, R. T., Yang, S., Rutherford, R. B., Krebsbach, P. H., Zhao, M. and Wang, D. (2004) Gene therapy approaches for bone regeneration. *Cells Tissues Organs* **176**, 95–108.
50. Rutherford, R. B., Moalli, M., Franceschi, R. T., Wang, D., Gu, K. and Krebsbach, P. H. (2002) Bone morphogenetic protein-transduced human fibroblasts convert to osteoblasts and form bone in vivo. *Tissue Eng.* **8**, 441–452.
51. Zhao, M., Zhao, Z., Koh, J. T., Jin, T. and Franceschi, R. T. (2005) Combinatorial gene therapy for bone regeneration: cooperative interactions between adenovirus vectors expressing bone morphogenetic proteins 2, 4, and 7. *J. Cell Biochem.* **95**, 1–16.
52. Yang, S., Wei, D., Wang, D., Phimphilai, M., Krebsbach, P. H. and Franceschi, R. T. (2003) In vitro and in vivo synergistic interactions between the

Runx2/Cbfa1 transcription factor and bone morphogenetic protein-2 in stimulating osteoblast differentiation. *J. Bone Miner. Res.* **18**, 705–715.
53. Hirata, K., Tsukazaki, T., Kadowaki, A., Furukawa, K., Shibata, Y., Moriishi, T., Okubo, Y., Bessho, K., Komori, T., Mizuno, A. et al. (2003) Transplantation of skin fibroblasts expressing BMP-2 promotes bone repair more effectively than those expressing Runx2. *Bone* **32**, 502–512.
54. Krebsbach, P. H., Gu, K., Franceschi, R. T. and Rutherford, R. B. (2000) Gene therapy-directed osteogenesis: BMP-7-transduced human fibroblasts form bone in vivo. *Hum. Gene Ther.* **11**, 1201–1210.
55. Byers, B. A., Pavlath, G. K., Murphy, T. J., Karsenty, G. and Garcia, A. J. (2002) Cell-type-dependent up-regulation of in vitro mineralization after overexpression of the osteoblast-specific transcription factor Runx2/Cbfal. *J. Bone Miner. Res.* **17**, 1931–1944.
56. Byers, B. A. and Garcia, A. J. (2004) Exogenous Runx2 expression enhances *in vitro* osteoblastic differentiation and mineralization in primary bone marrow stromal cells. *Tissue Eng.* **10**, 1623–1632.
57. Byers, B. A., Guldberg, R. E. and Garcia, A. J. (2004) Synergy between genetic and tissue engineering: Runx2 overexpression and in vitro construct development enhance in vivo mineralization. *Tissue Eng.* **10**, 1757–1766.
58. Gersbach, C. A., Byers, B. A., Pavlath, G. K. and Garcia, A. J. (2004) Runx2/Cbfa1 stimulates transdifferentiation of primary skeletal myoblasts into a mineralizing osteoblastic phenotype. *Exp. Cell Res.* **300**, 406–417.
59. Phillips, J. E., Gersbach, C. A., Wojtowicz, A. M. and Garcia, A. J. (2006) Glucocorticoid-induced osteogenesis is negatively regulated by Runx2/Cbfa1 serine phosphorylation. *J. Cell Sci.* **119**, 581–591.
60. Phillips, J. E., Guldberg, R. E. and Garcia, A. J. (2007). Dermal fibroblasts genetically modified to express Runx2/Cbfa1 are a mineralizing cell source for bone tissue engineering applications. **13**, 2029–2040.
61. Phillips, J. E., Guldberg, R. E. and Garcia, A. J. (2006). Mineralization capacity of Runx2/Cbfa1-expressing fibroblasts is scaffold-dependent. *Biomaterials.* **27**, 5535-5545.
62. Gersbach, C. A., Byers, B. A., Pavlath, G. K., Guldberg, R. E. and Garcia, A. J. (2004) Runx2/Cbfa1-genetically engineered skeletal myoblasts mineralize collagen scaffolds in vitro. *Biotechnol. Bioeng.* **88**, 369–378.
63. Sauer, G. R. and Wuthier, R. E. (1988) Fourier transform infrared characterization of mineral phases formed during induction of mineralization by collagenase-released matrix vesicles in vitro. *J. Biol. Chem.* **263**, 13718–13724.
64. Boskey, A. L., Guidon, P., Doty, S. B., Stiner, D., Leboy, P. and Binderman, I. (1996) The mechanism of beta-glycerophosphate action in mineralizing chick limb-bud mesenchymal cell cultures. *J. Bone Miner. Res.* **11**, 1694–1702.
65. Boyan, B. D., Schwartz, Z. and Boskey, A. L. (2000) The importance of mineral in bone and mineral research. *Bone* **27**, 341–342.

66. Bonewald, L. F., Harris, S. E., Rosser, J., Dallas, M. R., Dallas, S. L., Camacho, N. P., Boyan, B. and Boskey, A. (2003) von Kossa staining alone is not sufficient to confirm that mineralization in vitro represents bone formation. *Calcif. Tissue Int.* **72**, 537–547.
67. Bonewald, L. F., Dallas, S., Qiao, M., Camacho, N. P., Boyan, B. and Boskey, A. (1996) Comparison of mineralized matrix of fetal rat calvarial cells and MC-3T3-E1 osteoblast-like cells by Fourier transform infrared spectroscopy (FT-IR). *J. Bone Miner. Res.* **11**, S333.

21

In Vivo siRNA Delivery to the Mouse Hypothalamus Shows a Role of the Co-Chaperone XAP2 in Regulating *TRH* Transcription

Stéphanie Decherf, Zahra Hassani, and Barbara A. Demeneix

Summary

RNA interference mediated by small interfering RNAs (siRNAs) is a powerful tool for evaluating gene function *in vivo*. In particular it should be able to provide tissue-specific and developmental stage-specific knockdown of target genes in physiological contexts. However, there are few demonstrations of its use on neuronal specific genes *in vivo*. We recently developed a cationic lipid-based approach to study gene function in a neuronal context. In particular, we applied it to study how the novel partner for TRβ1, hepatitis virus B X-associated protein 2 (XAP2), a protein first identified as a co-chaperone protein, affects T3-transcriptional repression of the hypothalamic gene, *TRH*. The cationic lipid-based technique used, JetSI/DOPE, was previously shown to efficiently knockdown reporter gene mRNA *in vivo*. Using JetSI/DOPE to vectorize siRNA against XAP2 mRNA, we show that XAP2 is needed specifically for TRβ1-mediated (but not TRβ2) activation of hypothalamic TRH transcription. Thus, this cationic lipid-based siRNA strategy can effectively be used to reveal fine, tissue-specific and isoform-specific effects on neuronal gene transcription *in vivo*.

Key Words: TRH regulation; XAP2; gene transfer; RNAi; liposomal delivery; stereotaxic injections.

1. Introduction

Small interfering RNA (siRNA) duplexes function as sequence-specific mediators of RNA interference (RNAi), a post-transcriptional gene-silencing mechanism found in various animal species, including worms, flies, and

From: *Methods in Molecular Biology, vol. 433: Volume 1: Production and In Vivo Applications of Gene Transfer Vectors*
Edited by: J. M. Le Doux © Humana Press, Totowa, NJ

mammals reviewed in **refs. *1–4***. SiRNAs of 21–25 nucleotide RNA duplexes are produced from double-stranded RNAs (dsRNA), following digestion by an RNase III-like enzyme, Dicer ***(5)***. SiRNAs exert their gene-silencing activity through RNA-induced silencing complexes (RISCc) that degrade cognate mRNA ***(4,5)***; for review, see **ref. *6***.

The detection of RNAi activity in mammalian cells by long dsRNA has been hindered by the activation of a nonspecific pathway that represents a host response to viral infection. In this pathway, both dsRNA-dependent protein kinase (PKR) and 2′,5′-oligoadenylate synthetase are activated ***(7)***, resulting in a general inhibition of protein synthesis and nonspecific degradation of mRNA by RNaseL ***(8)***. However, direct delivery of 21-nucleotide siRNAs such as normally generated by dsRNA can bypass the activation of the nonspecific pathways leading to sequence-specific translation inhibition ***(9)***. This demonstration opened up immensely rich prospects for therapeutic applications as well as being a powerful tool for functional genomic investigations. However, whether applying siRNA technology to fundamental research or potential therapeutic ends, success hinges largely on the development of a delivery vehicle that can vectorize biologically competent siRNA to target cells. As in all in vivo gene delivery paradigms, there will be a need for optimization of delivery according to the route of administration, the tissue or cell type expressing the target gene, and whether long- or short-term effects are required. Furthermore, in the case of siRNA, one must consider two delivery mechanisms. Basically, on the one hand, one may use direct delivery of synthesized siRNA nucleotides with 3′ overhangs or, on the other, rely on introduction of a plasmid DNA encoding a hairpin construct (shRNA) that will be enzymatically degraded into siRNA in situ.

Whichever approach is chosen, the problem of delivery has to be considered and will again require optimization. This work only focuses on the siRNA delivery approach, while vectorization of shRNAs is currently undergoing optimization. The misadventures of gene therapy speak in favor of the former solution (i.e., direct delivery of siRNAs rather than plasmid DNA), where smaller, easily diffusible molecules have only to be delivered to the cytoplasm. Indeed, extracellular and intracellular diffusion as well as nuclear membrane crossing is known to severely hamper gene delivery. Furthermore, as the previous decade of work on gene delivery has amply shown, data on vectors obtained in in vitro settings will be of little use when transferring to the more stringent in vivo situations.

Thus, although very challenging from an experimental point of view, it is more useful in the long term to tackle the in vivo situation directly on the tissue targeted. A certain number of published in vivo RNAi experiments using siRNAs rely on the "hydrodynamic" method of delivery of naked nucleic acids in large volumes to hepatocytes ***(10,11)***. Although efficient, it seems difficult to transfer this technique to humans or to organs other than liver and skeletal muscles. Another recent study addressed delivery of naked siRNA to the peripheral nervous system for relief of chronic pain ***(12)***. However, large amounts of naked siRNA were required (400 μg/day) in perfusion over 6–7 days to achieve a 50% decrease in endogenous target gene mRNA levels. This result underscores the need to efficiently vectorize siRNA and calls for development of a synthetic vector strategy adaptable to small volumes. The central (CNS) and peripheral (PNS) nervous systems are particularly difficult targets for nucleic acid transfer and eventually gene therapy, mainly because of their restricted access due to the blood–brain barrier and their high degree of cellular diversity. Indeed, the brain is the tissue with the highest number of differentiated cell types. However, two factors make the CNS an exceedingly interesting target for gene delivery. First, CNS express a wide portion of genome, so that it is hardly feasible to knockout every gene of interest in a CNS-specific manner. Second, there are a number of endocrine regulations that target the CNS, and complex neuronal gene regulations are involved to reach an adapted and integrated response to various hormonal status.

Previous work in our laboratory has shown that intracranial delivery of siRNAs at picomolar amounts can inhibit exogenous gene expression (that of *Photinus pyralis*, carried by a plasmid with a *CMV* promoter) in newborn mice ***(13)***. Here we provide a methodology to address the problem of physiological regulation of a specific brain gene, using the example of hypothalamic thyrotropin-releasing hormone (*TRH*). *TRH* is a central regulator of the hypothalamo–hypophyseal–thyroid axis, with the final output being the thyroid hormones (TH) tetraiodothyronine (T_4) and triiodothyronine (T_3), T_3 being the biologically active form. TH homeostasis is vital during both development and maturity, and is maintained by negative feedback exerted by T_3 on TRH and thyroid-stimulating hormone (TSH) production in the hypothalamus and pituitary, respectively.

As for other target genes, T_3 modulates TRH transcription through its nuclear TH receptors (TRs). Among the TR subtypes, TRβ1 and TRβ2 (as opposed to TRα subtypes) are the key regulators of *TRH*. Both TRβ1 and TRβ2 are equally implicated in T_3-dependent repression of *TRH* ***(14–16)***, whereas TRβ1 has a more marked role in activation ***(17)***. This differential effect of TRβ1 and TRβ2 in regulating *TRH* transcription was assessed by two types of experiments. First, mutant mice invalidated for either one or both of the isoforms were studied, ***(16,18)*** and, second, *in vivo* effects of siRNAs that target differentially the two endogenous TRβ isoforms TRβ1 and TRβ2 on *TRH-luciferase* (*TRH-luc*) transgene transcription were examined ***(19)***.

We addressed the mechanisms of TRβ1 activation of TRH by investigating the role of a TRβ1-specific partner. One of the TRβ1 partners recently identified

(20) is the hepatitis virus B X-associated protein 2 (XAP2), also known as ARA9 or AIP *(21–23)*. XAP2 is known to modulate the transcriptional activity of the dioxin receptor (AhR) in vitro *(22)*, associating with hsp90 and regulating the intracellular localization of AhR *(24)*. However, no data have shown a physiological role for XAP2 or linked it to TRs.

We exploited the optimized JetSI/DOPE siRNA delivery technique in the newborn mouse's brain to examine whether it could be used to analyze the role of XAP2 in T_3-dependent regulation of *TRH* transcription. To this end, we examined the in vivo effects of siRNAs that target endogenous m*XAP2* on *TRH-luciferase* (*TRH-luc*) transgene transcription.

Our experiments showed that XAP2 exerts a functional role in modulating TRH transcription in vivo. By using a small inhibitory RNA (siRNA) knockdown approach, we showed that XAP2 is necessary for T_3-independent activation of TRH transcription mediated by TRβ1. Thus, our results show for the first time a functional and receptor isoform-specific in vivo role for XAP2. Thus, the JetSI™ siRNA delivery method is a particularly useful method for knockdown of specific mRNA in the brain, facilitating functional genomic studies without the need for producing complex lines of mutant mice.

2. Materials

2.1. Reagents and Plasmids

1. JetSI (40 mM): Resuspend JetSI (25 mg), provided as a lyophylized powder by Polyplus Transfection (Illkirsh, France), in 500 µl 100% ethanol to obtain a 40 mM solution.
2. Dioleoylphosphatidylethanolamine (DOPE; 80 mM): Resuspend 100 mg of DOPE (Sigma) in 300 µl chloroform and vortex vigorously until totally solubilized. Then, add 1360 µl of 100% ethanol to obtain an 80 mM solution of DOPE.
3. JetSI/DOPE: Mix 5 µl of 40 mM JetSI with 5 µl of 80 mM DOPE to obtain a 10 µl mix sample at a 1:2 molar ratio. Store at –20 °C in a screw-cap tube.
4. Glucose (25%): Dilute glucose (Sigma) to 25% in water, filter sterilize, and aliquot in single-use samples, and then store at –20 °C.
5. TRH-Luc plasmid (prepared by Plasmid Factory, Germany): TRH-Luc construct contains a rat TRH gene 5′ fragment extending from –544 to +84 base pairs (bp) cloned upstream of the *Photinus pyralis* luciferase-coding region.
6. pSG5-rTRb1 expression vector, this construct is described in **ref.** ***15***.
7. pSG5-rTRb2 expression vector, this construct is described in **ref.** ***17***. These vectors are available from B. Devergnes' laboratory in Lausanne, Switzerland.
8. siXAP2-1 is directed against mXAP2 nucleotides 489–509. siXAP2-1 is 21 bp dsRNAs designed using the HiPerformance Design Algorithm from Novartis AG (Qiagen). siXAP2-1 sequence is 5′-CACGTAGTCCTGTATCCTCTA-3′.

9. The second siRNA, siXAP2-2, is also designed using the HiPerformance Design Algorithm from Novartis AG and was obtained from Qiagen (SI00058562).
10. A scrambled sequence, siCT, is obtained from Qiagen. Its sequence is 5′-CGTCACGAGTACTCCGAAGTT-3′.
11. siRNAs are provided in a lyophilized form, stored at –20 °C in single-use samples at 20 mM concentration in siRNA buffer provided by Qiagen.

2.2. Stereotaxic Injections in Newborn Mice's Brains

1. One or two days old pups from male and female albinos mice (OF1) from Janvier (Le Genest St Isle, France) bred with ad libitum food and water.
2. Stereotaxic micro-injector (Narishige Instruments, Setagaya-Ku, Tokyo, Japan).
3. Thirty-microliter capillaries (Drummond Scientific, Polylaso, Strasbourg, France).

2.3. T3 Treatment

1. L-Triiodothyronine (T_3) is purchased from Sigma. Three milligrams of T_3 are dissolved in 200 ml of NaOH 0.1 M, and diluted in 0.9 % saline. This suspension at 3 mg/ml is stored at –20 °C for several months.
2. Hamilton 50- μl syringe.

2.4. Dissection

1. Petri dish, inverted and placed on ice in an ice-filled container.
2. Scalpel, two pairs of thin, curved forceps.
3. Eppendorf tubes containing 150μl of 4 °C ice-cold Reporter Lysis buffer 1× (Promega, Charbonniȇ-ses, France).
4. Manual potter for homogenization of tissue in luciferase buffer in Eppendorf tubes (e.g., battery-powered homogenizer, Kontes Vineland, New Jersey, USA).

2.5. Luciferase Assay

1. Luciferase Assay System kit (Promega).
2. Reporter Lysis 5× buffer from Promega.
3. Tube luminometer (LUMAT LB9507, Berthold Technologies, Bad Wildbad, Germany).
4. Tubes for luminometer.

3. Methods

In order to study gene function in an *in vivo* context using RNAi technology, one needs precise definition of the dose range and time course of the effect (siRNAs). Moreover, multiple controls are necessary to validate the specificity of the RNAi process.

In this nonviral in vivo gene transfer protocol, the main difficulty is in limiting inter-experiment variability. This implicates stringent conditions for the preparation of lipoplexes and polyplexes. Preparations must be done in less than 1 h before transfection (*see* **Note 1**).

3.1. Lipoplex Preparation

1. Keep all the solutions needed, except siRNAs, at room temperature for about half an hour before mix preparation: glucose 25%, L-PEI 100 mM, JetSI 40 mM, DOPE 80 mM, plasmid solutions. SiRNAs are kept on ice before pipetting.
2. Each newborn mouse receives 4 μl of lipoplex solution (*see* **Note 2**).
3. Lipoplexes preparation is carried out in three steps (*see* **Note 3**):

    a. DNA/siRNA formulation: Dilute the total plasmid DNA to complex in a 5% glucose solution to obtain a final concentration of 1 μg/ml for TRH-Luc and 100 ng/ml for pSG5-rTRβ1 or pSG5-rTRβ2 expression vector. Prepare a 800 nM siRNA solution in water. Mix equal volumes of DNA and siRNA solutions to obtain a concentration of 0.5 μg/ml for TRH-Luc, 50 ng/ml for pSG5-rTRβ1 or pSG5-rTRβ2 expression vector, and 400 nM siRNA. Note that in the following calculation, the amount of siRNA is negligible compared with that of plasmid DNA and so is ignored.
    b. Liposome formulation: Dilute a 10 μl sample of 40 mM JetSI and 80 mM DOPE fourfold in 100% ethanol to the respective concentrations of 10 mM and 20 mM. To prepare the transfection vector solution, we calculated the volume of 10 mM JetSI–20 mM DOPE to be used so as to obtain a ratio of 1.8 N/P DNA/lipid. The calculation can be done using the following formula, provided by Polyplus Transfection (http://www.polyplus-transfection.com/):

    $$\frac{x\ \mu\text{g of total DNA} * \text{N/P}}{5} = y\ \mu\text{l of the JetSI–DOPE mix}$$

    Dilute this JetSI–DOPE quantity in a 5% glucose solution of the same volume as the DNA/siRNA solution and thoroughly vortex. Leave the liposome solution at room temperature for 10 min.
    c. Lipoplexes formulation: Mix equal volumes of DNA/siRNA and liposome solutions and thoroughly vortex to get a 0.25 mg/ml TRH-luc, 25 ng/ml pSG5-rTRb1 or pSG5-rTRb2, and 200 nM siRNA lipoplex solution (*see* **Note 4**).

3.2. Stereotaxic Injections in the Newborn Mouse's Brain

1. Carry out transfections with 30-μl microcapillaries stretched to a final diameter of 1.5 μm and mounted on a micromanipulator. Check the regular conical shape of the micropipette with a binocular.
2. Anesthetize pups either 24 or 48 h old (weighing about 1.6–1.9 g) by hypothermia on ice before proceeding to the stereotaxic injections (*see* **Note 5**). For the

anesthesia, cover some ice with a piece of aluminium paper, and lay down a pup on it.

3. Lower a glass micropipette smoothly 2.5 mm through the skull, 1 mm lateral from the sagittal suture, and 2 mm anterior to lambda. Check the measures on the micromanipulator (*see* **Note 6**).
4. Inject slowly 2 μl of the 5% glucose solution containing the lipoplesces (*see* **Note 7**). Inject each pop bilaterally.
5. Keep the pups anesthetized on ice containing Petri dish during both injections.

3.3. T_3 Treatment

1. Dilute T_3 stock solution at 3 mg/ml 10 times in NaCl 9‰, to a 0.3 mg/ml final solution.
2. Treat each pup just after transfection with a 15 μl subcutaneous injection of T_3 (which corresponds to a treatment of 2.5 μg of T_3/g of body weight) or the same volume of saline solution (control) (*see* **Note 8**).

3.4. Dissection and Luciferase Assay

1. Decapitate injected newborn mice 20 h post transfection (i.e., at 48 h post natal). Remove the brain from the skull for precise dissection of the hypothalamus. Dissection is performed on an inverted ice-filled Petri dish so as to preserve the integrity of brain tissue and slow biochemical changes.
2. Place the brain ventral side up on the Petri dish, which is placed upside down in ice. The round shape of the hypothalamus is clearly visible just anterior to the cerebellum. With a scalpel, make a coronal cut (perpendicular to the anterior–posterior axis) at the anterior edge of the cerebellum. Remove and discard the cerebellum.
3. Then, make a coronal cut at the anterior edge of the hypothalamus at the optic chiasma level. Discard the anterior part of the brain.
4. Lay the obtained brain slice, containing the hypothalamic area, on its anterior surface. The round shape of the hypothalamus is clearly visible in the medio–ventral part of the brain slice. Cut parallel to the dorso–ventral axis at the lateral edges of the hypothalamus. Remove and discard the lateral cortex.
5. Then cut at 1 mm above the dorsal limit of hypothalamus. Remove and discard the dorsal cortex and thalamus.
6. Put the dissected hypothalamus in an Eppendorf tube containing 150 μl of 4 °C ice-cold Reporter Lysis buffer 1×.
7. Homogenize the hypothalamus in this solution with a manual potter (e.g., using a battery-powered homogenizer).
8. Perform Luciferase assay following the manufacturer's protocol (see results on **Fig. 1**) (*see* **Note 9**).

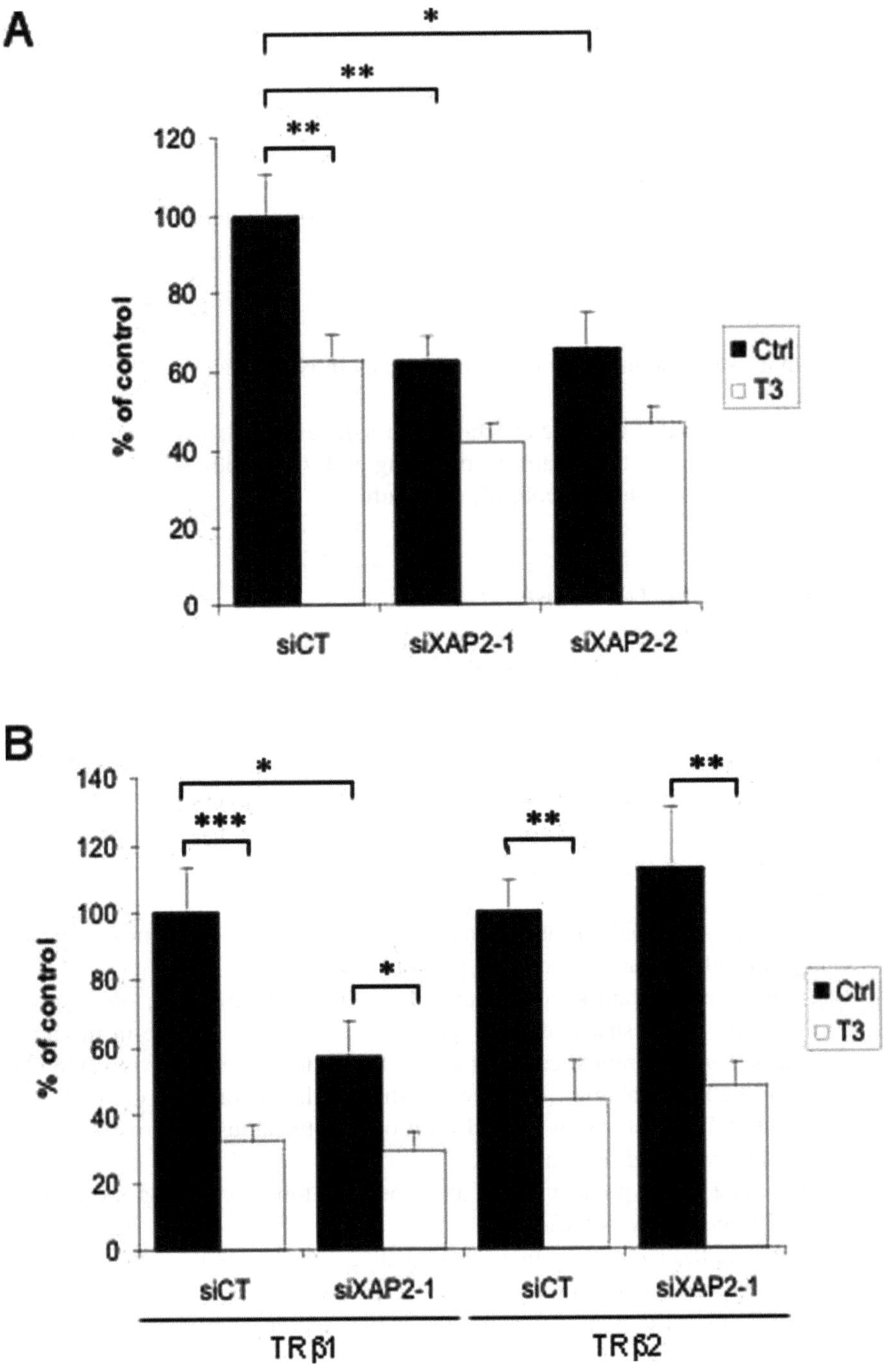
A
*
**
**
120
100
80
60
40
20
0
% of control
siCT
siXAP2-1
siXAP2-2
Ctrl
T3
B
*
***
*
**
**
140
120
100
80
60
40
20
0
% of control
siCT
siXAP2-1
siCT
siXAP2-1
TRβ1
TRβ2
Ctrl
T3

4. Notes

1. The lipoplexes size was measured in a time course experiment during the optimization process, from $t = 0$, time when the lipoplexes are prepared, until $t = 1$ h. The size of the complexes was found to be stable until 1 h. As longer times were not tested, we recommend to inject the lipoplexes within an hour after their formulation.
2. For *in vivo* gene delivery in the newborn or adult mouse's brain, the optimal final DNA concentration to be injected should not exceed 0.5 μg/μl.
3. The total volume of the lipoplex solution prepared for a series of injections should not exceed 60 μl. This allows better homogenization of the complexes. If a total volume of more than 60 μl is required for the experiment, multiple samples of <60 μl should be prepared and combined before use. For example, for a 150 μl solution, three 50 μl identical solutions should be prepared and mixed after preparation of the complexes. For gene delivery in the brain using JetSI–DOPE, optimal transfection implies an optimal charge ratio between negative charges (carried by DNA phosphates) and positive charges (carried by L-PEI) of 1.8 (N/P = 1.8). The molar ratio between JetSI and DOPE is optimal at 1:2; that is why they are kept in a mixed preparation at 40 and 80 mM, respectively.
4. Just after lipoplexes formulation by vortexing, carefully check if the solution does not contain white flakes of precipitated DNA. The final solution can also seem opaque or cloudy. If it does, start the formulation process again until the solution obtained is clear and transparent, which means that the lipoplexes solution is homogenous.

Fig. 1. In the mouse hypothalamus, hepatitis virus B X-associated protein 2 (XAP2) small inhibitory RNA (siRNA) affects only TRβ1-mediated activation of thyrotropin-releasing hormonegene transcription. (**A**) Two different XAP2 siRNAs affect the activation but not the T_3-dependent repression of TRH transcription. TRH-luc transcription was measured in hypothyroid 2-day-old mice, 20 h after hypothalamic injection of 1 μg reporter construct, 100 ng of TRβ1 expression vector, and 200 nM of siXAP2-1, siXAP2-2 or scrambled siRNA (siCT) complexed with JetSI/DOPE (*see* **Subheading 3**). After transfection, mice were treated with T_3 (2.5 μg/g of body weight, open bars) or saline (filled bars). Values are means ± SEM, $n \geq 12$; $*p < 0.05$; $**p < 0.01$. The experiment was repeated twice with the same results. (**B**) XAP2 affects transcription mediated by TRβ1 but not by TRβ2. TRH-luc transcription was measured in hypothyroid 2-day-old mice, 20 h after hypothalamic injection of 1 μg reporter construct, 100 ng of TRβ1 or TRβ2 expression vector, and 200 nM of siXAP2-1 or scrambled siRNA (siCT) complexed with JetSI/DOPE. After transfection, mice were treated with T_3 (2.5 μg/g of body weight, open bars) or saline (filled bars). Values are means ± SEM, $n \geq 6$; $*p < 0.05$; $**p < 0.01$. TRβ1 and TRβ2 results are from different sets of experiments. The experiment was repeated twice showing the same results.

5. For the anesthesia, cover some ice with a piece of aluminium paper, and lay down a pup on it. It will go into hypothermia within 5 min. Inject the pup when it becomes immobile. Do not leave the pup on ice for too long, otherwise it can die from hypothermia. If the color of the pup's body fades, take it off the ice and allow it to warm back to room temperature.
6. Stereotaxic coordinates given here were checked by brain dissection and measurement.
7. When lowering the micropipette in the brain, maintain the pup's head under the micromanipulator with a slight pressure. Then inject slowly the lipoplexes (over about 20–30 s). Before lifting the micropipette out of the skull, reduce the slight pressure of your fingers on the skull. Once the micropipette is out, check the head is dry, with no drop coming out of the injection site. If it is the case, do not inject again at this site, mark the pup to identify it as being badly injected, and continue the experiment. At most, only a small drop of blood should be seen, marking the injection point.
8. For the subcutaneous hormone injection, fold the back skin of the pup and insert the tip of the needle in the fold. At 1 day old, the pup's skin is very thin, so that the needle can be seen under the skin. Slowly inject the T3 solution, and carefully withdraw the needle to avoid pulling out the injected solution, with no liquid escaping from the injection site. If it is the case, or if the pup bleeds, do not inject again at this site, mark the pup to identify it, and continue the experiment. The hormonal injection is done subcutaneously only because of the size of the pup. It does not allow intraperitoneal injection, which could be lethal.
9. It is possible to investigate the effect of siRNAs on two genes at a time by using the Dual Luciferase Assay System (Promega), which allows analysis of two different reporter genes expression in the same sample.

References

1. Gonczy P, Echeverri C, Oegema K, Coulson A, Jones SJ, Copley RR, Duperon J, Oegema J, Brehm M, Cassin E, Hannak E, Kirkham M, Pichler S, Flohrs K, Goessen A, Leidel S, Alleaume AM, Martin C, Ozlu N, Bork P, Hyman AA (2000) Functional genomic analysis of cell division in C. elegans using RNAi of genes on chromosome III. *Nature* **408**, 331–336.
2. Fire A, Xu S, Montgomery MK, Kostas SA, Driver SE, Mello CC (1998) Potent and specific genetic interference by double-stranded RNA in Caenorhabditis elegans. *Nature* **391**, 806–811.
3. Hannon GJ (2002) RNA interference. *Nature* **418**, 244–251.
4. Hammond SM, Bernstein E, Beach D, Hannon GJ (2000) An RNA-directed nuclease mediates post-transcriptional gene silencing in Drosophila cells. *Nature* **404**, 293–296.
5. Bernstein E, Caudy AA, Hammond SM, Hannon GJ (2001) Role for a bidentate ribonuclease in the initiation step of RNA interference. *Nature* **409**, 363–366.

6. Dykxhoorn DM, Novina CD, Sharp PA (2003) Killing the messenger: short RNAs that silence gene expression. *Nat Rev Mol Cell Biol* **4**, 457–467.
7. Taylor DR, Shi ST, Romano PR, Barber GN, Lai MM (1999) Inhibition of the interferon-inducible protein kinase PKR by HCV E2 protein. *Science* **285**, 107–110.
8. Zhou A, Paranjape J, Brown TL, Nie H, Naik S, Dong B, Chang A, Trapp B, Fairchild R, Colmenares C, Silverman RH (1997) Interferon action and apoptosis are defective in mice devoid of 2´,5´-oligoadenylate-dependent RNase L. *EMBO J* **16**, 6355–6363.
9. Elbashir SM, Harborth J, Lendeckel W, Yalcin A, Weber K, Tuschl T (2001) Duplexes of 21-nucleotide RNAs mediate RNA interference in cultured mammalian cells. *Nature* **411**, 494–498.
10. McCaffrey AP, Meuse L, Pham TT, Conklin DS, Hannon GJ, Kay MA (2002) RNA interference in adult mice. *Nature* **418**, 38–39.
11. Song E, Lee SK, Wang J, Ince N, Ouyang N, Min J, Chen J, Shankar P, Lieberman J (2003) RNA interference targeting Fas protects mice from fulminant hepatitis. *Nat Med* **9**, 347–351.
12. Dorn G, Patel S, Wotherspoon G, Hemmings-Mieszczak M, Barclay J, Natt FJ, Martin P, Bevan S, Fox A, Ganju P, Wishart W, Hall J (2004) siRNA relieves chronic neuropathic pain. *Nucleic Acids Res* **32**, e49.
13. Hassani Z, Lemkine GF, Erbacher P, Palmier K, Alfama G, Giovannangeli C, Behr JP, Demeneix BA (2005) Lipid-mediated siRNA delivery down-regulates exogenous gene expression in the mouse brain at picomolar levels. *J Gene Med* **7**, 198–207.
14. Guissouma H, Ghorbel MT, Seugnet I, Ouatas T, Demeneix BA (1998) Physiological regulation of hypothalamic TRH transcription in vivo is T3 receptor isoform specific. *FASEB J* **12**, 1755–1764.
15. Guissouma H, Dupre SM, Becker N, Jeannin E, Seugnet I, Desvergne B, Demeneix BA (2002) Feedback on hypothalamic TRH transcription is dependent on thyroid hormone receptor N terminus. *Mol Endocrinol* **16**, 1652–1666.
16. Abel ED, Ahima RS, Boers ME, Elmquist JK, Wondisford FE (2001) Critical role for thyroid hormone receptor beta2 in the regulation of paraventricular thyrotropin-releasing hormone neurons. *J Clin Invest* **107**, 1017–1023.
17. Dupre SM, Guissouma H, Flamant F, Seugnet I, Scanlan TS, Baxter JD, Samarut J, Demeneix BA, Becker N (2004) Both thyroid hormone receptor (TR)beta 1 and TR beta 2 isoforms contribute to the regulation of hypothalamic thyrotropin-releasing hormone. *Endocrinology* **145**, 2337–2345.
18. Forrest D, Hanebuth E, Smeyne RJ, Everds N, Stewart CL, Wehner JM, Curran T (1996) Recessive resistance to thyroid hormone in mice lacking thyroid hormone receptor beta: evidence for tissue-specific modulation of receptor function. *EMBO J* **15**, 3006–3015.
19. Guissouma H, Froidevaux MS, Hassani Z, Demeneix BA (2006) In vivo siRNA delivery to the mouse hypothalamus confirms distinct roles of TR beta isoforms in regulating TRH transcription. *Neurosci Lett* **406**, 240–243.

20. Froidevaux MS, Berg P, Seugnet I, Decherf S, Becker N, Sachs LM, Bilesimo P, Nygard M, Pongratz I, Demeneix BA (2006) The co-chaperone XAP2 is required for activation of hypothalamic thyrotropin-releasing hormone transcription in vivo. *EMBO Rep* **7**, 1035–1039.
21. Carver LA, Bradfield CA (1997) Ligand-dependent interaction of the aryl hydrocarbon receptor with a novel immunophilin homolog in vivo. *J Biol Chem* **272**, 11452–11456.
22. Ma Q, Whitlock JP, Jr. (1997) A novel cytoplasmic protein that interacts with the Ah receptor, contains tetratricopeptide repeat motifs, and augments the transcriptional response to 2,3,7,8-tetrachlorodibenzo-p-dioxin. *J Biol Chem* **272**, 8878–8884.
23. Meyer BK, Pray-Grant MG, Vanden Heuvel JP, Perdew GH (1998) Hepatitis B virus X-associated protein 2 is a subunit of the unliganded aryl hydrocarbon receptor core complex and exhibits transcriptional enhancer activity. *Mol Cell Biol* **18**, 978–988.
24. Kazlauskas A, Sundstrom S, Poellinger L, Pongratz I (2001) The hsp90 chaperone complex regulates intracellular localization of the dioxin receptor. *Mol Cell Biol* **21**, 2594–2607.

22

Efficient Retroviral Gene Transfer to Epidermal Stem Cells

Pedro Lei and Stelios T. Andreadis

Summary

Skin is an attractive target for gene modification to treat skin diseases, wound healing, or even systemic disorders. Although retroviral transduction results in permanent genetic modification, differentiation and eventually loss of the transduced cells from the epidermis and in temporary transgene expression. Therefore, it is important to develop methods that promote gene transfer to epidermal stem cells, which self-renew and regenerate the epidermis for extended periods of time. Here we describe an efficient protocol that results in high levels of retroviral gene transfer to human epidermal stem cells by immobilizing retrovirus on a recombinant fibronectin (rFN) fragment. In contrast to the traditional method, transduction on rFN promotes gene transfer to epidermal stem cells and prevents loss of clonogenic potential due to exposure of cells to retroviral supernatant. Notably, transduction on rFN does not require addition of toxic polycations such as polybrene. Overall this method provides a simple, fast, and efficient means to modify human epidermal stem cells for cutaneous gene therapy and for biological studies that require stable genetic modification.

Key Words: Fibronectin; retrovirus; gene therapy; human keratinocytes; epidermal stem cells.

1. Introduction

The skin is the largest organ and outermost layer of the body. Therefore, it has great potential for gene modification and serves as "bioreactor" to deliver therapeutic proteins in vivo. In fact some investigators have used

From: *Methods in Molecular Biology, vol. 433: Volume 1: Production and In Vivo Applications of Gene Transfer Vectors*
Edited by: J. M. Le Doux © Humana Press, Totowa, NJ

modified skin cells to treat a variety of skin-related diseases such as epidermolysis bullosa and lamellar ichthyosis ***(1–5)*** and other systemic disorders such as hemophilia ***(6,7)***, adenosine deaminase (ADA)-deficiency ***(8)***, and leptin deficiency ***(9)***. Additionally, epidermal keratinocytes have been modified to overexpress growth factors including platelet-derived growth factor-A (PDGF-A) ***(10)***, insulin-like growth factor (IGF)-I ***(11)***, keratinocyte growth factor (KGF) ***(12)***, and vascular endothelial growth factor (VEGF) ***(13,14)*** and examine their potential in enhancing wound healing.

Despite these promising results, long-term transgene expression from the skin remains inefficient. One possible explanation is that transduced cells are lost in vivo due to fast turnover of the skin (~1 month). Hence the only way to ensure sustained transgene expression is to deliver therapeutic genes to the keratinocyte stem cell compartment, which has unlimited proliferative potential, and therefore can repopulate the epidermis while retaining transgene expression.

Here we describe a retroviral gene transfer protocol that utilizes recombinant fibronectin (rFN) fragment (RetroNectin) to promote gene transfer to keratinocytes ***(15)***. This protocol yields up to 90% transduced cells. To demonstrate transduction to the stem cells, we employ a clonal assay that identifies stem cells by the ability to form holoclones—colonies that are large enough to demonstrate high clonogenic potential. Using this assay, we show that at least 50% transduced keratinocytes display high proliferative potential. By contrast, less than 20% of transduced cells show high proliferative potential when using traditional transduction protocol, i.e., transduction on tissue culture plastics (TTCP) without fibronectin. Notably, cells transduced on fibronectin yield higher number of holoclones than cells transduced on tissue culture plastic (97 vs. 61 colonies per 800 cells plated) (*see* **Fig. 1**). These data demonstrate that fibronectin can facilitate gene transfer to epidermal stem cells, which have the potential for achieving long-term transgene expression in vivo.

1.1. Retronectin (CH296)

Retronectin, also termed as CH296, is commercially available through Takara Mirus (Madison, MI). It is a rFN fragment that consists of three components: cell-binding domain (C-domain), heparin-binding domain (H-domain), and CS-1 sequence. Cells have been shown to bind to the C-domain and CS-1 via the integrin receptors $\alpha5\beta1$ and $\alpha4\beta1$, respectively ***(16)***. On the contrary, retroviruses are shown to bind to the H-domain of retronectin ***(16)***, via the virus-surface heparan sulfate proteoglycans ***(17)***.

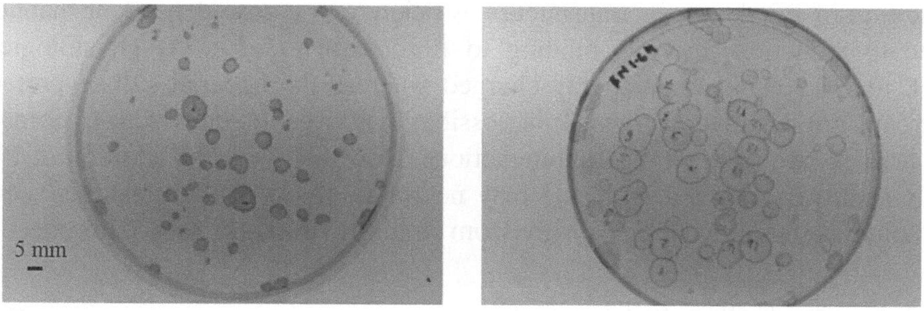

(a) (b)

Fig. 1. Adherence of keratinocytes to recombinant fibronectin (rFN) increased the number of holoclones. Human primary keratinocytes were transduced either on (**A**) TCP or on (**B**) rFN. When the cells reached 80–90% confluence, they were trypsinized and replated at clonal densities (800 cells per 100-mm tissue culture dish) on mitomycin-C-treated 3T3/J2 feeder layers. After 14 days in culture, cells were washed, fixed, and counterstained with hematoxylin. Colonies greater than 5 mm in diameter were visually identified and GFP^+ holoclones were quantified using a fluorescence microscope. Representative plates with keratinocyte colonies are shown.

1.2. Fibronectin-Mediated Gene Transfer

1.2.1. Transduction on Fibronectin with Immobilized Retrovirus

Early studies used the fibronectin fragment (CH296) to capture retrovirus and facilitate gene transfer to suspension cells such as hematopoietic stem cells and T lymphocytes ***(18–23)***. Briefly fibronectin was first coated on a plastic surface. A fresh stock of retrovirus was then added and the virus was allowed to diffuse and bind to fibronectin. After removal of unbound viruses and other transduction inhibitors such as chondroitin sulfate ***(24)***, target cells were added to initiate gene transfer.

In this setup, the level of active virus captured on the surface depends on several parameters, namely the diffusion and decay of virus in solution. By examination of these rate parameters, we found that virus immobilization is a strong function of both time and temperature. At high temperature (37 °C) due to the slow diffusion and fast decay of virus ***(25–27)***, the optimum level of active virus was captured on the surface in about 7 h. On the contrary, at lower temperature (25 °C), as virus decay is slower, higher levels of active virus could be achieved with longer incubation times (24 h) ***(26)***.

To further improve virus binding, we examined the virus–fibronectin interaction and found that virus binds to the H-domain via viral surface heparan sulfate proteoglycans. Surprisingly, in contrast to traditional transduction protocols, polycations such as polybrene (PB) was not required for transduction. In fact, addition of PB decreases transduction efficiency in a dose-dependent manner, due to reduction of virus binding to fibronectin ***(17,26)***. At physiological pH, the H-domain is positively charged whereas the heparan sulfate proteoglycans are negative charged. It is possible that virus interacts with H-domain at least in part by electrostatic interactions. However, with the addition of PB, this positively charged molecule may neutralize the negatively charged virus in solution, thus preventing the virus from binding to the H-domain of fibronectin.

1.2.2. Transduction on Fibronectin

In transduction on fibronectin with immobilized retrovirus (TFIR), virus is loaded on the substrate followed by addition of target cells. However, transduction on fibronectin (TFN) can also take place with target cells plated on fibronectin before viruses are added. The difference is that in TFN, cells are already in the proliferation cycle when they encounter the virus, which is necessary for retroviral transduction. Despite the difference in the cell status, TFIR and TFN show no significant difference in transduction efficiency ***(28)***. However, the kinetics of these processes varies dramatically. Whereas TFIR requires at least 7 h of virus loading to achieve maximum gene transfer, TFN takes less than 2 h to obtain similar transduction efficiency ***(26,28)***. Additionally, because target cells are present during virus loading, gene transfer kinetics of TFN depends on the origin of the virus. In particular, whereas virus originated from human packaging cells yields similar gene transfer levels for 2–8 h loading, virus from mouse packaging cells reached a maximum at short times (0.5–2 h) but decreased at longer times ***(15,28)***. The reason for such a decrease with time is that conditioned medium from virus producer cells contains toxic metabolic bi-products that affect epidermal cell viability. Finally, in contrast to TFIR, addition of PB did not affect the level of gene transfer in this system, possibly because the virus can directly interact with cells on fibronectin rather than binding to fibronectin before encountering the cells ***(17,28)***.

1.2.3. Combined Transduction Protocol (TFIR/TFN)

The gene transfer efficiency of either TFIR or TFN is 40–60%. To obtain higher transduction levels, one should combine these two protocols in sequential order, i.e., TFIR followed by TFN. Notably, the level of gene transfer in this protocol is sensitive to target cell density. At very low cell densities (2500–5000 cells/cm^2), transduction efficiency is high but the cells show very slow proliferation rates. As cell densities increase, however, transduction efficiencies decrease monotonically. Hence the optimal transduction target cell density is found to be 12,500 cells/cm^2, the density at which gene transfer remains high without compromising cell growth rates. Using these settings, one can easily obtain about 80% of transduced keratinocytes. Thus this protocol provides a convenient way to obtain high levels of transduced primary cells without drug selection and at the same time preserves the "stemness" of epidermal cells during gene modification in vitro.

1.3. Clonal Assay

To evaluate the efficiency of gene transfer to the stem cell compartment, we employed a clonal assay, which identifies stem cells by their ability to form holoclones, i.e., colonies of large enough size to demonstrate high clonogenic potential ***(15)***. Following the work of Barrandon and Green, we defined a holoclone as a colony of keratinocytes that was larger than 5.0 mm in diameter (~40,000 cells) ***(29)***. After modifying keratinocytes on fibronectin, we replated the cells at clonal densities (800 or 1600 cells/100-mm dish) and allowed them to grow and form isolated colonies. In 14 days, colonies larger than 5.0 mm in diameter were picked and evaluated for transgene expression. As each holoclone is composed of the progeny of a stem cell, the fraction of transgene-expressing holoclones should reflect the fraction of transgene-expressing stem cells.

2. Materials

2.1. Cell Culture

1. PA317-LXCGN cells (Steven Greenberg, Roswell Park Cancer Institute, Buffalo NY).
2. Phoenix-Ampho cells [American Type Culture Collection (ATCC), Manassas, VA].
3. 3T3/J2 (ATCC).
4. Dulbecco's modified Eagle's medium (DMEM; Gibco BRL, Grand Island, NY).
5. Antibiotic-Antimycotic (Gibco BRL).
6. Heat-inactivated calf serum [HICS: Bovine Calf Serum (Hyclone, Logan, UT) incubated at 55 °C for 30 min with mixing every 5 min].
7. Fetal bovine serum (FBS; Gibco BRL).
8. Bovine serum (BS; Gibco BRL).
9. Tissue culture flasks (75-cm^2; BD Falcon, Bedford, MA).
10. Mitomycin-C (Sigma, St. Louis, MO).
11. Phosphate-Buffer Saline (PBS).
12. Keratinocyte Culture Medium (KCM): DMEM/Ham's F-12 (Gibco) (3:1 ratio), FBS (10%), Antibiotic-Antimycotic (1%), Adenine (Sigma) (24 mg/l), cholera toxin (*Vibrio cholerae*, Type Inaba 569 B; Calbiochem, La Jolla, CA) (10^{-10} M), Hydrocortisone (Calbiochem) (0.4 µg/ml), insulin (Sigma) (5 µg/ml), transferrin, (Boehringer Mannheim, Indianapolis, IN) (5 µg/ml), triiodo-L-thyronine (Sigma) (2×10^{-9} M).

13. Tissue culture dishes (60-mm tissue culture dishes; Becton Dickinson Labware, Franklin Lakes, NJ).
14. Conical tubes (15- and 50-ml; Laboratory Product Sales, Rochester, NY).
15. Trypsin (0.25%) containing 0.5 mM EDTA (Gibco BRL).
16. Serological pipettes (5-ml; Corning, NY).
17. Versene: 0.48 mM EDTA in PBS.
18. Epidermal growth factor (EGF; BD Biosciences, Mountain View, CA).
19. Keratinocyte-Serum Free Medium (Keratinocyte-SFM, Gibco BRL).
20. Tissue Culture Roller Drum (TC-8, Brunswick Scientific Co. Inc., Edison, NJ).

2.2. Recombinant Retrovirus Production

1. Phoenix-Ampho cells.
2. PA317-LXCGN cells.
3. Tissue culture flasks (75- and 150-cm^2).
4. DMEM.
5. Microcentrifuge tubes (1.5-ml).
6. FUGENE 6 (Roche Diagnostics, Indianapolis, IN).
7. pBMN-I-GFP plasmid (Gary Nolan, Stanford University, Palo Alto, CA).
8. 500-ml bottle top cellulose acetate filters (0.45-μm pore sized; Corning).
9. Antibiotic-Antimycotic.
10. FBS.

2.3. Retroviral Mediated Gene Modification of Human Primary Keratinocytes on Recombinant Fibronectin

1. Non-tissue culture plates (24 wells; BD Falcon).
2. Retronectin (rFN; Takara Mirus).
3. Human primary keratinoctyes.
4. Keratinocyte-SFM (Gibco BRL).

2.4. Measurement of Transduction Efficiency by Flow Cytometry

1. PBS.
2. Trypsin.
3. DMEM + 1% Antibiotic-Antimycotic + 10% FBS.
4. Propidium iodine (PI; Molecular Probes, Grand Island, NY).

2.5. Clonal Analysis

1. 3T3/J2 cells.
2. Mitomycin-C.
3. KCM.
4. EGF.
5. PBS.

6. Paraformaldehyde (Sigma).
7. Harris' Alum Hematoxylin (EMD Chemicals Inc., Gibbstown, NJ).
8. Kodak Electrophoresis Documentation and Analysis System 290 (Kodak EDAS 290; Kodak Company, Rochester, NY).
9. ImageJ (National Institutes of Health, Bethesda, MD USA).

3. Methods

3.1. Cell Culture

3.1.1. Maintenance of Virus Producer Cells

PA317-LXCGN virus-producing cells are cultured in DMEM + 1% Antibiotic-Antimycotic + 10% FBS, Phoenix-Ampho virus-packaging cells are cultured in DMEM + 1% Antibiotic-Antimycotic + 10% HICS, and 3T3/J2 feeder layer cells are cultured in DMEM + 1% Antibiotic-Antimycotic + 10% BS. All cells are cultured for no more than 12 passages for consistent results.

3.1.2. Isolation and Maintenance of Human Primary Keratinocytes

1. The day before isolation, plate 3T3/J2 feeder cells at $1.2–1.8 \times 10^6$ cells in a 75-cm^2 tissue culture flask (*see* **Note 1**).
2. On the day of isolation, treat 3T3/J2 cells with 15 ml mitomycin-C (15 μg/ml) for 2–3 h (*see* **Note 2**).
3. Wash the cells extensively with 5 ml PBS for three times (*see* **Note 3**).
4. Add 15 ml KCM to the flask and incubate for at least 1 h before use.
5. Obtain specimens of fresh newborn human foreskins from the hospital (*see* **Note 4**).
6. Add 1–3 ml PBS to a 100-mm tissue culture dish.
7. Place the tissue, with the epidermis down, on the Petri dish.
8. Trim the tissue to remove fat and connective tissue with a pair of sterile sharp rounded scissors (*see* **Note 5**).
9. Transfer the tissue into a 50-ml conical tube.
10. Rinse the tissue with 30 ml PBS for eight times (*see* **Note 6**).
11. Add 3 ml trypsin to a 60-mm tissue culture dish.
12. Transfer the tissue to the tissue culture dish.
13. Mince the tissue with a pair of sterile sharp rounded scissors for 10–15 min (*see* **Note 7**).
14. Add 3 ml trypsin to a 15-ml conical tube.
15. Pre-wet a 5-ml serological pipette with trypsin.
16. Transfer all the minced tissue into the 15-ml conical tube from **step 14**.
17. Wash the tissue culture dish with 3 ml versene.
18. Transfer the versene into the conical tube.
19. Repeat **steps 17** and **18** once. The conical tube should now contain 12 ml solution.

20. Leave the conical tube on a cell culture roller drum (1.0–1.5 rpm) at 37 °C.
21. After 30 min, remove the conical tube from the rotator and allow the tissue to settle down by gravity.
22. Carefully transfer the supernatant into a clean 15-ml tube.
23. Centrifuge the supernatant at 140×*g* for 5 min.
24. Resuspend the cells with 1 ml conditioned KCM (*see* **step 4**) and add to the feeder layer in the T-75 flask.
25. Add 6 ml trypsin and 6 ml versene to the minced tissue and return the conical tube on the rotator at 37 °C.
26. Repeat **steps 21–25** for three more times.
27. Three days after isolation, replace cells with KCM + 10 ng/ml EGF.
28. Check cells daily.
29. Before cell colonies start to merge, subculture cells at 1/10 dilution either on mitomycin-C-treated 3T3/J2 feeder layers (*see* **steps 1–4**) or in keratinocyte-SFM (*see* **Note 8**).

3.2. Recombinant Retrovirus Production

3.2.1. Virus Production by Transient Transfection

1. Plate retrovirus packaging cells (Phoenix-Ampho) at 1.875×10^6 cells per 75-cm^2 tissue culture flask (*see* **Note 9**).
2. On the next day, add 100 μl DMEM with no antibiotics or serum into a sterile 1.5-ml microcentrifuge tube.
3. Add 22.5 μl FUGENE 6 to the microcentrifuge tube and incubate at room temperature for 5 min.
4. Add 7.5 μg plasmid (pBMN-I-GFP) to the FUGENE 6 mixture.
5. Incubate at room temperature for 45 min.
6. Add the FUGENE 6/DNA complex dropwise to the Phoenix-Ampho cells.
7. After overnight incubation, replace cells with fresh medium (DMEM + 1% Antibiotic-Antimycotic + 10% HICS).
8. Harvest virus-containing supernatant in 24 h, filter through 0.45-μm pore sized filters, aliquot, and store at –80 °C until use (*see* **Note 10**).
9. Replenish cells with fresh medium.
10. Repeat **step 8**.

3.2.2. Virus Production from Permanent Producer Cells

1. Culture virus-producing cells (PA317-LXCGN) in 150-cm^2 tissue culture flask until they reach confluence.
2. Replenish cells with fresh medium (DMEM + 1% Antibiotic-Antimycotic + 10% FBS).
3. Harvest virus-containing supernatant in 24 h, filter through 0.45-μm pore sized filters, aliquot, and store at –80 °C until use.
4. Repeat **steps 2–3** once.

3.3. Retroviral Mediated Gene Modification of Human Primary Keratinocytes on Recombinant Fibronectin

1. Coat the wells of a 24-well non-tissue culture treated plate with 500 μl rFN (10 μg/ml). Leave the plate at 4 °C overnight (*see* **Note 11**).
2. Load 250 μl virus at 37 °C and with 10% CO_2 for 6–7 h (*see* **Note 12**).
3. Add 25,000 keratinocytes in 0.5 ml keratinocyte-SFM and incubate at 37 °C and with 10% CO_2 overnight (*see* **Note 13**).
4. On the next day, load 250 μl of fresh virus at 37 °C and with 10% CO_2 for 2 h (*see* **Note 14**).
5. Replace the virus with 500 μl fresh keratinocyte-SFM.
6. Thereafter, change medium every 2 days until they reach 80–90% confluence.
7. Determine the transduction efficiency by flow cytometry (as described below in **Subheading 3.4.**).

3.4. Measurement of Transduction Efficiency by Flow Cytometry

1. When cells reach confluence, wash once with PBS.
2. Detach cells from the surface with trypsin.
3. Inactivate trypsin with medium containing serum (e.g., DMEM + 1% Antibiotic-Antimycotic + 10% FBS).
4. Centrifuge cells at 140×*g* for 5 min.
5. Resuspend cells in PBS containing 1 μg/ml PI. Leave cells on ice for 15 min.
6. Determine the fraction of GFP^+ cells and GFP^- cells by flow cytometry (*see* **Note 15**).
7. Calculate transduction efficiency (TE):

$$TE = \frac{\%\text{GPF}^{+}\text{cells}}{\%\text{GPF}^{-}\text{cells} + \%\text{GPF}^{+}\text{cells}} \times 100\%$$

3.5. Clonal Analysis

1. When cells reach confluence, take a small sample of cells and determine the transduction efficiency by flow cytometry (*see* **Subheading 3.4.**).
2. Plate cells at clonal density (800–1600 cells) on mitomycin-C-treated 3T3/J2 cells in a 100-mm tissue culture dish (for mitomycin-C treatment, refer to **Subheading 3.1.2., steps 1–4**).
3. Replenish cells with fresh medium (KCM + 10 ng/ml EGF) every 3 days.
4. Fourteen days after cell seeding, remove culture medium and wash cells once with PBS.
5. Fix cells with 4% paraformaldehyde in PBS for 10 min at room temperature.
6. Stain cells with hematoxylin and photograph the entire dish area at a fixed distance using Kodak EDAS 290 imaging and documentation system.

7. Determine the area of each clone using ImageJ and use it to calculate the effective diameter. Colonies with diameter greater than 5 mm are defined as holoclones (*see* **Note 16**).
8. Determine the fraction of GFP$^+$ holoclones using a fluorescence microscope (*see* **Note 17**).

4. Notes

1. Depending on the time the cells are plated, the seeding density varies. If cells are plated early in the morning, 1.2×10^6 cells/flask is sufficient. On the contrary, if cells are plated late in the day, 1.8×10^6 cells/flask is required. This seeding density will ensure >95% confluence on the next day.
2. Mitomycin-C inhibits proliferation of 3T3/J2 cells, which would otherwise outgrow keratinocytes. However, mitomycin-C-treated fibroblasts can still produce soluble factors that promote keratinocyte proliferation and clonogenic potential.
3. It is important to wash away all the mitomycin-C to avoid inhibition of keratinocyte proliferation.
4. Immediately after surgery, human foreskins are collected in 50-ml conical tubes containing 30 ml of KCM and transported to the laboratory on ice. Although it is best to isolate cells on the same day, we have successfully isolated cells from tissues that were stored at 4 °C for up to 2 days post-harvest. To avoid disease transmission from human tissue, wear gloves at all times. After use, surgical instruments such as scissors should be treated with bleach, washed thoroughly with soap and water, and autoclaved. Disposable items such as cell culture dishes should be discarded in proper biohazard bags.
5. It is important to remove *all* connective tissue to avoid fibroblast contamination and obtain a more homogenous keratinocyte culture.
6. This washing step is required to remove contaminants from the tissue.
7. It is important to cut the tissue into as small pieces as possible to increase the yield of keratinocyte isolation.
8. Do not let all the colonies merge with each other as keratinocytes differentiate resulting in loss of proliferative potential. In addition, do not allow cells to go beyond passage 5 because they undergo terminal differentiation.
9. It is important to plate cells at a density that will give 50–70 % confluence at the day of transfection.
10. Filtration through 0.45-μm pore size filters removes cell debris from the virus-containing supernatant. Do not use filters with smaller pore size (i.e., 0.22 μm) as they also capture retroviral particles (0.1 μm diameter) resulting in significant loss of virus titer. Alternatively, virus-containing supernatant can be centrifuged at $200\times g$ for 5 min to remove cell debris.
11. Non-tissue culture-treated plate is required for consistent rFN coating.
12. Because of decay of retroviruses, the optimum loading time at 37 °C is about 7 h ***(26)***.

13. The optimum transduction efficient is obtained at 12,500 cells/cm^2. At higher cell densities, the transduction efficiency decreases ***(15)***.
14. The optimum transduction for this step is 2 h. Transduction for longer times (especially with retroviruses of producer cells of mouse origin) reduces transduction efficiency, probably due to toxic effects of the virus supernatant on keratinocytes ***(15)***.
15. Only viable (PI negative) cells are used to evaluate the transduction efficiency.
16. Cells that are capable of forming colonies >5 mm in diameter (~40,000 cells) are defined as holoclones—colonies of large enough size to demonstrate high clonogenic potential ***(29)***.
17. Transduction on rFN increased both the fraction of GFP$^+$ holoclones and the number of holoclones, suggesting that rFN may preserve the stemness of epidermal cells ***(15)***.

References

1. Choate, K. A., Kinsella, T. M., Williams, M. L., Nolan, G. P. and Khavari, P. A. (1996) Transglutaminase 1 delivery to lamellar ichthyosis keratinocytes. *Hum Gene Ther* **7**, 2247–2253.
2. Choate, K. A., Medalie, D. A., Morgan, J. R. and Khavari, P. A. (1996) Corrective gene transfer in the human skin disorder lamellar ichthyosis. *Nat Med* **2**, 1263–1267.
3. Freiberg, R. A., Choate, K. A., Deng, H., Alperin, E. S., Shapiro, L. J. and Khavari, P. A. (1997) A model of corrective gene transfer in X-linked ichthyosis. *Hum Mol Genet* **6**, 927–933.
4. Dellambra, E., Vailly, J., Pellegrini, G., Bondanza, S., Golisano, O., Macchia, C., Zambruno, G., Meneguzzi, G. and De Luca, M. (1998) Corrective transduction of human epidermal stem cells in laminin-5-dependent junctional epidermolysis bullosa. *Hum Gene Ther* **9**, 1359–1370.
5. Dellambra, E., Pellegrini, G., Guerra, L., Ferrari, G., Zambruno, G., Mavilio, F. and De Luca, M. (2000) Toward epidermal stem cell-mediated ex vivo gene therapy of junctional epidermolysis bullosa. *Hum Gene Ther* **11**, 2283–2287.
6. Alexander, M. Y., Bidichandani, S. I., Cousins, F. M., Robinson, C. J., Duffie, E. and Akhurst, R. J. (1995) Circulating human factor IX produced in keratin-promoter transgenic mice: a feasibility study for gene therapy of haemophilia B. *Hum Mol Genet* **4**, 993–999.
7. Page, S. M. and Brownlee, G. G. (1997) An ex vivo keratinocyte model for gene therapy of hemophilia B. *J Invest Dermatol* **109**, 139–145.
8. Fenjves, E. S., Schwartz, P. M., Blaese, R. M. and Taichman, L. B. (1997) Keratinocyte gene therapy for adenosine deaminase deficiency: a model approach for inherited metabolic disorders. *Hum Gene Ther* **8**, 911–917.
9. Larcher, F., Del Rio, M., Serrano, F., Segovia, J. C., Ramirez, A., Meana, A., Page, A., Abad, J. L., Gonzalez, M. A., Bueren, J., Bernad, A. and Jorcano, J. L. (2001) A cutaneous gene therapy approach to human leptin deficiencies: correction

of the murine ob/ob phenotype using leptin-targeted keratinocyte grafts. *FASEB J* **15**, 1529–1538.
10. Eming, S. A., Medalie, D. A., Tompkins, R. G., Yarmush, M. L. and Morgan, J. R. (1998) Genetically modified human keratinocytes overexpressing PDGF-A enhance the performance of a composite skin graft. *Hum Gene Ther* **9**, 529–539.
11. Eming, S. A., Snow, R. G., Yarmush, M. L. and Morgan, J. R. (1996) Targeted expression of insulin-like growth factor to human keratinocytes: modification of the autocrine control of keratinocyte proliferation. *J Invest Dermatol* **107**, 113–120.
12. Andreadis, S. T., Hamoen, K. E., Yarmush, M. L. and Morgan, J. R. (2001) Keratinocyte growth factor induces hyperproliferation and delays differentiation in a skin equivalent model system. *FASEB J* **15**, 898–906.
13. Supp, D. M., Supp, A. P., Bell, S. M. and Boyce, S. T. (2000) Enhanced vascularization of cultured skin substitutes genetically modified to overexpress vascular endothelial growth factor. *J Invest Dermatol* **114**, 5–13.
14. Supp, D. M. and Boyce, S. T. (2002) Overexpression of vascular endothelial growth factor accelerates early vascularization and improves healing of genetically modified cultured skin substitutes. *J Burn Care Rehabil* **23**, 10–20.
15. Bajaj, B. G., Lei, P. and Andreadis, S. T. (2005) Efficient gene transfer to human epidermal keratinocytes on fibronectin: in vitro evidence for transduction of epidermal stem cells. *Mol Ther* **11**, 969–979.
16. Hanenberg, H., Xiao, X. L., Dilloo, D., Hashino, K., Kato, I. and Williams, D. A. (1996) Colocalization of retrovirus and target cells on specific fibronectin fragments increases genetic transduction of mammalian cells. *Nat Med* **2**, 876–882.
17. Lei, P., Bajaj, B. and Andreadis, S. T. (2002) Retrovirus-associated heparan sulfate mediates immobilization and gene transfer on recombinant fibronectin. *J Virol* **76**, 8722–8728.
18. Pollok, K. E., Hanenberg, H., Noblitt, T. W., Schroeder, W. L., Kato, I., Emanuel, D. and Williams, D. A. (1998) High-efficiency gene transfer into normal and adenosine deaminase-deficient T lymphocytes is mediated by transduction on recombinant fibronectin fragments. *J Virol* **72**, 4882–4892.
19. Moritz, T., Dutt, P., Xiao, X., Carstanjen, D., Vik, T., Hanenberg, H. and Williams, D. A. (1996) Fibronectin improves transduction of reconstituting hematopoietic stem cells by retroviral vectors: evidence of direct viral binding to chymotryptic carboxy-terminal fragments. *Blood* **88**, 855–862.
20. Conneally, E., Eaves, C. J. and Humphries, R. K. (1998) Efficient retroviral-mediated gene transfer to human cord blood stem cells with in vivo repopulating potential. *Blood* **91**, 3487–3493.
21. Dao, M. A., Hashino, K., Kato, I. and Nolta, J. A. (1998) Adhesion to fibronectin maintains regenerative capacity during ex vivo culture and transduction of human hematopoietic stem and progenitor cells. *Blood* **92**, 4612–4621.
22. Kiem, H. P., Andrews, R. G., Morris, J., Peterson, L., Heyward, S., Allen, J. M., Rasko, J. E., Potter, J. and Miller, A. D. (1998) Improved gene transfer into baboon marrow repopulating cells using recombinant human fibronectin fragment

CH-296 in combination with interleukin-6, stem cell factor, FLT-3 ligand, and megakaryocyte growth and development factor. *Blood* **92**, 1878–1886.

23. Dardalhon, V., Noraz, N., Pollok, K., Rebouissou, C., Boyer, M., Bakker, A. Q., Spits, H. and Taylor, N. (1999) Green fluorescent protein as a selectable marker of fibronectin-facilitated retroviral gene transfer in primary human T lymphocytes. *Hum Gene Ther* **10**, 5–14.
24. Le Doux, J. M., Morgan, J. R. and Yarmush, M. L. (1996) Proteoglycans secreted by packaging cell lines inhibit retrovirus infection. *J Virol* **70**, 6468–6473.
25. Chuck, A. S., Clarke, M. F. and Palsson, B. O. (1996) Retroviral infection is limited by Brownian motion. *Hum Gene Ther* **7**, 1527–1534.
26. Bajaj, B., Lei, P. and Andreadis, S. T. (2001) High efficiencies of gene transfer with immobilized recombinant retrovirus: kinetics and optimization. *Biotechnol Prog* **17**, 587–596.
27. Andreadis, S., Lavery, T., Davis, H. E., Le Doux, J. M., Yarmush, M. L. and Morgan, J. R. (2000) Toward a more accurate quantitation of the activity of recombinant retroviruses: alternatives to titer and multiplicity of infection. *J Virol* **74**, 3431–3439.
28. Bajaj, B., Behshad, S. and Andreadis, S. T. (2002) Retroviral gene transfer to human epidermal keratinocytes correlates with integrin expression and is significantly enhanced on fibronectin. *Hum Gene Ther* **13**, 1821–1831.
29. Barrandon, Y. and Green, H. (1987) Three clonal types of keratinocyte with different capacities for multiplication. *Proc Natl Acad Sci USA* **84**, 2302–2306.

Index

MIX
Papier aus verantwortungsvollen Quellen
Paper from responsible sources
FSC® C105338

If you have any concerns about our products,
you can contact us on
ProductSafety@springernature.com

In case Publisher is established outside the EU,
the EU authorized representative is:
Springer Nature Customer Service Center GmbH
Europaplatz 3, 69115 Heidelberg, Germany

Printed by Libri Plureos GmbH
in Hamburg, Germany